International Conference "Advanced Problems of Electrotechnology" (APET-2020)

IOP Conference Series: Materials Science and Engineering
Volume 950

Yekaterinburg, Russia
1 - 2 October 2020

ISBN: 978-1-7138-3198-3
ISSN: 1757-8981

Printed from e-media with permission by:

Curran Associates, Inc.
57 Morehouse Lane
Red Hook, NY 12571

Some format issues inherent in the e-media version may also appear in this print version.

Institute of Physics
Dirac House, Temple Back
Bristol BS1 6BE UK

Phone: 44 1 17 929 7481
Fax: 44 1 17 920 0979

techtracking@iop.org

Additional copies of this publication are available from:

Curran Associates, Inc.
57 Morehouse Lane
Red Hook, NY 12571 USA
Phone: 845-758-0400
Fax: 845-758-2633
Email: curran@proceedings.com
Web: www.proceedings.com

TABLE OF CONTENTS

1. Preface

International conference "Advanced Problems of Electrotechnology APET-2020" was held October 1-2, 2020 in the Ural Federal University named after the first President of Russia B.N.Yeltsin.

This conference presents papers that largely reflect problems that scientists and specialists in the field of modern electrotechnology solve, as well as in related fields of technology. This applies, first of all, to electrothermal equipment, magnetohydrodynamics andelectrotechnologies, power supplies and power supply systems of this equipment.

Conference was held with the support of the Department «Electrical engineering» of the Ural Power Engineering Institute.

2. Lists of committees/participants

Organizing Committee

Chairmen
Sarapulov F. Professor, Ural Federal University

Co-Chairmans
Frizen V. Ural Federal University
Sarapulov S. Ural Federal University

Members of Committee

Aliferov A.I. Novosibirsk State Technical University
Baake E. University of Hannover, Germany
Doležel I. Institute of Thermomechanics of the CAS, Czech Republic
Galunin S. St. Petersburg Electrotechnical University LETI
Jakovičs A. University of Latvia, Latvia
Karban P. University of West Bohemia, Czech Republic
Khripchenko S.Y Institute of Continuous Media Mechanics of the UB of RAS
Kolesnichenko I. Institute of Continuous Media Mechanics of the UB of RAS
Konyaev A. Ural Federal University
Kuvaldin A. National Research University Moscow Power Engineering Institute
Nacke B. University of Hannover, Germany
Pleshivtseva Y. Samara State Technical University
Timofeev V. Siberian Federal University
Bytchkov S. Ural Federal University
Tarasov F. Ural Federal University
Bolotin K. Ural Federal University
Shvydky E. Ural Federal University
Smolyanov I. Ural Federal University
Prakht V. Ural Federal University
Kazakbaev V. Ural Federal University

IOP Conf. Series: Materials Science and Engineering **950** (2020) 011002 doi:10.1088/1757-899X/950/1/011002

Peer review declaration

All papers published in this volume *of IOP Conference Series: Materials Science and Engineering* have been peer reviewed through processes administered by the Editors. Reviews were conducted by expert referees to the professional and scientific standards expected of a proceedings journal published by IOP Publishing.

- **Type of peer review: Single-blind**
- **Conference submission management system:**
- **Number of submissions received: 40**
- **Number of submissions sent for review: 40**
- **Number of submissions accepted: 34**
- **Acceptance Rate (Number of Submissions Accepted / Number of Submissions Received X 100): 85**
- **Average number of reviews per paper: 1**
- **Total number of reviewers involved: 40**
- **Any additional info on review process:**
- **Contact person for queries: e.l.shvydky@urfu.ru**

IOP Conf. Series: Materials Science and Engineering **950** (2020) 012001 doi:10.1088/1757-899X/950/1/012001

Adjustable Electric Drive Based On Radiation-Resistant Induction Motors

V.I. Denisenko[1], A.M. Ziuzev[1], K.A. Kondakov[1], A.N. Moiseychenkov[1], V.V Lytkin[1], K.N. Dvoeglazov[2], D.A. Lukonin[3]
[1] Ural Federal University named after the first President of Russia B.N.Yeltsin
[2] Bochvar High-Tech Research Institute of Inorganic Materials
[3] LLC RusAtom-Engineering

Abstract. The object under consideration is an adjustable electric drive based on radiation-resistant DAR induction motors with a power of 0.75 kW and 2.2 kW, with a nominal stator field rotation speed of 1000 rpm and 1500 rpm and an environmental protection method IP 68, designed to use in the equipment for hydrometallurgical processing of spent nuclear fuel (SNF) of the processing module of the pilot demonstration energy complex (PM PDEC). Bench tests of DAR prototypes confirmed that the developed motors in accordance with electromagnetic parameters, energy and starting characteristics fully comply with serial asynchronous motors of similar power and speed and satisfy the requirements of the technical specifications. The control system with the Altivar ATV212/ATV320 frequency converter ensures stable operation of the DAR in the range of 1:50 speed control in the "sensorless" version, the functions and interface of the control system meet the specified requirements. Recommendations are given on expanding the tasks of automation of the technological process, which can be provided by means of the developed control system based on software implementation. An assessment of the thermal state during liquid cooling of engines during operation from the control system.

1. Introduction

For the first time in the Russian Federation in the innovation and implementation center "Electromechanical and Electrotechnological Complexes" (IEC "Electromekhtechnokom") Ural Federal University named after B.N. Yeltsin, together with VNIINM JSC, Uralelectromash CJSC and Rusatom Engineering LLC, as part of the Proryv project area, developed and manufactured radiation-resistant asynchronous electric motors with a control system with a service life of at least 3 years. This will allow us to abandon the traditionally problematic transmission of torque through the biological protection of the premises and place the engines directly in the equipment installation chambers with a high radiation dose level. This will reduce the cost of equipment, construction and installation costs and operating costs. The engines are designed for the technological equipment of the pilot demonstration energy complex (PDEC) for the processing of nuclear waste.

In accordance with the statement of work (SOR), the engines must be capable of long-term operation in PDEC processing modules with a high level of radiation (up to 10^8 rad), in a chemically aggressive environment with nitric acid vapor, the temperature of which reaches + 80° C. Fulfillment of these requirements was implemented by the design of the DAR according to the method of protection against environmental influences - IP 68, which implies complete tightness of the housing structure,

the use of radiation-resistant materials in the design of engines. In accordance with the statement of work, the control system must provide a range of regulation of the engine speed of not less than 1:50. In addition, in accordance with the statement of work, a prerequisite is the use of water-cooled engines to prevent overheating of the medium in PDEC processing modules.
DAR prototypes were made on the basis of the electromagnetic core of serial asynchronous motors of similar power and speed of CJSC Uralelectomash. Other design modifications of asynchronous motors with an asymmetric magnetic core [1,2] or with ring windings [6,7] or an electromagnetic core with ventilation-induction motors of a sensorless control system [3-5,8,9] can be used as an electromagnetic core.

The development of the control system and the manufacture of the control cabinet were carried out at the Electromekhtechnokom IT Center by the departments of Electrical Machines and Electric Drive and Automation of Industrial Installations. Testing of DAR prototypes, including the control system, was carried out in the research laboratory of the Department of Electrical Machines.

To regulate the speed of a new generation of radiation-resistant asynchronous DAR electric motors, the frequency control method was chosen as the most energy-efficient method among the known [11,12]. The engine control system is designed on the basis of a typical serial frequency converter (FC) of the Altivar family of Schneider Electric. To implement operator control functions, solve technological problems of plant automation, driven by the developed electric motor, and provide external access to control the system, it includes an industrial programmable logic controller (PLC) and an operator panel (OP) with touch control. When choosing a set of appropriate equipment, they were guided by its functional characteristics and minimal cost. The Schneider Electric equipment range includes all elements of modern automation systems, has available software and Russian-language operating manuals necessary for design and maintenance, the equipment itself has positively established itself in Russia, and is used in many projects in various industries [10,13].

The problem under considerations was the requirements for the engine control system for equipment for hydrometallurgical processing, namely, providing a control range of 1:50 without a speed sensor, it should be noted that in the characteristics of the inverter for general purposes, such an indicator in its direct representation is absent. The boundary values of the output frequency of the supply voltage are indicated, for example, 0.5 ... 200 Hz, and it does not say what quality indicators (statism of mechanical characteristics, bandwidth in speed) are able to provide an electric motor speed control system. In this regard, the task of obtaining experimental adjusting mechanical characteristics of motors with a selected frequency converter and control system was set in the work [14]. Obviously, to obtain the indicated characteristics, special testing equipment is needed to load an electric machine powered by a frequency converter with a calibrated resistance moment value. The generally accepted solution for such tests is the use of electromechanical stands, in which two electric machines are combined by a common shaft, one of which, for example, the test one, operates in the motor mode, and the second - load, works in the generator mode, creating a braking torque, the value of which is determined by the calculated or evaluated by a torque sensor integrated in the coupling. If it is necessary to evaluate the dynamic properties of the electric drive system, the test list is supplemented by removing the system reactions to stepwise or harmonic input influences, the so-called “control response” and “disturbance response”. It should be noted that at this stage of operation, there were no special requirements for the dynamic performance of the electric drive, therefore, the corresponding items are not included in the test program.

2. Methodology and test results of the electric drive to ensure a given range of speed control

Tests of prototypes of radiation-resistant electric motors, including the control system, were conducted in the laboratory of the Department of EM UralENIN UrFU on an electric machine unit consisting of rigidly connected test engine and a load electric machine (see Figure 1). As a load machine, the 1PH8101 induction motor was used, powered by a frequency converter with an active input rectifier, which provides energy recovery to the network, which makes it possible to obtain a generator (i.e.,

braking) mode of operation of the machine. The drive was controlled from the control cabinet (see Figure 2).
The engine torque was measured using a TRD-3K type torque sensor integrated in the coupling (nominal torque 3 kgf· m, accuracy class 0.03, manufacturer Dacell, Korea). The rotation speed was measured using an encoder loading machine (2048 pulses per revolution). Electrical quantities (current and voltage) were measured using LEM sensors (at the inverter output: CV3-1000, LA 25-NP, two wattmeter circuit; at the inverter input: LV25-P / SP5, LA 25-NP, two wattmeter circuit). To record signals from sensors, the ADC QMBox17-48 was used.

Figure 1. Test bench

Figure 2. Control cabinet for DAR-2.2

The test results showed that the control system with the Altivar ATV212 IF can be recognized as satisfying the requirements of the technical specifications, as ensuring stable operation of the ED in the specified range of speed regulation in the "sensorless" version (without a speed sensor on the shaft), which is illustrated by the mechanical characteristics shown in Fig. 3 and 4.

The problem of the importance in evaluating the performance of an electric drive was considered because of the nature of the load of the production mechanism. In particular, mechanisms with load characteristics of the "fan" type (a group of centrifugal mechanisms), in which the moment of resistance created by them on the motor shaft, is determined by a power-law dependence on speed n: $M_l = f(n^k)$, where $k > 2$, reduce the load on the engine when switching to a reduced speed, which, of course, reduces the requirements for the accuracy of maintaining a given speed.
The clarification of the requirements for the linearity of mechanical characteristics (varying in speed depending on the value of the resistance moment), expressed in absolute units or in fractions of a given value, it may turn out that the obtained characteristics of developed engines with the proposed control system cannot be considered satisfactory. In this case, to increase the overload capacity of the drive with the inverter and expand the speed control range in the "sensorless" version, it is recommended to use the Altivar ATV320 inverter with a power one step higher than the power of the ED.

Comparison between the experimental characteristics shown in Fig. 4 and fig. 5 outlines that Altivar ATV320 provides more stringent mechanical characteristics of the engine. It should be noted that the functions and interface of the control system (controller and operator panel) provide input of the speed task and displays the main variables available in Altivar.

To solve technological problems with the use of DAR and IF, additional information is required on the load characteristics of the working bodies of the mechanisms driven by the proposed electric drive. In this case, the tasks of automation of the technological process can be solved by software using the developed control system when completing the technological equipment with a set of appropriate sensors and actuators and / or devices.

IOP Conf. Series: Materials Science and Engineering **950** (2020) 012001 doi:10.1088/1757-899X/950/1/012001

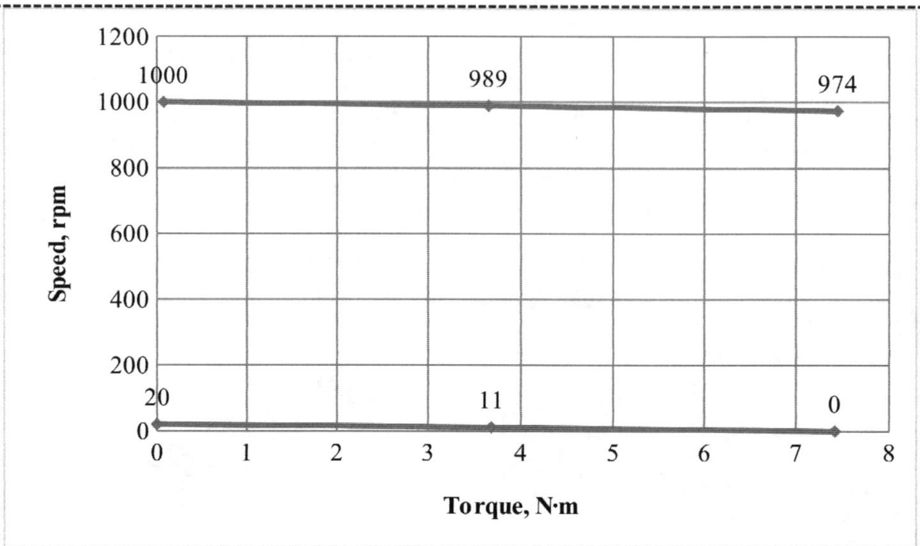

Figure 3. Mechanical characteristics of the DAR-0.75 engine when powered by an Altivar ATV212 frequency converter

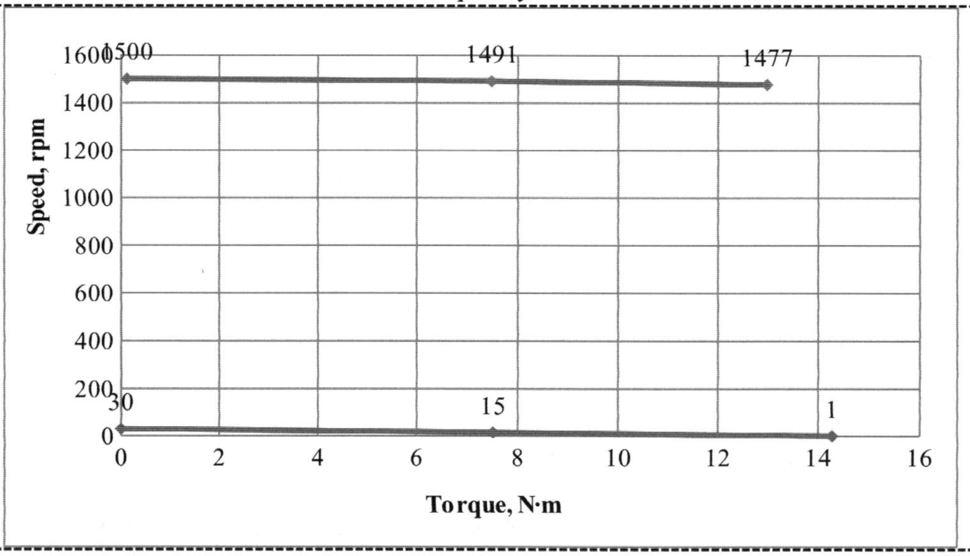

Figure 4. Mechanical characteristics of the DAR-2,2 engine when powered by an Altivar ATV212 frequency converter

3. The results of thermal tests of DAR in the electric drive

Thermal tests of DAR as part of an electric drive were carried out in the research laboratory of the Department of Electric Machines of UrFU:

1) for the DAR 2.2 engine when powered from a mains sinusoidal voltage of industrial frequency;
2) for the DAR 0.75 engine powered by a Schneider Electric Altivar 212 frequency converter.

When conducting a heating test ED DAR-2.2, the temperature of the incoming cooling water is 20 °C. The phase voltage and current in the stator winding corresponded to the nominal load mode. The results are presented in Figure 6.

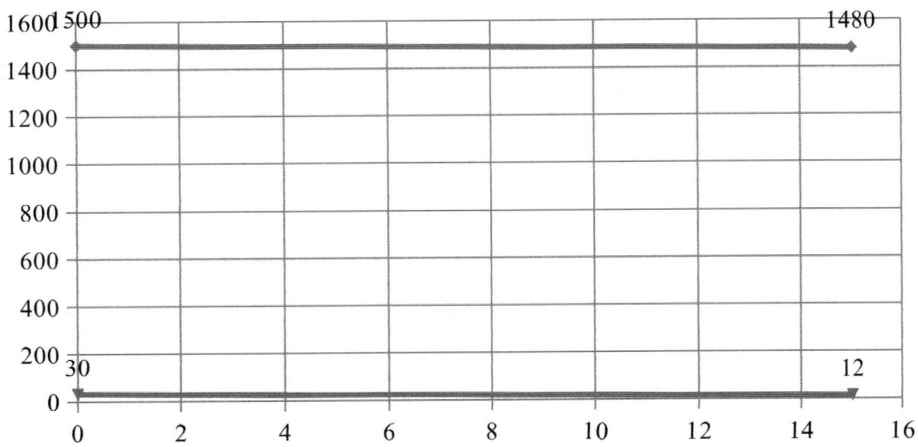

Figure 5. Mechanical characteristics of the DAR-2,2 engine when powered by an Altivar ATV320 frequency converter

Figure 7 shows a thermogram of an experimental sample of DAR-2.2 during heating tests at the test bench of the research laboratory of the Department of Electrical Machines of Ural Federal University. As can be seen from the above thermogram, the temperature of the most heated point on the bearing shield does not exceed the upper values on the color temperature scale on the right. Similar tests were performed for DAR-0.75.
Based on the results of thermal tests, a verification of the thermal calculation method developed to assess the thermal state of the DAR was performed. The technique was compiled using the equivalent thermal circuit method [2,3].
The solution of the ETS system of equations is made in matrix form for 15 nodes in a MathCad environment with 2 reference nodes that determine the ambient temperature and water temperature in the inlet pipe of the cooler. The calculations were performed at ambient and water temperatures and water flow rates in the cooler given in the thermal test protocols. In addition, the calculations used data on steel losses and mechanical losses found during bench tests at CJSC Uralelectromash. Electric losses in the stator and rotor windings were determined by the method of successive approximations by adjusting the active resistance of the windings according to the calculated values of their temperature.

When assessing the thermal state of the engine, we used the dependence of the heat transfer coefficient on the flow rate of cooling water, found as a result of calculating the liquid cooler by a numerical method based on the developed housing design.

Comparison of the steady-state values of the calculated temperatures in the nominal load mode of the experimental DAR samples with the experimental data showed that the developed method gives overestimated temperatures compared with the experiment. In comparison with the established value of the average temperature of the stator winding DAR-2.2, equal to 49.5°C, the excess was 15%, and with respect to the average temperature 49.5°C of the stator winding DAR-0.75, the excess was 5%, which is acceptable in technical calculations. In general, heating of the DAR stator windings does not exceed the values specified in the technical specifications, taking into account the increase in ambient temperature to + 80°C.

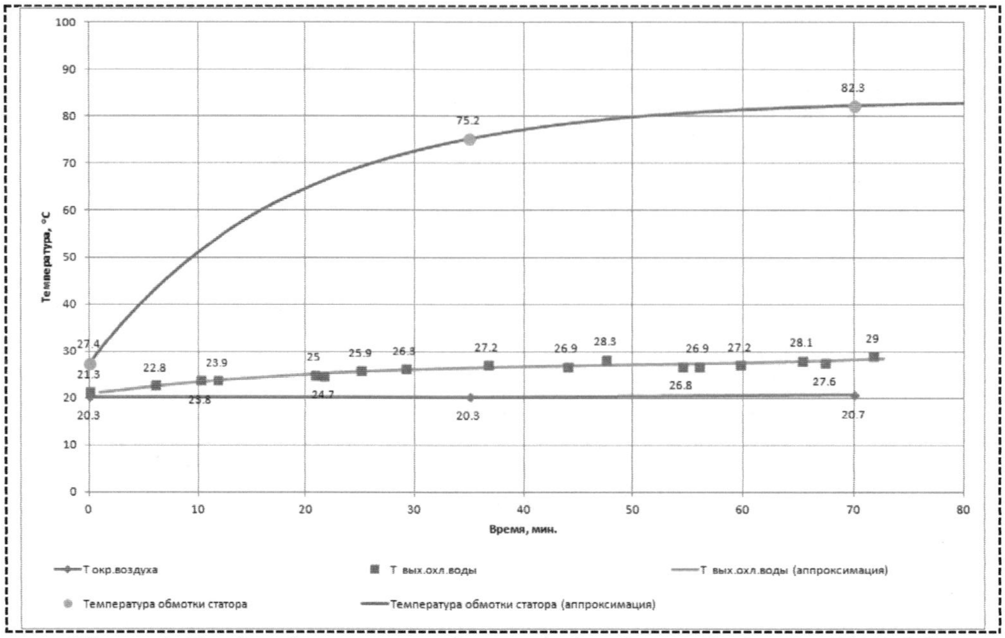

Figure 6. Graphs of temperature changes during the test for heating DAR-2,2

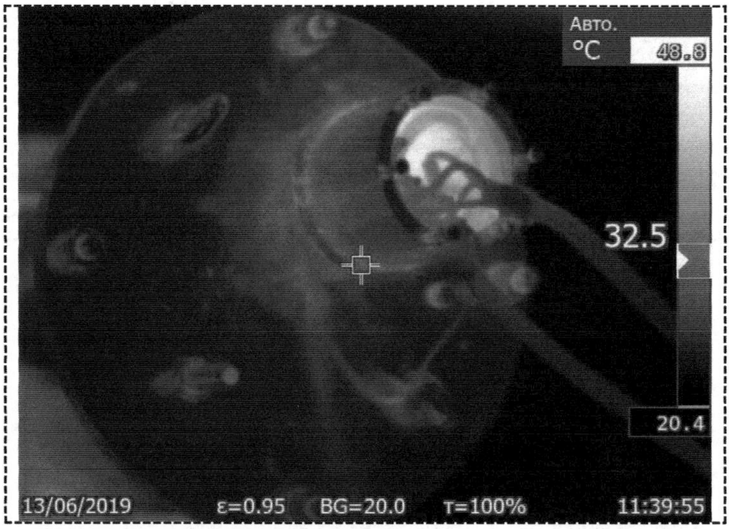

Figure 7. Thermogram of the DAR-2,2 engine. Bearing side view with terminal box

4. Conclusion

Bench tests of DAR prototypes carried out in the testing laboratory of CJSC Uralelectromash and at the Department of EM UralENIN UrFU confirmed that the developed engines with a control system for electromagnetic parameters, thermal, energy, starting and control characteristics fully comply with the technical specifications. They provide a predetermined level of speed regulation and can work in ODEC processing modules with a high level of heating in an aggressive environment.

References

[1] Patent for invention RU No. 094401, IPC H 02 K 1/4, 17/2 Stator of a reversible induction motor / Goldin R.G., Denisenko V.I., Plastun A.T., Pulnikov A.A. Priority 03/01/1996

[2] Baurzhan Bakubaev, Viktor Denisenko, Vladimir Nedzelskiy. Developing a highly reliable asynchronous motor with an asymmetric magnetic core for specialoperating conditions / Selected, Peer reviewed papers from the International Conference for Young Scientists "ELECTRICAL ENGENEERING. ELECTROTECHNOLOGY. ENERGY", June 9 – 12, 2015, Novosibirsk, Russia. Applied Mechanics and Materials. Vol. 792, pp. 90 – 94, Sep. 2015. ISBN – 13: 978-3-03835-548-9. DOI: 10.4028/www.scientific.net / AMM.772.90.

[3] Gayfutdinov, A. Radiation-Resistant Valve-Inductor-Type Motor as a Part of Technological Installation for Recycling Nuclear Industrial Waste / A. Gayfutdinov, A. Plastun, M. Baranov, A. Beketov // Applied mechanics and materials. – 2014. – 698. – pp. 111-115

[4] Patent RU No. 138724 U1. Radiation resistant induction motor of low power. / M.V. Baranov, A.R. Beketov, V.I. Denisenko, A.T. Plastun. Application No. 2014141369/07 of 09/10/2013

[5] Beketov, A.R. Designing and testing a system of controlling a switched reluctance motor with ceramic insulation, using no pickups / Denisenko, V.I., Moiseichenkov, A.N., Plastun, A.T., Lukonin, D.A. , Shaikhiev, A.R., Sanin V.K. / 2015 International Siberian Conference on Control and Communications (SIBCON). Proceedings. –Omsk: Omsk State Technical University. Russia, Omsk, May 21–23, 2015. IEEE Catalog Number: CFP15794-CDR. ISBN: 978-1-4799-7102-2. INSPEC Accession Number: 15287751 DOI:10.1109/SIBCON.2015.7147149.

[6] RF patent No. 2121207. Anchor of a multiphase electric machine / Plastun A.T. / filing application 1996-06-09, publication of the patent: 10.27.1998

[7] Tikhonova O.V., Kulakov N.N., Malygin I.V., Plastun A.T. Features of modeling an induction motor with ring windings in the ANSYS Maxwell software package, Sat. Proceedings of the IX All-Russian Scientific Conference of Young Scientists "Science. Technologies. Innovations", NSTU, December 2015, RSCI.

[8] RF patent 2547682. A method of obtaining an electrical insulating coating on the surface of an electrically conductive material / Baranov M.V., Denisenko V.I., Lukonin D.A., Nikulin S.L., Plastun A.T. / priority from 10/09/2013, registration 03/16/2015

[9] Baranov, M.V. Nano-oxide materials for increasing the operational characteristics of AC machines / M.V. Baranov, A.R. Beketov, V.I. Denisenko, A.T. Plastun, A.R. Gayfutdinov D.A. Lukonin, Sh.D. Gudaev // Proceedings of the Sverdlovsk Research Institute of Chemical Engineering. - 2013 .-- 20 (84). - from. pp. 68-72

[10] CERN [Electronic resource] Access mode: https://radiation-damage.web.cern.ch/, dated 05.24.2019.

[11] Sokolovsky G.G. Frequency-controlled AC electric drives: a textbook for students. higher textbook. institutions / G.G. Sokolovsky. - M.: Publishing Center "Academy", 2006. – 272p.

[12] Vinogradov A.B. Vector control of AC electric drives / GOUVPO "Ivanovo State Power Engineering University named after V.I. Lenin ". _ Ivanovo, 2008 - 298 p.

[13] Schneider Electric Russia [Electronic resource] Access mode:https://www.se.com/ru/ru//, dated 15.03.2020.

[14] Shonin Oleg Borisovich, Novozhilov Nikita Gennadievich, Kryltsov Sergey Borisovich Observer of the rotor speed in the scalar control system of an asynchronous electric drive // ES and K. 2016. No. 2 (31). URL: https://cyberleninka.ru/article/n/nablyudatel-chastoty-vrascheniya-rotora-v-skalyarnoy-sisteme-upravleniya-asinhronnym-elektroprivodom (accessed: 03.03.2020

IOP Conf. Series: Materials Science and Engineering **950** (2020) 012002 doi:10.1088/1757-899X/950/1/012002

An electrovortex flow around two fully submerged electrodes

A Chudnovsky[1], Yu Ivochkin[2], A Jakovics[3], S Pavlovs[3], I Teplyakov[2] and D Vinogradov[2]

[1] JSC LATVO, Ganibu Dambis street 53, Riga, Latvia
[2] Joint Institute for High Temperatures, Izorskaya street 13/2, Moscow, Russia
[3] University of Latvia, Jelgavas street 3, Riga, Latvia

[1] eivf19@gmail.com

Abstract. Electrovortex flow (EVF) of liquid metal in container with submerged electrodes and bifilar current supply relates to problem of melt stirring in electrometallurgical aggregates like welding, electroslag remelting, direct current (DC) electrical arc furnaces (EAF) and others. Maximum values of electromagnetic (EM) force and flow velocity are developing in the region between electrodes. Sophisticated 3D flow patterns are forming around them. In presenting paper, the examples of measured electrical potential and 3D self-magnetic field of supplied DC in comparing with numerical calculation results as well as with quasi-2D model of EM fields are discussed. Several numerically computed 3D EVF patterns are presented for two regimes: i) laminar and ii) turbulent, which is obtained with LES (Large Eddy Simulation) approach. There is good qualitative and quantitative correspondence between experimentally obtained data and numerical computations results as well as with previously published results.

1. Introduction

EVF between submerged electrodes have been studied in 1970–1980 for purposes of physical modelling of electrometallurgical technologies. In considered model, a spatially distributed electrical current with density J interacts with self-magnetic fields B and generates EM force $F = J \times B$. This force field is non-potential and drives conductive liquid circulation in the confined container [1].

Due to essentially 3D pattern, the main investigation methods were concerned with EM force field estimations, prognoses of flow patterns character and flow observation on a model's free surface. The measurements of velocity along symmetry axis were performed with fiber optical sensor in [2]. A review of various flow patterns between two electrodes and basic flow properties are presented in [3].

The development of innovative technologies is the cause of renewed interest to EVF at present time.

Current paper considers laboratory scale 3D system of relatively simple common configuration with two planes of symmetries and bifilar DC supply.

3D self-magnetic EM field is researched with the following approaches:

• measurements at experimental setup;

• analytical calculations using developed quasi 2D model;

• numerical computations with application of commercial package *ANSYS Maxwell*.
3D EVF patterns are analyzed with the following approaches:
• numerical study of flow laminar and turbulent regimes using *ANSYS FLUENT* package; LES model of turbulence is used;
• flow patterns obtained during observation of liquid metal free surface.

2. Scheme of experimental setup and computational model

The scheme of model is shown in figure 1(a). The photo of experimental setup is presented in figure 1(b) – mirror of liquid metal reflects surrounding fragments of laboratory.
Cold (melting point is 10.5°*C*) liquid metal (*Galinstan*) is placed in a cylindrical container (diameter D = 220 mm) of low depth (H = 20 mm). Top surface of liquid is free.
Two copper cylindrical electrodes (diameter d = 20 mm) are built in bottom-up through the container bottom and are fully submerged into liquid – flat top ends of electrodes are at the same level with liquid free surface. Electrode axes are located on a plane of bath symmetry and on equal opposite distances from the bath axis. The distance between electrode axes $2x_e$ = 60 mm.
Bifilar power supply scheme is used so that electrical current circuit is closed through a liquid metal between cylindrical walls of electrodes. Total current I can be changed in the interval from 100 A till 900 A.
Electrical conductivity for liquid metal (*Galistan*) and copper are $\sigma_m\ 3 \div 3.3 \cdot 10^6\, S/m$ and $\sigma_e\ 5.8 \cdot 10^7\, S/m$.

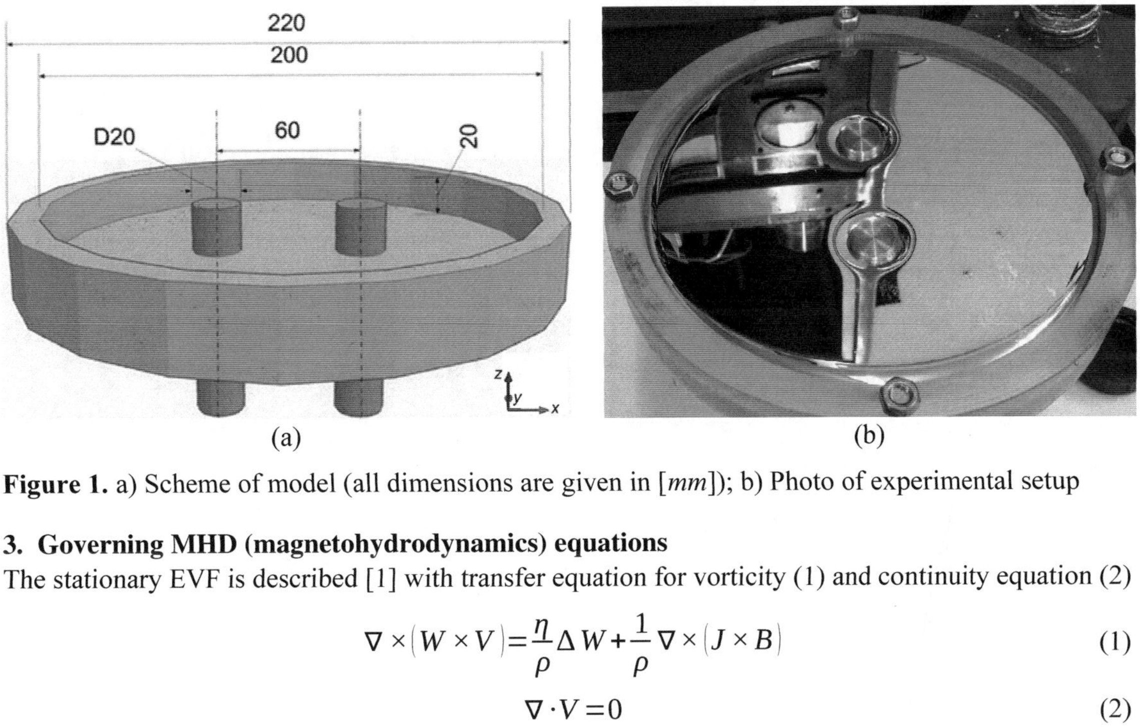

(a) (b)

Figure 1. a) Scheme of model (all dimensions are given in [*mm*]); b) Photo of experimental setup

3. Governing MHD (magnetohydrodynamics) equations

The stationary EVF is described [1] with transfer equation for vorticity (1) and continuity equation (2)

$$\nabla \times (W \times V) = \frac{\eta}{\rho} \Delta W + \frac{1}{\rho} \nabla \times (J \times B) \tag{1}$$

$$\nabla \cdot V = 0 \tag{2}$$

where V is velocity and $W = \nabla \times V$ is vorticity for liquid metal. Dynamic viscosity η and density ρ are constants.
EM field is described with *Maxwell* equations (3)–(6) and *Ohm*'s low (7)

$$\nabla \times B = \mu_0 J \tag{3}$$

$$\nabla \cdot B = 0 \quad (4)$$

$$\nabla \times E = 0 \quad (5)$$

$$\nabla \cdot E = 0 \quad (6)$$

$$J = \sigma E \quad (7)$$

where E is electrical field intensity; μ_0 is magnetic constant and electrical conductivity σ for liquid metal σ_m and electrodes σ_e are constants.

The following assumptions are taken into account:

• there is no influence of fluid flow on EM fields – $V \times B \ll J/\sigma$ (electrodynamics approximation);
• free surface deformation is not essential and don't influence the distribution of EM field;
• thermogravity convection is neglected in comparison with strong EM stirring.

The Cartesian coordinate system x, y, z with orts I_x, I_y, I_z are used. The origin is placed at the bath bottom centre so, that z-axis directs to the free surface, x-axis – to the right electrode and y-axis – depthward from an electrodes plane. There are two plane of symmetries geometry of model: y = 0 is the electrodes' plane and x = 0 is the orthogonal plane.

4. Analytical solution for EM field using quasi 2D approximation

The approximations for the case of $\sigma_e >> \sigma_m$ are following:

• z-component of current density in liquid is equal to zero;
• considered functions depend on coordinates x, y:

$$J_x = \partial\varphi/\partial x = -\partial\psi_1/dy,\ J_y = \partial\varphi/\partial y = \partial\psi_1/\partial x,\ J_z = 0 \quad (8)$$

$$\partial\varphi/\partial z = \partial\psi_1/\partial z = 0 \quad (9)$$

where $\varphi = \varphi(x, y)$ is electrical potential and $\psi_1 = \psi_1(x, y)$ is electrical stream function in planes $z = const$ inside liquid layer $0 \le z \le H$.

Complex potential $Q = Q(u) = \varphi + i\psi_1$ of complex variable $u = x + iy$ [2] for model, shown in figure 1(a), can be written as follows:

$$Q = \frac{A}{2\pi} \ln \frac{(U-1)}{(U+1)} \quad (10)$$

$$\varphi = \frac{A}{2\pi} \ln \sqrt{¿¿¿¿} \quad (11)$$

$$\psi_1 = \frac{A}{2\pi}\left(arctg\frac{y}{x+1} - arctg\frac{y}{x-1}\right) \quad (12)$$

where φ_0 is the calibration point potential.

Then from equation (3) and expressions (8),(9) we can get related expressions for self-magnetic field of the current:

$$B_x = (b + cz)\frac{\partial\psi_1}{\partial x} \quad (13)$$

$$B_y = (b + cz)\frac{\partial\psi_1}{dy} \quad (14)$$

$$B_z = \psi_1(c-1) \quad (15)$$

and corresponding expressions for EM force vector and its curl:

$$F = J \times B = F_x I_x + F_y I_y + F_z I_z = (1-c)\nabla \frac{\psi_1^2}{2} - (b+cz)\,|\,\nabla \psi_1\,|^2 I_z \qquad (16)$$

$$\nabla \times F = \frac{-\partial F_z}{\partial y} I_x + \frac{\partial F_z}{\partial x} I_y \qquad (17)$$

Here all sizes are related to x_e, the constant A in equations (10)–(12) is proportional to total current, constants b and c in equations (13)–(15) will be checked experimentally. As it seen from equation (16), only z-component of EM force is non-potential; and horizontal gradients of F_z define the curl of EM force (17). Note that the maximum values of J, B, F and $\nabla \times F$ in liquid metal can be found in the electrode plane $y = 0$ near side wall of electrodes and between them [2, 3]. Thus main flow pattern is generated in central region as well. Consequently, the influence of external boundary to EM field in central area of bath is neglected and therefore external boundary is not included into complex potential in equation (10).

5. Comparison of measured and computed EM field

EM field for 3D model (figure 1(a)) has been numerical computed with ANSYS Maxwell package.
At experimental setup (figure 1(b)) the electric potential has been measured for two planes:

- for electrodes plane $y=0$ (electrodes axes lie in this plane) between electrodes for coordinates $-20\,mm < x < 20\,mm$ and $0 < z < 20\,mm$;
- for orthogonal plane $x=0$ and coordinates $-20\,mm < x < 20\,mm$ and $0 < z < 20\,mm$;

The step between measurements points is equal 1 *mm* for all coordinates *x*, *y* and *z*.

5.1. Electrical potential

Comparison of measured (figure 2(a)) and calculated (figure 2(b)) results for electrical potential distribution shows, that analytical solution using quasi 2D approximation is enough for precise estimations of EM field at middle horizontal plane $z = 10\,mm$ between electrodes.

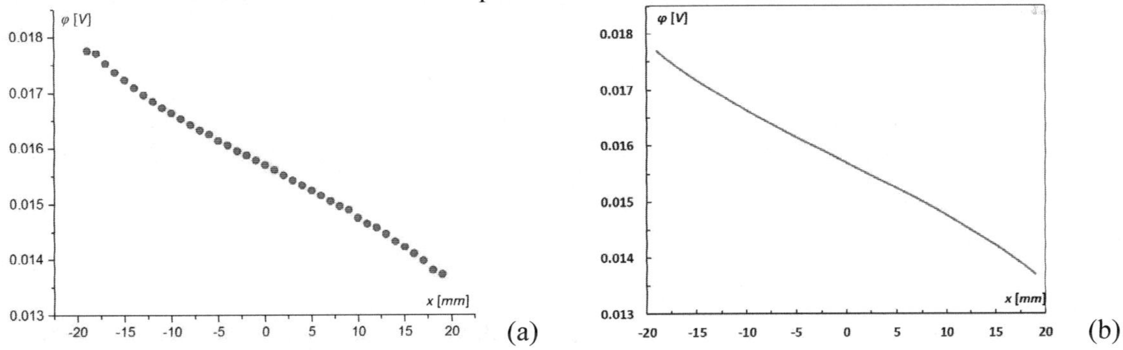

Figure 2. Distribution of electrical potential between electrodes at plane $z = 10\,mm$:
a) measurements results; b) results of calculation (red line) with quasi 2D approximation – equation (11) and constant values $\varphi_0 = 0.0157$ and $A = 5000$

5.2. Induction of magnetic field

Measured (figure 3 (a)) and numerically computed (figure 3 (b)) results for y-component of magnetic field induction are of good qualitative correspondence. Magnetic field is stronger near electrodes in zone of bath bottom ($z = 0$).

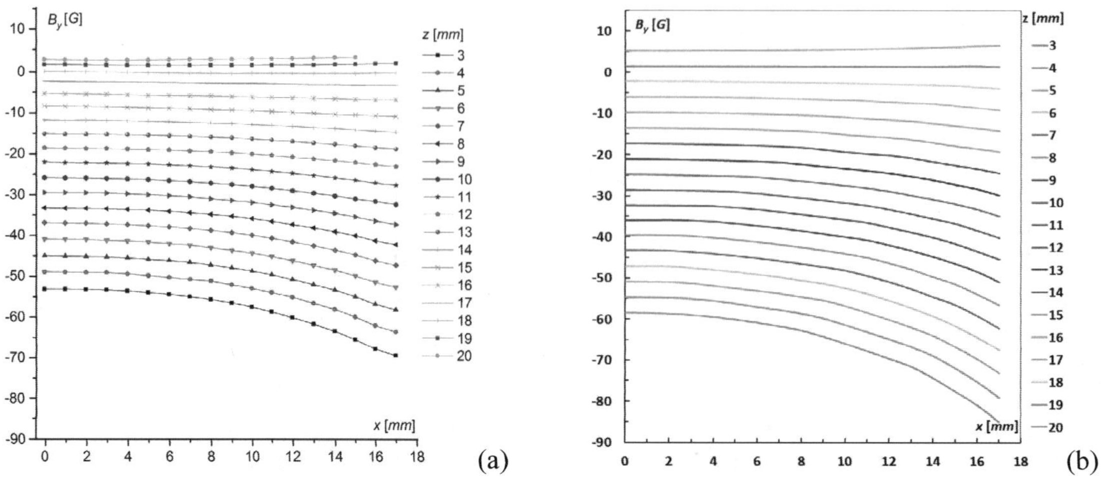

Figure 3. Distribution of magnetic field induction *By(x, z)* at electrodes plane *y* = 0 (electrodes axes lie in this plane): a) measurements results; b) results of numerical computations with *ANSYS Maxwell*

The quantitative distinctions may be concerned with the following reason:

• measured electrical conductivity of *Galistan* σ_m in experimental setup is for 10% smaller than declared in material certificate and thus is smaller than σ_m values, which is used in numerical computations.

6. EVF patterns

6.1. Computational results

Liquid metal EVF for 3D model (figure 1(a)) has been numerical computed with *ANSYS FLUENT* package. The following flow regimes are considered:

• ***Turbulent regime*** is obtained with applications of LES model. Estimated with expression (18) value of *Reynolds* number is $\Re \; 10^4$. As characteristic dimension the liquid metal height H is chosen, maximum values of velocity is $v_{max} \; 0.18\, m/s$. *Galistan* viscosity is $\eta = 2.2 \cdot 10^{-3} kg/(m \cdot s)$ and density is $\rho = 6.361 \cdot 10^3 kg/m^3$. Electrical current supplied over electrodes is $I_{\square} = 1000\, A$.

• ***Laminar regime*** of EVF is computed for dynamic viscosity $\eta_l = 0.22\, kg/(m \cdot s)$, which is chosen for two orders greater than *Galistan* viscosity. Estimated value of *Reynolds* number is $\Re \; 40$, maximum values of velocity is $v_{max} \; 0.07\, m/s$.

$$\Re = \frac{v_{max} H}{\eta / \rho} \tag{18}$$

Laminar regime can be described with *Stokes* approximation (linear equation (1) without convective term) for extremely slow flows. Further development of laminar flow needs description with nonlinear equation (1) until turbulent pulsations appear. In numerical computations, the laminar flow regime of liquid metal can be reached with decrease of supplied electrical current from $I_{\square} = 1000\, A$ to value $I_l = 10\, A$ (100 times less). For this case non-dimensional parameter, estimated with expression (19), decreases from $S_{\square} = 4.2 \cdot 10^7$ (turbulent regime) to value $S_l = 4.2 \cdot 10^3$ (laminar regime), that is four orders less. The same result can be obtained with liquid viscosity increasing for two orders without variations of electrical current.

$$S = \frac{\mu_0 \rho I^2}{4 \pi^2 \eta^2} \tag{19}$$

IOP Conf. Series: Materials Science and Engineering **950** (2020) 012002 doi:10.1088/1757-899X/950/1/012002

Computational results are illustrated as follows:
- streamlines for laminar flow regime correspond to quasi-steady state (flow time $t \;\; 30\,min$) – figures 4(a), 5(a), 6(a);
- streamlines for turbulent flow regime are time-averaged for flow time period $t \;\; 100\,s$ – figures 4(b), 5(b), 6(b).

3D patterns of EVF are presented for the following characteristic planes:
- at electrodes plane $y = 0$ (figure 4) maximum values of melt velocity are obtained alongside electrodes walls. At free surface between electrodes, there are liquid metal counter-moving jets, coming from electrodes to vertical axis z;
- at plane $x = 0$, which is orthogonal to electrodes plane $y = 0$ (figure 5), maximum values of melt velocity are obtained at free surface, where are obtained divergent jets from axis z;
- at horizontal middle plane $z = 10\ mm$ (figure 6) are obtained four contours of melt circulation, maximum values of melt velocity are reached around electrodes and in zone between electrodes.

Melt flows, shown in three characteristic planes, mentioned above, are forming the following integrated 3D pattern of EVF. Melt circulation is driven by EM force, which has maximum value between two electrodes. Upstream jets alongside walls of submerged electrodes transform into counter-moving jets on a free surface with further circulation both in electrodes plane and in orthogonal plane. Closing of all streams of melt is forming four contours circulation, which is observed in horizontal middle plane.

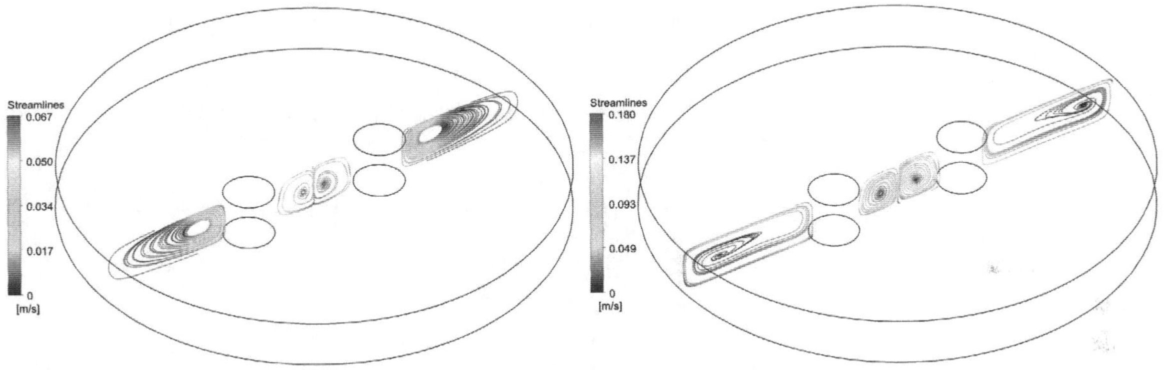

(a) (b)

Figure 4. Streamlines at plane $y = 0$: a) laminar flow; b) turbulent flow.

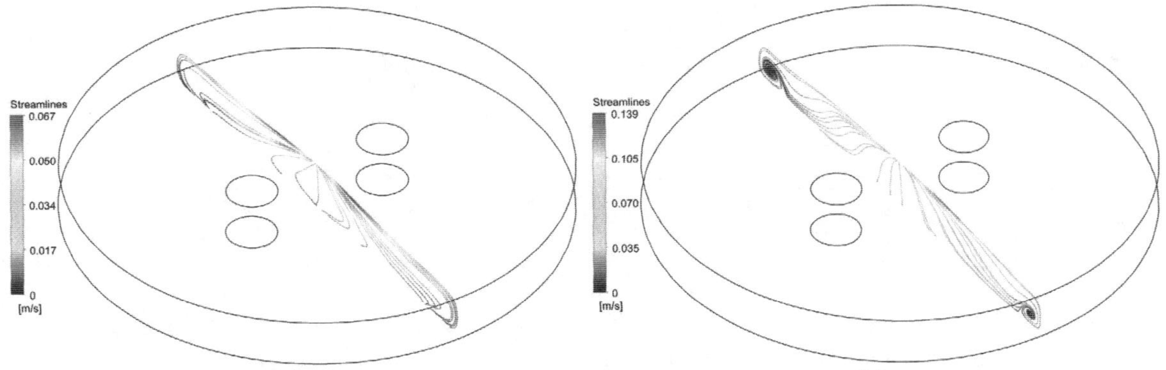

(a) (b)

Figure 5. Streamlines at plane $x = 0$: a) laminar flow; b) turbulent flow

IOP Conf. Series: Materials Science and Engineering **950** (2020) 012002 doi:10.1088/1757-899X/950/1/012002

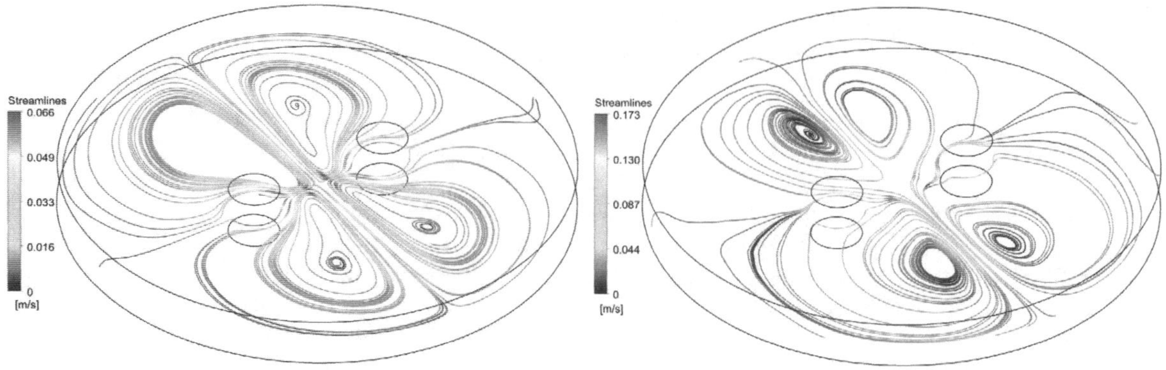

(a) (b)

Figure 6: Streamlines at middle horizontal plane z = 10 *mm*: a) laminar flow; b) turbulent flow

EVF patterns computed for turbulent and laminar regime are qualitatively similar. Turbulent flow patterns consist of large vortices of intensive circulations, but there are also lots small vortices especially for instantaneous distribution of streamlines. For laminar flow, circulation vortices are smooth and slow without small vortices.

6.2. Qualitative comparison of computed and experimental EVF patterns

Figure 7 shows good qualitative correspondence of numerically computed (a) and experimentally observed (b) four contours circulation.

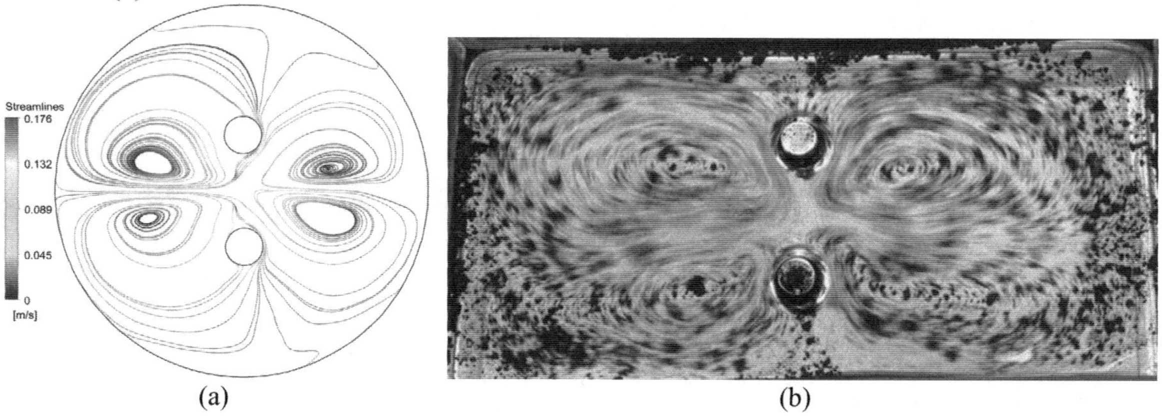

(a) (b)

Figure 7. EVF turbulent regime of liquid metal (top view): a) computed streamlines in horizontal middle plane for model with small thickness of liquid metal; b) photo of experimentally observed circulation at the free surface [3].

7. Conclusion

• Measured values of electrical potential and magnetic field induction in central part of liquid metal bath both qualitatively and quantitatively correspond on reasonable level to results of 3D numerical computation as well as correspond to calculations using derived quasi 2D analytical solution.

• Numerically computed electrovortex flow 3D patterns have characteristic four contours of circulation in horizontal plane, which qualitatively correspond to similar experimentally obtained photos of melt circulation.

• Numerical results for turbulent regime make it possible to research of flow characteristics for large values of total supplied electrical current. Computed patterns for slow laminar regime allow studying the details of 3D flow with relation to its driving sources – 3D field of EM force vorticity.

Acknowledgements
This work was funded by European Regional Development Fund under contract "Development of numerical modelling approaches to study complex multiphysical interactions in electromagnetic liquid metal technologies" (No. 1.1.1.1/18/A/108)

References

[1] Bojarevich V, Freiberg J, Silova E and Scherbinin E 1989 *Electrically induced vortical flows* (Cluwer Academic Press, Springer Netherlands).
[2] Dement'ev S, Zilin V, Ivochkin Yu, Oksman A and Chudnovsky A 1988 Magnetohydrodynamics **24**
[3] Chudnovsky A 2017 Magnetohydrodynamics **53**

IOP Conf. Series: Materials Science and Engineering **950** (2020) 012003 doi:10.1088/1757-899X/950/1/012003

Calculation of the performance of the electromagnetic magnetic fluid separator non-magnetic materials

Yu B Kazakov, V A Filippov

Ivanovo State Power University, 34 Rabfacovskaya st., Ivanovo, 153003, Russian Federation

Abstract. Magnetic fluid separation is a promising area in the field of disposal and recycling of non-magnetic materials. Performance and separation accuracy are important features of any separator. The methods for calculating the productivity offered at this moment have a number of assumptions that can significantly affect the result (constant magnetic permeability of the magnetic fluid, linear change in tension along the height of the gap, etc.), and are applicable only to specific models of separators. The aim of the study is to develop a methodology for calculating the performance of an electromagnetic magnetic fluid separator of non-magnetic materials, taking into account the distribution of the magnetic field in the gap and the influence of the hydrodynamic properties of the magnetic fluid on the movement of particles in the separation zone. The proposed approach allows us to calculate the mass and volumetric capacities of the magnetic fluid separator, the dependences of the capacities for each fraction separately, and the total on various parameters of the separator (for example, on the angle of inclination of the pole pieces). The obtained results can be used to assess the optimality of the adopted parameters when designing a magnetic fluid separator. Also, the results obtained will help optimize the installations already in operation.

1. Introduction.

Every year, the volumes of electronic waste in the world increase and in 2016 more than 40 million tons were produced. In the future, the situation will only worsen it is assumed that at the beginning of the next decade, at least 50 million tons of electronic waste will be released in the world posing a threat to human health and the environment [1].

Unsuitable for further use devices and materials may contain valuable natural resources, such as gold, silver, copper, platinum, etc. Their competent processing can provide reduction of environmental pollution, conservation of natural resources and return to the production of non-ferrous, precious metals and other materials. So, for example, in alluvial deposits, the gold content is up to 0.1 - 0.5 g / t of ore, in primary ores 1 - 5 g / t. And in the secondary raw materials, the gold content reaches 100 g / t and even 100 kg / t. The proportion of precious metals that can be obtained from recycled materials reaches 20 - 50%. So, in 2016, the cost of such materials not extracted from discarded electronics amounted to $ 55 billion [2].

During secondary processing, the performance of the technological scheme is determined by the productivity of each stage. The performance of magnetic fluid separation determines the power of the previous plants for grinding, magnetic separation and the required capacity for the following operations [3,4].

Despite this, the issue of performance is often resolved by empirical studies of an already created device. So, in [5] the dependences of the separation efficiency on the productivity of the experimental setup are given (figure 1).

Separation efficiency is evaluated by Zemlyansky's criterion:

$$E_Z = \varepsilon_1 + \varepsilon_2 - 100,$$

where ε_1 is the extraction of light fractions into tails, %, ε_2 is the extraction of heavy fractions into concentrate, %.

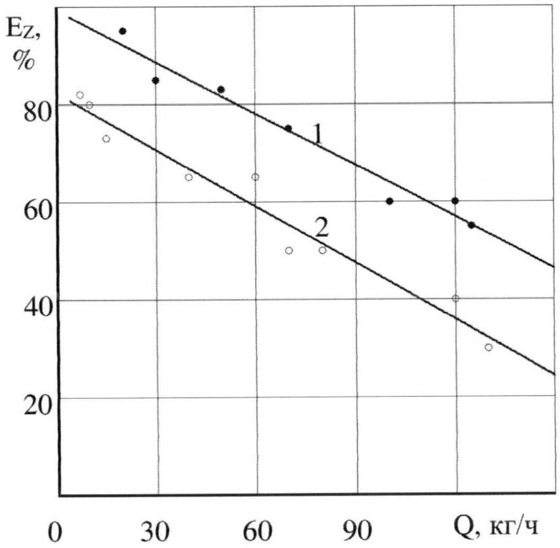

Figure 1. Dependencies of separation efficiency according to the Zemlyansky (E_Z) criterion on productivity: 1 - supply of material along the entire length of the magnetic fluid layer, 2 - supply of material to the beginning of the magnetic fluid layer [5].

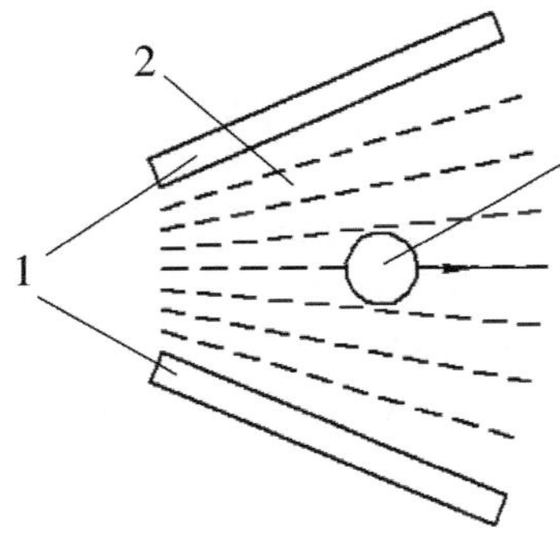

Figure 2. Magneto-liquid separator with diverging poles: 1 - pole tips, 2 - magnetic fluid, 3 - non-magnetic body [6].

In [6], the issue of the performance of a magnetic fluid separator with poles diverging in the horizontal plane is considered (Figure 2). This design provides additional strength, which helps to move the light fraction in the horizontal direction.

The following expressions are used to calculate the performance of the separator.

The final velocity of the non-magnetic particle in the horizontal direction:

$$\upsilon_{kx} = k_3\sqrt{\frac{d}{\rho_{kx}}\left(\mu_0\chi H\frac{\partial H}{\partial x} + \rho_p g\sin\alpha\right)},$$

where k_3 is the empirical coefficient, d is the particle diameter, ρ_{kx} is the horizontal component of the quasi-density of the magnetic fluid, and χ is the magnetic susceptibility of the magnetic fluid.

The final velocity of a non-magnetic particle in the vertical direction:

$$\upsilon_{ky} = k_1 k_2\sqrt{d\frac{\rho_p - \rho_{ky}}{\rho_{ky}}},$$

where k_1, k_2 are empirical coefficients, ρ_{ky} is the vertical component of the quasi-density of the magnetic fluid, and ρ_p is the particle density.

Required separation zone length:

$$L = k_4 h\sqrt{\frac{\rho_{ky}\mu_0\chi H\frac{\partial H}{\partial x} + \rho_p g\sin\alpha}{(\rho_p - \rho_{ky})\rho_{kx}}},$$

where k_4 is the empirical coefficient.

IOP Conf. Series: Materials Science and Engineering **950** (2020) 012003 doi:10.1088/1757-899X/950/1/012003

Separator capacity:

$Q = \rho \cdot \Delta h \cdot \theta \cdot b \cdot v_{к} / \beta$,

where Δh is the thickness of the layer of material being transported, θ is the degree of loosening of the fraction, the working width of the separator channel, β is the fraction of the light fraction in the feedstock.

The above methods for determining the performance are suitable exclusively for specific types of magnetic fluid separators and include a few empirical coefficients, which makes it difficult to use them in the development of new devices. Also, several assumptions are applied that may affect the accuracy of determining this parameter.

The use of numerical modeling of physical fields to calculate the performance of magnetic-liquid separators can give higher accuracy and the possibility of applying a new technique in the design of new ones.

2. Research Methods.

To calculate the particle trajectory, when moving in the working gap of the magnetic fluid separator (MFS), the following model was adopted. Four forces act on a non-magnetic particle with density and volume V in a magnetic fluid with density and magnetization M (Fig. 3): gravity force $\overline{F}_{T} = \rho \cdot \overline{g} \cdot V$, Archimedes force $\overline{F}_{A} = -\rho_{Ж} \cdot \overline{g} \cdot V$, buoyancy from the side of the magnetic fluid

$$\overline{F}_{M} = -\mu_0 \cdot M \cdot \nabla H \cdot V, \tag{1}$$

where ∇H is the gradient of the magnetic field, and the resistance force $\overline{F}_{V}$, due to the viscosity of the magnetic fluid (MF), determined by the Stokes formula

$$\overline{F}_{V} = -3\pi \cdot \eta \cdot d \cdot \overline{v}, \tag{2}$$

where: η-dynamic viscosity of the MF; $\overline{v}$ - particle velocity.

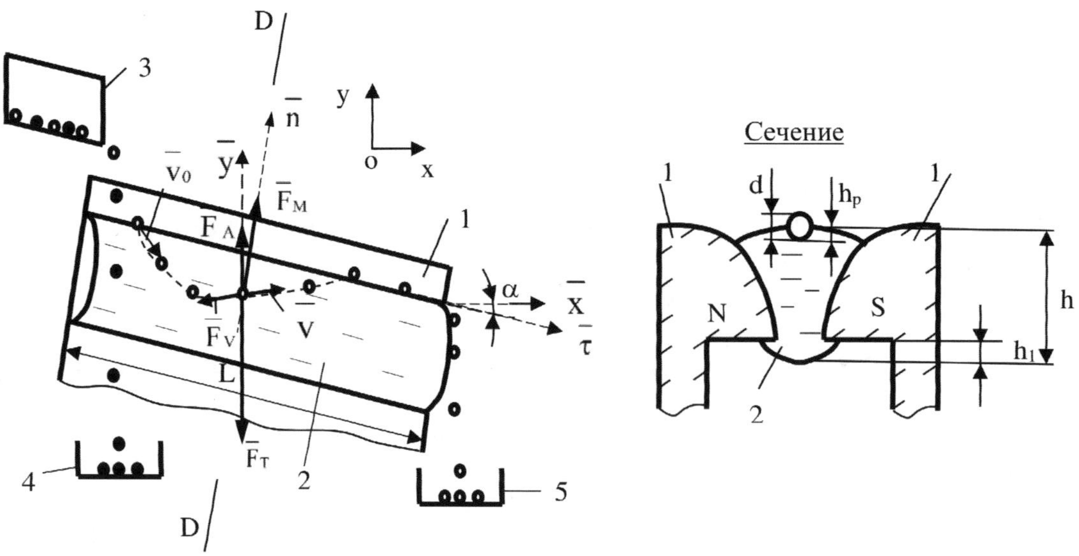

Figure 3. The movement of non-magnetic particles in a magneto-liquid separator: 1 - pole pieces forming a working gap; 2 - magnetic fluid; 3 - device for supplying non-magnetic particles; 4 - container for heavy particles; 5 - container for light particles.

The equation of motion of a particle can be obtained based on Newton's second law.

$$\overline{F}_T + \overline{F}_A + \overline{F}_M + \overline{F}_V = m \cdot \frac{d\overline{v}}{dt}, \tag{3}$$

where: $m = \rho V$ is the mass of the particle. We assume that the particles are spherical and have a diameter d. After substituting all the values into Newton's formula, the equation of motion of a nonmagnetic particle in the working gap of the MFS will have the form

$$\rho V \cdot \frac{d\overline{v}}{dt} + 3\pi \cdot \eta \cdot d \cdot \overline{v} + \left[-\mu_0 M \cdot |\nabla H| \cdot V + (\rho - \rho_Ж) \cdot \overline{g} V \right] = 0. \tag{4}$$

We accept the following assumptions:

- MF has a linear rheological characteristic and constant dynamic viscosity (η = const);
- the working gap of the MFS is designed correctly, as a result of which the third term on the left side of equation (4) is constant (it represents the resulting force that acts on the particle in statics).

We write equation (4) in the form:

$$a \frac{d\overline{v}}{dt} + b \cdot \overline{v} + c = 0, \tag{5}$$

where a, b, c are constant values independent of the magnitude of the velocity.
In projections on the axis, equation (4) has the form:

$$\rho g V(-\overline{e}_y) - \rho_{MF} g V(-\overline{e}_y) - \mu_0 M \cdot \left[|\nabla H| \sin\alpha(-\overline{e}_x) + |\nabla H| \cos\alpha(-\overline{e}_y) \right] V -$$

$$-3\pi\eta d(v_x \overline{e}_x + v_y \overline{e}_y) = \rho V \frac{d}{dt}(v_x \overline{e}_x + v_y \overline{e}_y)$$

$$\rho V \frac{dv_x}{dt} + 3\pi \cdot \eta \cdot d \cdot v_x - \mu_0 M |\nabla H| \sin\alpha V = 0 \tag{6}$$

$$\rho V \frac{dv_y}{dt} + 3\pi \cdot \eta \cdot d \cdot v_y + \left[-\mu_0 M |\nabla H| \cos\alpha + (\rho - \rho_{MF}) g \right] V = 0 \tag{7}$$

As a result of integration of equations (6) and (7), we obtain expressions for the particle velocity along the axes

$$v_x = \left(v_{0x} + \frac{c_x}{b} \right) \cdot e^{-\frac{b \cdot t}{a}} - \frac{c_x}{b}. \tag{8}$$

$$v_y = \left(v_{0y} + \frac{c_y}{b} \right) \cdot e^{-\frac{b \cdot t}{a}} - \frac{c_y}{b}, \tag{9}$$

where: $a = \rho V$; $b = 3\pi \cdot \eta \cdot d \cdot k_b$;

$c_y = k_c \left(-\mu_0 M \cdot |\nabla H| \cdot V \cos\alpha - \rho_{MF} \cdot gV \right) + \rho \cdot gV$;

$c_x = -k_c \cdot \mu_0 M \cdot |\nabla H| \cdot V \sin\alpha$,

where k_b, k_c – coefficients taking into account the degree of immersion of the particle in the MF;

$$k_b = \frac{S_I}{S} = \frac{h_p}{d}, k_c = \frac{V_I}{V} = \frac{6(d\frac{h_p^2}{2} - \frac{h_p^3}{3})}{d^3}, \tag{10}$$

where h_p is the height of the submerged part of the particle, d is its diameter.
The following parameters are set in a computer program for calculating the particle trajectory:

- particle density ρ;

- fluid density ρ_{MF};
- the angle of the poles (surface of the magnetic fluid) α;
- the length of the active part of the poles L;
- the height of the column of magnetic fluid h;
- the height of the minimum working gap relative to the lower surface of the magnetic fluid h_l (the boundary of the change in the direction of the magnetic force FM);
- magnetization curve of magnetic fluid;
- the value of the dynamic viscosity of the breast η;
- particle diameter d;
- the initial position of the particle X_0, Y_0;
- projections of the initial velocity vector v_{0X}, v_{0Y}.

When calculating the trajectory by known coordinates at the previous iteration, it is determined whether the particle is in magnetic fluid. If not, then the change in speed is determined under the action of only gravity. If the particle is in the MF, then the degree of its immersion in the MF is determined, the coefficients kb and kc are determined by (10), and the velocity is calculated using the above formulas (7-8). The known speeds determine the change in the coordinates of the particles at the considered step of the change in time. The coordinates are written to the array and are ultimately displayed on the graphs (Figures 4 - 5) [7].

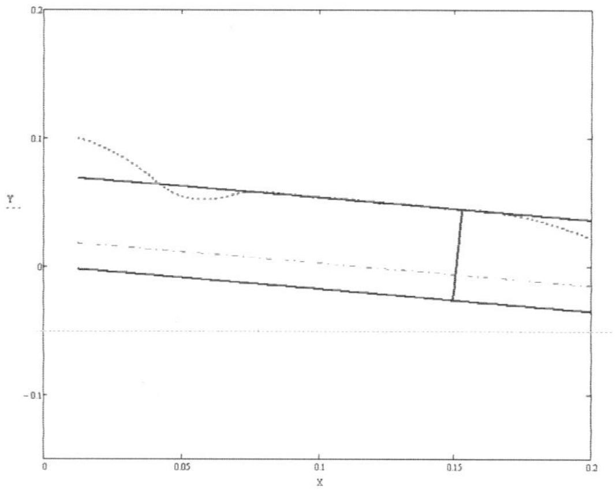

Figure 4. The calculated value of the trajectory of an aluminum particle with a diameter of 5.6 mm at an angle of inclination of the poles of 10 °. Dotted line - particle trajectory, solid line - MF borders, dash-dot line - boundary of the change in the direction of action of the magnetic force.

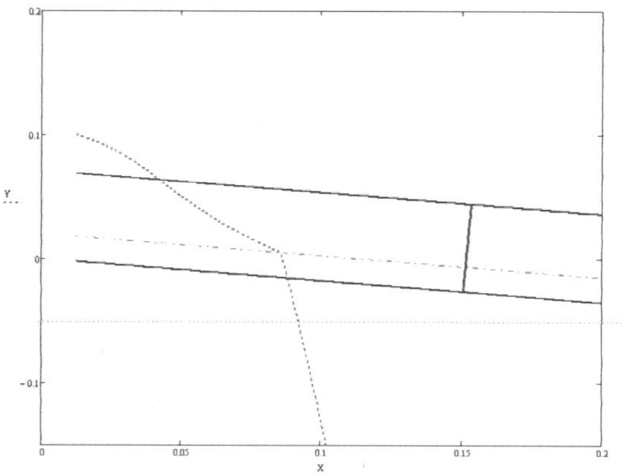

Figure 5. The calculated value of the trajectory of motion of a copper particle with a diameter of 4.9 mm at an angle of inclination of the poles of 10 °. Dotted line - particle trajectory, solid line - MF borders, dash-dot line - boundary of the change in the direction of action of the magnetic force.

The performance of the separator when moving the shared material in a monolayer can be calculated by the following formula:

$$Q = \rho d \cdot b \cdot v / k,$$

where d is the average particle diameter of the material to be separated, b is the width of the layer of the material to be separated, v is the average speed of the mixture to be shared towards the unloading device, and k is the fill factor of the material [8]. Since the movement of the separated mixture in the magnetic fluid separator occurs in two directions: in the horizontal plane for the light fraction, and in the vertical for the heavy fraction, two capacities should be considered - in the light and heavy fractions.

Thus,

$Q_L = \rho_L \cdot d_L \cdot b \cdot v_{xav} / k_L$ - productivity for a light fraction;

$Q_H = \rho_H \cdot d_H \cdot \delta v_{yav} / k_H$ - performance for the heavy fraction, where dl and dt are the diameters of light and heavy particles, respectively, b is the arc length of the upper surface of the magnetic fluid, δ is the minimum clearance of the magnetic fluid separator (figure 6), v_{xav} and v_{yav} are the average speeds along the corresponding axes, k_L and k_H are the filling factors of the light and heavy fractions, respectively [5,6]. To obtain volumetric performance, you must use the following formula:

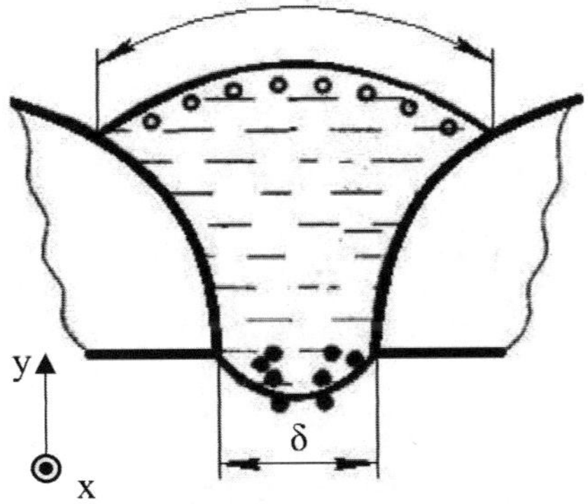

Figure 6. Gap of MFS

$Q_V = Q/\rho$, where ρ is the density of the corresponding fraction. Using the models described above, particle velocities in the separation zone of the separator can be obtained.

Since the particle velocity during its movement in the separation zone is not constant, it is necessary to calculate the average velocity:

$$v_{av} = \frac{\int_0^T v dt}{T},$$

where T is the time of movement in the separation zone.

Numerically average speed:

$$v_{av} = \frac{\sum_1^N v_i \Delta t}{T},$$

where v_i is the velocity at the i-th iteration, N is the iteration number at which the particle exits the separation zone. If $T = N \cdot \Delta t$, then

$$v_{av} = \frac{\sum_1^N v_i}{N}.$$

3. The results of the study.

As an example, the performance of a magnetic fluid separator was calculated for different angles of inclination of the pole pieces. The following magnetic fluid parameters were taken for calculation: viscosity 7.5 Pa · s, density 1.15 g/cm^3, saturation magnetization 25 kA/m [9]. The separated mixture is two-component with particles with a diameter of 1.5 mm, and densities of 7.26 g/cm^3 (tin) and 2.7 g/cm^3 (aluminum) for heavy and light fractions, respectively. The parameters of the separator correspond to the laboratory setup: the length of the separation zone is 15 cm, the height of the separation zone is 7 cm, the arc length of the upper surface of the magnetic fluid is 2.5 cm, the minimum gap is 1 cm.

As a result of the calculation, the speeds and residence time in each section were obtained, based on which the dependences of the mass (figure 7) and volumetric (figure 8) capacities on the angle of inclination of the pole pieces were calculated (the calculation results are summarized in table 1).

Table 1. Dependences of the mass and volumetric performance of the magnetic fluid separator on the angle of the poles.

α, deg	**0**	**10**	**20**	**30**	**40**	**45**
Q_H, kg/h	75,48	78,74	83,15	88,94	-	-
Q_L, kg/h	-	17,85	42,09	53,3	55,83	55,95
Q_Σ, kg/h	75,48	96,59	125,24	142,24	55,83	55,95
Q_{VH}, l/h	10.40	10.85	11.45	12.25	0.00	0.00
Q_{VL}, l/h	0.00	6.61	15.59	19.74	20.68	20.72
$Q_{V\Sigma}$, l/h	10.40	17.46	27.04	31.99	20.68	20.72

4. Discussion and conclusions.

In contrast to the methods considered earlier, the developed technique uses the full picture of the field distribution in the gap; in calculating the productivity, instantaneous values of particle velocities are used, which undoubtedly increases the accuracy of the calculation. Also, the developed technique can be applied to various types of magnetic fluid separators.

The dependences obtained make it possible to select the optimal parameters of the magneto-liquid separator to ensure maximum performance and energy efficiency of the separation process. They also help to evaluate the adequacy of some parameters of the magneto-liquid separator. So, for example, from the graphs (figure 7,figure 8), it can be seen that at an angle of inclination of the pole tips of more than 40 degrees, the heavy fraction begins to fall into the light and this position of the poles is unacceptable. Thus, using the program developed based on this algorithm [10], it is possible to design a magnetic fluid separator that maximally meets the requirements of the separation process in the processing of secondary materials.

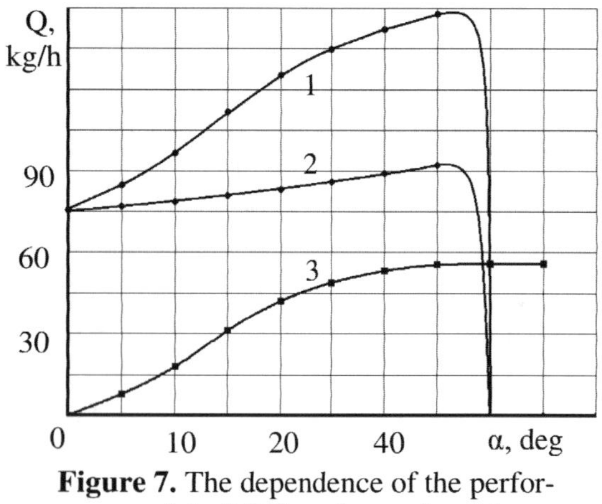

Figure 7. The dependence of the performance of the MFS on the angle of the poles: 1 - total, 2 - for the heavy fraction, 3 - for the light fraction.

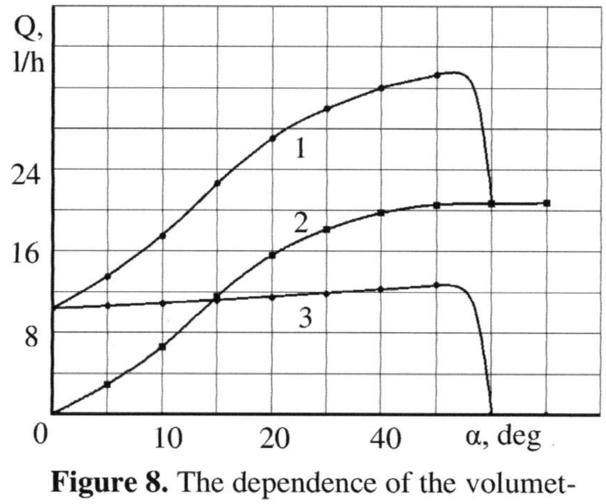

Figure 8. The dependence of the volumetric capacity of the MFS on the angle of the poles: 1 - total, 2 - for the light fraction, 3 - for the heavy fraction.

References

[1] K P Balde, V Forti, V Grey, R Kyur, P Stegmann. *Global'nyy monitoring elektronnykh otkhodov,* 2017 god, Universitet Organizatsii Ob"edinennykh Natsiy (UOON), Mezhdunarodnyy soyuz elektrosvyazi (MSE) i Mezhdunarodnaya assotsiatsiya po tverdym otkhodam (MATO), Bonn/Zheneva/Vena.

[2] Barkan M, Chinenkova M. *Tekhnologiya izvlecheniya dragotsennykh metallov iz otkhodov elektrotekhniki* [Elektronnyy resurs] http://www.waste.ru/module s/section/item.php?itemid=233

[3] L.S Strizhko, Yu.Yu. Kostyukhin, G.V. Kruzhkova, E.A. Ivanova. *Izvlechenie tsvetnykh i blago-rodnykh metalloviz elektronnogo loma: ekonomicheskie pokazateli i strategiya tsenoobrazovaniya* // Izvestiya vuzov. Tsvetnaya metallurgiya № 3. 2013. pp. 28 - 33
[4] A N Telyakov, L V Ikonin. *Rezul'taty ispytaniya tekh-nologii pererabotki radioelektronnogo loma*// Zapiski Gornogo instituta. T.169. - Sankt-Peterburg. 2006. pp193 – 195.
[5] Gulyaikhin E V, Solodenko A B, Bochkarev G R. *Separatsiya mineral'nogo syr'ya v psevdoutya-zhelennykh sredakh.* – Novosibirsk: Nauka, 1984. – 140 p.
[6] N D Kravchenko, V I Karamzin. *Magnitnaya separatsiya otkhodov tsvetnykh metallov.* M.: Met-allurgiya, 1986, - 129 p.
[7] Kazakov Yu B, Stradomskiy Yu I, Filippov V A. *Matematicheskaya model' dvizheniya nemagnit-nykh chastits v rabochem zazore gidrostaticheskogo magnitozhidkostnogo separatora.* // Sbornik nauchn. trudov 17-oy Mezhdun. Plesskoy nauchn. konf. po nanodispersnym magnitnym zhidkostyam. – Ples. sentyabr' 2016. pp. 178-186
[8] Pelevin A E. *Magnitnye i elektricheskie metody obogashcheniya.* Magnitnye metody obogash-cheniya: uchebnik / Pelevin A. E.; Ural. gos. gornyy un-t. – Ekaterinburg: Izd-vo UGGU, 2018. – 296 p.
[9] Kolchanov N V, Kolesnichenko E V. *Vyazkost' magnitnykh kolloidnykh zhidkostey pri razlichnykh temperaturakh i ob"emnykh dolyakh tverdykh chastits* // Vestnik Permskogo universiteta. Fizika. 2017. № 4 (38). S. 37–45. doi: 10.17072/1994-3598-2017-4-37-44
[10] Svidetel'stvo o gosudarstvennoy registratsii programm dlya EVM №2019616505 *«Raschet traektoriy nemagnitnykh chastits v zone razdeleniya magnitozhid-kostnogo separatora»* / Filippov V A.// FGBOU VO IGEU – Zayavka № 2019615527 ot 15 maya 2019 g. Zaregistriro-vano v Reestre programm dlya EVM 23 maya 2019 g

Development brushless direct current electric motor of the Pilger mill drive for the technology of seamless pipes manufacturing based on the magnetic system with claw-poles

Bakhtiyor Kosimov[1,2] *Sergey* Gandzha[1] *Dilshod* Aminov[1,2], Azamdzhon Davlatov[1], *Dilovar* Gulov[1], Ilhom Abdulloev[1]

[1]South Ural State University, Chelyabinsk, Russia
[2]Tajikistan Power Energy Institute, Khushoniyon, Tajikistan

Abstract. Currently, metallurgical industry contains a large number of morally and physically outdated electrical equipment that requires modernization. It requires a large capital expenditure to replace it. For this reason, it is slow to implement, long-term use for the purpose of payback, and slow to withdraw from manufacturing. In this regard, the new equipment must be innovative and in many respects ahead of the existing level of development of the industry. The article suggests such a solution. To replace the existing collector motor of the pilger mill drive for the manufacturing of seamless pipes, which has been operating since 1928 and is in a critical condition, a large-sized low-speed brushless motor with a large built-in inertial mass is proposed, which has a permanent magnet magnetic system and claw-poles. Such a technical solution for drives of this class has not been used in the world. The motor eliminates the existing 120-ton flywheel, improves reliability due to contactless current supply, and reduces operating costs by eliminating excitation losses. The motor design allows its Assembly without additional tooling, which is very important for large motors with permanent magnets. To develop a unique motor, a design system consisting of a synthesis subsystem and an analysis subsystem was created. The synthesis subsystem implements multi-level single-criteria optimization. As a result of its operation, the optimal geometry is determined based on the selected criterion. The analysis system confirms the calculation reliability using simplified optimization methods and finally removes technical risks before manufacturing an industrial design. It is based on the well-developed Ansys Electronics Desktop and Ansys Icepak systems for CAE electric machines. The analysis has several stages, which includes electromagnetic and thermal analysis. It is based on the well-developed for CAE systems electric machines Ansys Electronics Desktop and Ansys Icepak. The analysis has several stages, which includes electromagnetic and thermal analysis. The effectiveness of the design system was tested on a real project that was performed for the Chelyabinsk rolling mill. This approach is recommended for the design of other similar types of brushless direct current drive.
Key words: pilger mill, CAE systems, synthesis, analysis, thermal analysis, brushless direct current motor, claw-poles, magnetic system, tangential magnets, two-dimensional model, three-dimensional model.

IOP Conf. Series: Materials Science and Engineering **950** (2020) 012004 doi:10.1088/1757-899X/950/1/012004

1 Introduction

The Ural region is traditionally one of the centers of heavy industry. Historically, large-scale mechanical engineering and metallurgy have developed here for more than two centuries. These trends are still present today. A feature of the development of this industry is inertia. First of all, this is due to the large capital costs of modernizing production. Scientific developments and innovative solutions are much slower to be implemented in this area than in other industries. This can explain the fact that many heavy industry enterprises in Russia still

Corresponding author: kosimov.energy@mail.ru

use pre-war and post-war equipment. In this regard, it is very important that technical solutions for this industry are significantly ahead of time and innovative. They will be slowly implemented and then work for a long time until moral and physical wear and tear.

One such example is the current pilger mill drive at the Chelyabinsk pipe rolling plant for the production of seamless pipes of various diameters that are used in the oil and gas sector. The drive has been operating since 1928. The main parameters of the synchronous collector motor from Siemens are shown below, and its General appearance is shown in Fig. 1:

rated power of 2.75 MW;
rated voltage power supply 6 kV DC;
rated speed of 35 rpm.

Fig. 1. The pilger mill electric motor of the Chelyabinsk pipe rolling plant

The electric drive rotates a massive flywheel with a diameter of 9 m and a weight of 120 tons.
Continuous electric motor operation has led to physical and moral wear and tear. Numerous defects were found that could cause an accident.
As obvious problems can be noted the appearance of microcracks on the motor shaft due to shock load, which can lead to the destruction of the shaft.
Taking into account the 3-shift operation of the drive, special attention should be paid to solving issues of energy efficiency and energy saving of the new electric motor and its drive.
The key task of the enterprise at present is to replace the existing collector electric motor with a more modern, reliable, energy-efficient contactless electric motor.

2 Requirements for a new drive

According to the terms of the customer, when replacing the electric motor in question, the main components and assemblies of the existing drive must remain unchanged, namely:

two-roll cage with variable cross-section gauge;
feeding mechanism.

Therefore, only the electric motor and the massive flywheel must be replaced.
The customer's requirements for new equipment are listed below:

the new unit should not exceed the dimensions of the manufacturing room facility where it should be placed;
to increase reliability and reduce operating costs the new drive must have a contactless current supply;

in order to be able to roll pipes of different diameters, the drive speed adjustment must be within the range of 30-45 rpm, while ensuring the stability of the selected speed;

it is necessary to reduce operating losses by increasing the efficiency of the drive;

the time it takes to replace the old drive with a new drive should not cause continuous production to stop for a long time;

the proposed electric motor and its control system should have the lowest possible cost.

3 Selecting the basic motor option for the pilger mill drive

Analysis of trends in the development of modern electrical engineering allows us to choose a brushless electric drive as the most prospective one [1-5]. This is due to the fact that the power and reliability of brushless machines are steadily increasing, including in aggressive environments. At the same time, the price of such drives is reduced and becomes affordable for medium-sized enterprises. This is facilitated by the development of production of high-power high-coercivity permanent magnets and power electronics for high currents and voltages. Currently, the development of controlled current frequency converters at a power of tens and hundreds of megawatts does not present practical problems. A well-established vector control, which is effective from the point of view of mass and energy indicators. Various designs with efficient magnetic systems for the inductor have appeared. These positive trends allow us to design and produce valve engines and generators for tens and hundreds of megawatts.

But practice shows that powerful and large-sized brushless direct current (BLDC) machines are produced with electromagnetic excitation without permanent magnets. This is due to the technology of manufacturing machines with permanent magnets, which have a large diameter. When assembling an electric machine, when the inductor is inserted into the armature, very strong magnetic forces occur. When using powerful magnets, these forces can reach several tons. The rotor is attracted to the stator walls, and assembly becomes almost impossible. Theoretically, it is possible to design equipment for the assembly of such electric machines, but calculations show that the cost of these devices can be ten times higher than the cost of the motor itself. The same problems occur when disassembling the inductor for routine maintenance and repair work.

Investigations to solve this technological problem led to the selection and analysis of the brushless machine with claw-poles [6-8]. The design of an inductor with claw-poles with a permanent magnet is shown in Fig. 2

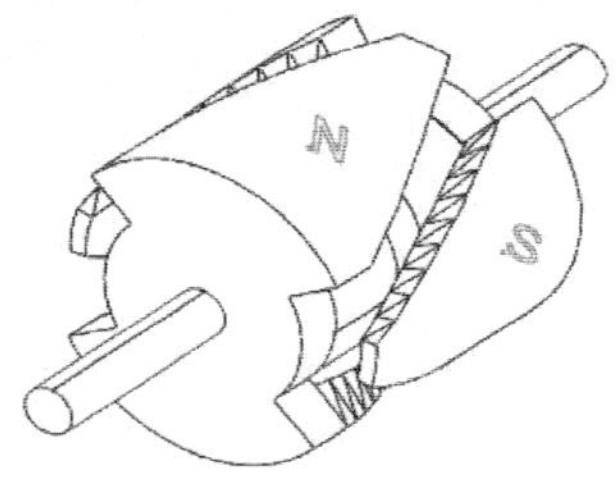

Fig. 2. General view of the rotor with claw-poles and a permanent magnet

It should be noted that the magnetic system with claw-poles cannot be considered effective. It has a large scattering flux that saturates the magnetic core, but does not participate in energy conversion.

With this disadvantage, this design has a great advantage. The inductor can be assembled directly in the electric machine itself in parts using the following technology:

1. Initially, a shaft with the lower clip of claw-poles is inserted into the anchor. This part of the magnetic system at this stage of assembly does not contain permanent magnets and, for this reason, will not be attracted to the walls of the anchor;
2. At the second stage of assembly, a permanent magnet is mounted in the inductor. Depending on its size, it can be glued together in separate parts, or it can be mounted as a whole;
3. At the third stage, the upper cage with claw-poles is inserted into the inductor. Under the action of the electro-magnetic forces of a permanent magnet, it must be drawn into the inductor;
4. At the last final stage, the bearing mounting shield completes the assembly.

The build technology is shown in fig. 3

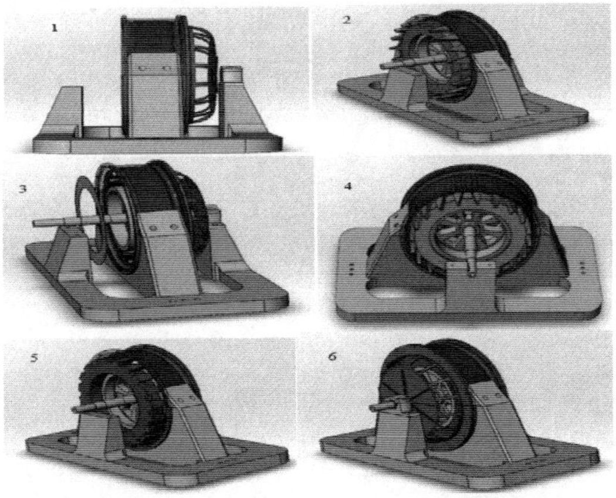

Fig.3. Technology motor Assembly (1-installation of the stator Assembly, 2-installation of the first half of the claw-rotor, 3,4 - installation of a permanent magnet, 5-installation of the second half of the claw-rotor, 6-final Assembly of the motor).

This technology of Assembly of the claw-poles brushless motor (CPBM) does not require expensive equipment and special devices and is offered for the first time.

By optimizing the geometry and choosing the correct size of the circled permanent magnet, the magnetic system can be made efficient enough with a large working magnetic flux.

Magnetic systems with claw-poles are well studied [9-13], but this applies to electric drives of low and medium power. The development of multi-megawatt brushless machines with a bore diameter of several meters is unique for engineering practice. In this regard, the task of theoretical research and practical implementation of such a design in real production has become urgent.

4 Statement of the CPBM synthesis and analysis problem

A special feature of the pilger mill drive is the low speed of the shaft rotation at high impact loads. Operating conditions do not allow the use of a step-up gearbox for the motor due to high dynamic moments, so the motor has large dimensions. It is impossible to apply the similarity theory in this case, when an optimally designed low-power motor proportionally increases in size to a given power. In electrical engineering, this technique is not used. This is due to sharply nonlinear relationships between the linear dimensions of the engine, electromagnetic loads and its electromagnetic power.

The design of large-capacity and large-size motors is associated with high capital costs. It is impossible to lead a project along the traditional path: creating and researching a mock-up model, a prototype, or a serial product. Modern development of computer technology and software allows you to go through these stages on a digital double before making a real motor. To do this, you need to create a design system for synthesizing the optimal geometry and analyzing the resulting electronic prototype.

For synthesis, it is necessary to choose the most suitable method of nonlinear programming for this mathematical model.

For better analysis of electrical machines of this class, you can use well-developed and proven CAE systems based on the finite element method, such as Ansys Electronics Desktop. Such a design system for designing the CPBM was created. The functional diagram of the design system is shown in the following figure:

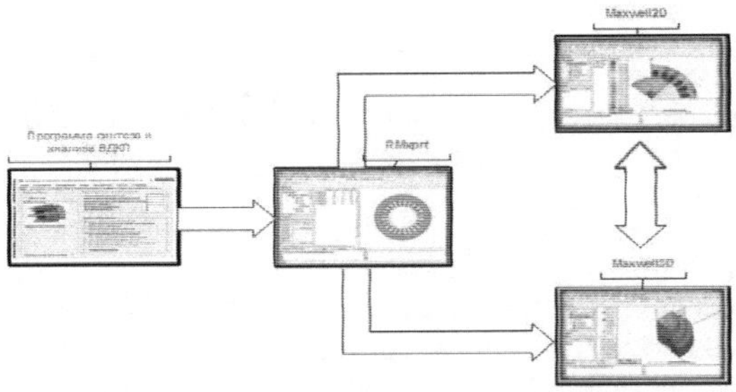

Fig. 4. Functional diagram of the design system for a brushless machine with the claw-like magnetic system

Division of the project system into two parts: synthesis and analysis is explained by the following reasons. Geometry optimization involves a large number of iterations when changing independent variables. Depending on the selected method, the number of mathematical model calculation cycles can be up to several thousand. It is clear that optimal design methods can only work with a very simplified model with a large number of restrictions [14,15]. It is impossible to leave this model as the final one, even after the optimization procedure, because of the high risks of capital expenditures for the development of design and technological documentation and sample production. A thorough analysis of the results obtained on the basis of more accurate methods of digital duplicates is necessary. This task is performed by the analysis system.

The length of the article does not allow us to describe both systems in detail. The work of the synthesis system is presented in a number of already published articles [16]. Briefly note that the synthesis system is based on the method of substitution schemes. The optimal design problem is formulated as a multi-level single-criteria problem. For the algorithm for iterating through independent variables, a method is chosen that combines the Gauss-Seidel coordinate descent method when moving to the optimum with the Fibonacci method when choosing a step. The algorithm has a built-in method of penalty functions when searching for an optimum on the boundary.

We will describe the construction of the analysis system and the results of its work in more detail, since several innovative solutions were adopted during its development.

As it was noted, the Ansys Electronics Desktop software package was selected for the analysis system out of numerous possible options [17-20]. It contains a large number of well-developed methods of typical designs of electric machines. CPBM is included in the list of these machines. The software package allows you to make a preliminary analysis of this machine in RMxprt mode. This mode makes the calculations based on the method of equivalent circuits. To do this, just fill in table forms with the machine geometry, materials, and parameters of the nominal mode, which greatly facilitates the analysis of the machine. However, it should be noted that in the latest version of the program, you can only analyze the generator mode of operation. For the pilger mill drive requires an analysis of the motor operation. Due to the specifics of the drive operation, it may differ significantly from the generator mode, which leads to technical risks when introducing the drive into production.

The software package allows you to deploy the task in a 3D model. The program does this automatically when you select Create Maxwell 3D Design mode. It is not possible to expand the problem of analyzing a gate machine into a 2D model, since the machine geometry does not have flat symmetry.

At this stage of creation of the system, the developers encountered a very big problem. Test calculations showed that when using the Ansys Electronics Desktop program for a 3D model in the mode of solving the transient dynamic problem for a machine with a claw-like magnetic system, the waiting time for the results of calculating the main characteristics was from 8 to 15 hours. It should be noted that a powerful University supercomputer was used to solve the problem. The procedure of parallelizing the problem solution to several processor cores did not significantly improve the result. One of the probable reasons was the fact that the formation of the design grid came from an air gap of several millimeters, and the dimensions of the entire engine were more than 6 meters. The grid generator divided the calculated area into a very large number of elements. At the same time, for the convergence of the process, the grid was repeatedly rebuilt.

Of the existing standard designs of brushless machines in the Ansys Electronics Desktop software package, the closest to a claw-poles brushless motor was the design of a BLDC type brushless machine with tangential poles. This brushless machine has a similar armature. The inductors of these machines differ, but there are no electric or magnetic losses in the inductor. The above suggests that in cases of choosing an identical magnetic flux in the inductor of a brushless machine with a tangential excitation system, and a completely identical armature, you can get almost a twin machine for analyzing the main characteristics.

At the same time, BLDC analysis can be performed in a two-dimensional problem statement, which reduces the calculation time to 25-45 minutes.

When writing the program of synthesis and analysis of the projected electric machine with claw-poles, the justified principles of replacement were made at the stage of synthesis and analysis, the description of which is discussed below.

5 Description of the software package for designing brushless machines with claw-poles (CPBM)

Implementation of the task starts with the synthesis. At this stage, the optimal geometry of the machine and the winding data of a real CPBM are determined. The obtained results of optimization calculations based on a pre-prepared script are transmitted to the Ansys Electronics Desktop program. The correctness of the calculated data is checked in the RMxprt mode, which is based on this construction. At the same time, the accuracy of solving the optimization problem is checked for the main parameters (efficiency, losses, flow in the gap, moment, induced by EMF in phases).

If the marked basic parameters differ in the synthesis and analysis subsystem, the procedure is repeated with the correction of these data in the synthesis subsystem, since the analysis subsystem is considered a reference.

After optimization for the main parameters without leaving the RMxprt mode, the real machine with claw-poles CPBM is replaced by a machine analogous to BLDC. For replacement, the data obtained from the claw-poles model is transmitted to the model of a brushless machine with a tangential excitation system. In this case, the internal diameter of the armature of the BLDC machine is selected so that the main magnetic flux passing from the rotor to the stator coincides with the magnetic flux of the CPBM armature.

IOP Conf. Series: Materials Science and Engineering **950** (2020) 012004 doi:10.1088/1757-899X/950/1/012004

At the same time, we check the identity of the characteristics of both machines.

After all tasks are completed, the machine automatically switches from RMxprt mode to Maxwell 2D Design mode. In this mode, the machine geometry is constructed and analysis is performed using the finite element method. The calculation time of the electromagnetic field parameters is 20-40 minutes.

At the final stage, a three-dimensional (3D) model of the claw-poles machine CPBM is analyzed with the corresponding real parameters. The calculation time is several hours, but it is quite acceptable for a one-time calculation.

Below is shown one of the tabs of the developed optimization program implemented in the programming language Delphi for an example of a brushless motor with a claw-like magnetic system with the following parameters: 2.5 MW, 600 V, 40 rpm to drive the pilger mill.

Figure 4 shows the results of the first stage (the synthesis program). The program is an author's development and is written in the Delphi programming language. It determines the optimal geometry of the machine according to the customer's technical specification.

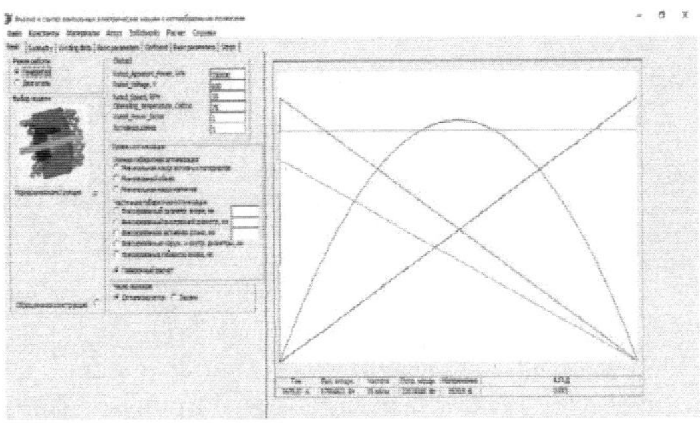

Fig. 4. Program for optimizing the geometry of a brushless machine with a claw-like magnetic system

To check the correctness of the parameters calculated at this stage, they are passed to the Ansys Electronics Desktop program in RMxprt mode using a pre-prepared script.

The results of calculating the claw-poles brushless machine in RMxprt mode are shown in Fig. 5.

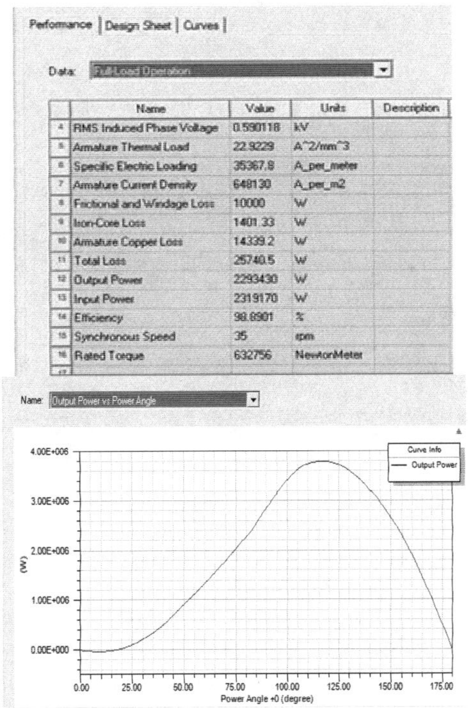

Fig. 5. Results of calculation of the claw-poles brushless machine in RMxprt mode

IOP Conf. Series: Materials Science and Engineering **950** (2020) 012004 doi:10.1088/1757-899X/950/1/012004

To obtain the results of calculating the data of the brushless motor with the tangential inductor (BLDC), we take the same parameters (stator dimensions and winding data) of the corresponding machine with claw-poles (CPBM) as the initial data. In this case, the magnetic flux of the twin machine is selected using the internal diameter equal to the magnetic flux of the machine with claw-poles (CPBM).

The analysis results of the calculated data of the brushless machine with a tangential system in RMxprt mode are shown in Fig. 6.

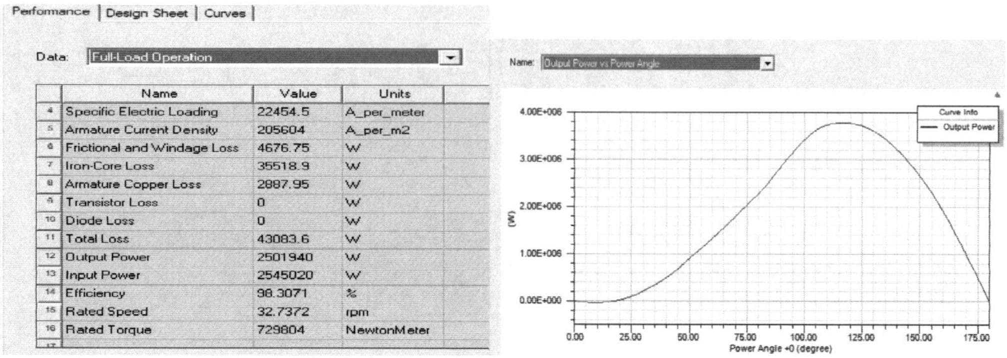

	Name	Value	Units
4	Specific Electric Loading	22454.5	A_per_meter
5	Armature Current Density	205604	A_per_m2
6	Frictional and Windage Loss	4676.75	W
7	Iron-Core Loss	35518.9	W
8	Armature Copper Loss	2887.95	W
9	Transistor Loss	0	W
10	Diode Loss	0	W
11	Total Loss	43083.6	W
12	Output Power	2501940	W
13	Input Power	2545020	W
14	Efficiency	98.3071	%
15	Rated Speed	32.7372	rpm
16	Rated Torque	729804	NewtonMeter

Fig. 6. Analysis result of calculated data of the brushless machine with a tangential system in RMxprt mode

Follow-up actions are performed in Maxwell 2D Design. To perform this action, the design of a brushless DC motor with permanent magnets (BLDC) is deployed in a flat two-dimensional model. This model has an analogy with a machine with claw-poles, i.e. it has the same anchor and magnetic flux. Simulating an electronic circuit for performance analysis is performed in the Ansys Citrix Circuit.

Simulation results of the BLDC motor with tangential magnets in Maxwell 2D Design, as well as the diagrams of its simulator are shown in Fig. 7.

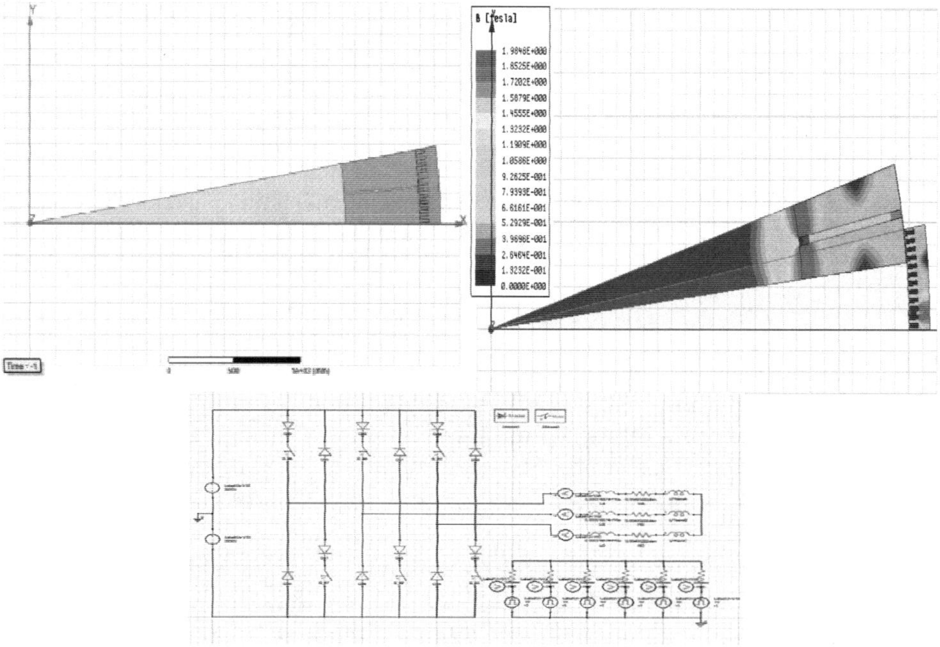

Fig. 7. BLDC motor simulation with tangential magnets in Maxwell 2D Design

The calculated curve of the electromagnetic moment is shown in Fig. 8

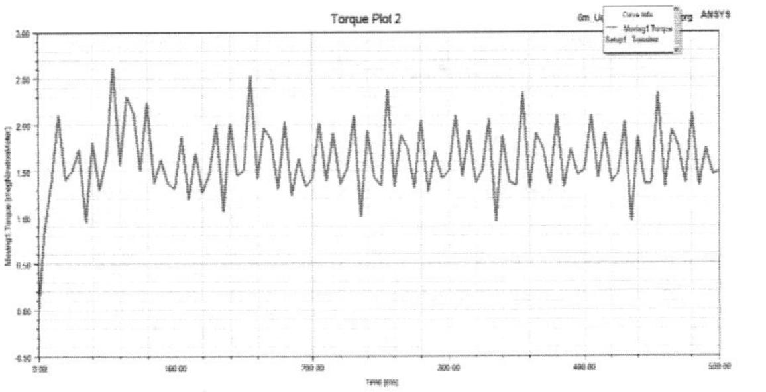

Fig.8. Electromagnetic torque curve

At the final stage, after the final determination of the optimal geometry and winding data, the transition to the RMxprt mode for the machine with claw-poles takes place. From this mode, using the options of the Ansys Electronics Desktop program, a real three-dimensional model of the machine is built for the final analysis and construction of the main characteristics.
Three-dimensional model in Maxwell 3D Design mode is shown in figure 9.

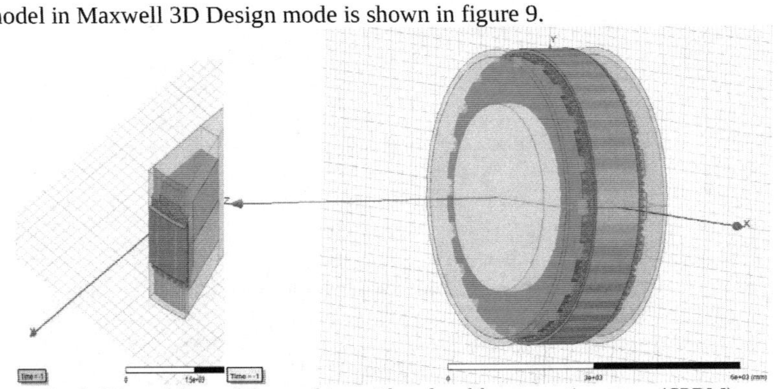

Fig. 9. 3D modeling of the machine with a claw-like magnetic system (CPBM)
Electromagnetic moment curve of the machine with the claw-like magnetic system calculated for a three-dimensional model in Maxwell 3D Design mode is shown in Fig. 10.

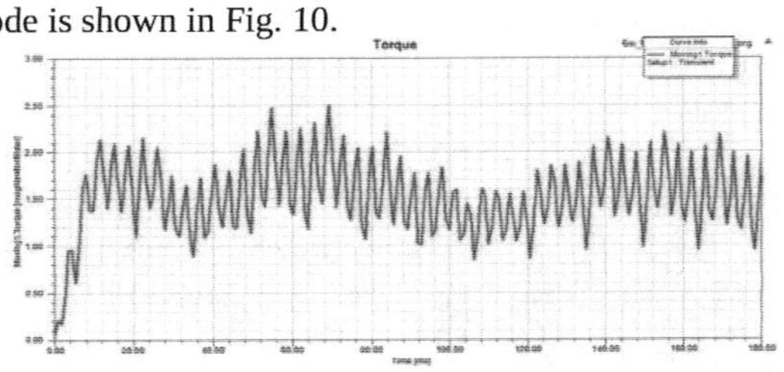

Fig. 10. Electromagnetic moment curve of the machine with a claw-like magnetic system in a three-dimensional model from the Maxwell 3D Design mode

Figure 10 shows that the calculated torque curve of a claw-like magnetic system (CPBM) motor is almost identical to the same curve for the motor BLDC (see figure 8).

Fig. 11 shows the design system stages in the form of the following algorithmic scheme for a clearer understanding of its functioning.

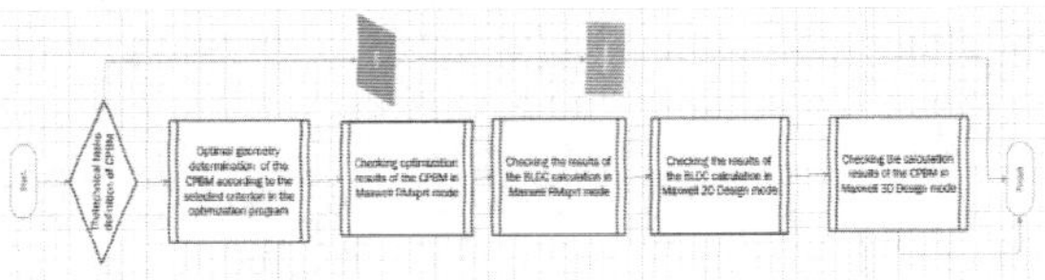

Fig. 11. Algorithmic scheme of the project system functioning.

Thus, using this design system, it is possible to perform synthesis and multi-stage analysis of the electric machine with a claw-like magnetic system.

IOP Conf. Series: Materials Science and Engineering **950** (2020) 012004 doi:10.1088/1757-899X/950/1/012004

6 Thermal analysis of the state CPBM using the Ansys Icepak software package

Calculation of the thermal state of high-power brushless DC motors is a very important step in the analysis. Its evaluation requires the use of precise modern programs based on the finite element method. These calculations are labor-intensive, but they significantly reduce the technical risks before manufacturing a prototype [18-20]. The following positive trend should be noted when creating modern CAE systems. It is now possible to combine different programs into a single model to solve a related problem through the Ansys Workbench shell. Let's use this opportunity and create a calculated thermal model based on the following calculation scheme. Initially, we will perform CPBM electromagnetic state calculation in the program Ansys Electronics Desktop. Based on the results of this calculation, we will more accurately determine the losses in steel and copper of the armature winding. Next, we will pass the data to the Workbench shell. It is a link to the heat calculation program. In the next step, we will pass the data directly to the Icepak thermal field calculation program.
Thus, the related magnetodynamic and thermodynamic tasks are solved. This calculation scheme is shown in Fig. 12

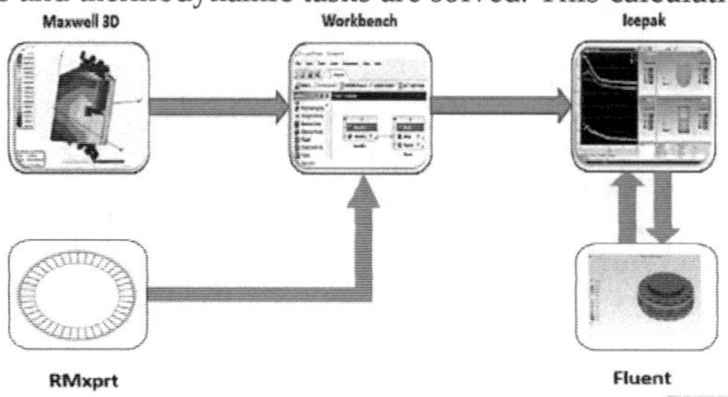

Fig. 12 Thermal calculation scheme

In the Icepak program, the final configuration of the thermal task is performed:
- materials are refined taking into account their thermodynamic properties;
- set the size of the area where the heat exchange is performed;
- set the speed of the cooling agent and the nature of its movement (laminar, turbulent);
- the calculation grid is set up.

Thermal calculation is performed in an iterative way, during which the partition grid is refined into finite elements and the convergence of the computational process is determined.

As a result of the calculation, the temperature in each element that the motor model is divided into is determined. Thus, we can calculate the temperature field in contrast to the integral temperature, which is determined by the method of equivalent thermal substitution schemes. The temperature distribution for the 3D model is shown in figure 13. The figure shows the maximum calculated temperature of 359 degrees Kelvin. When converted to degrees Celsius, this value will be 86 C^0.

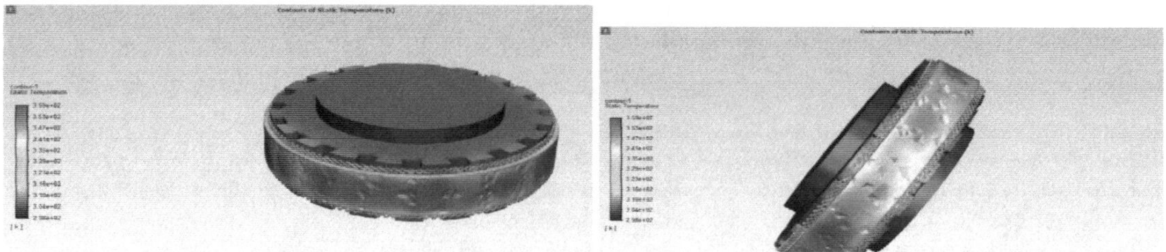

Fig. 13. Temperature field of CPBM calculated in the Ansys Icepak software package

Thus, thermal calculation by the finite element method gives an accurate picture of the CPBM heating. Positive simulation results finally remove technical risks for development results of CPBM that allows you to move onto the final stage – the creation of three-dimensional solid models and developing design documentation for the creation of a prototype.

Analysis of results. Modern applications of the CAE system allow you to calculate and analyze the parameters and characteristics of electrical machines.

It should be noted that these programs require very high performance and large computer resources. Not every engineer and researcher can afford such opportunities. One of the main problems during the calculation is the duration of the calculation process.

Practice shows that the developer of electric machines does not always need to complicate the calculation model. To obtain accurate results of three-dimensional objects can be represented in the form of two-dimensional models that do not require large computational resources.

This approach was applied to the brushless machine with a claw-like magnetic system. After the synthesis stage, where the simplified model was optimized with a substitution scheme, the results of optimization data were analyzed in the Ansys Electronics Desktop software environment, where the problem of calculating the magnetic field for a two-dimensional model (Maxwell 2D Design) and a three-dimensional model (Maxwell 3D Design) was solved sequentially.

Differences in the main parameters (efficiency, electromagnetic moment, speed, magnetic flows, rotation EMF) of the design system functional scheme for the brushless machine with a claw-like magnetic system are presented below:

the error between the calculated parameters obtained in the Delphi and RMxprt program for the machine with claw-poles was from 5 to 9 %;

the error between the results in RMxprt for the machine with claw-poles and the RMxprt results for a brushless machine with tangential magnets was 3-5 %;

the error between the results of the calculation in Maxwell 2D mode for the brushless motor with tangential magnets and the results of the calculation in Maxwell 3D for the machine with claw-poles was 5-7 %;

the error between the synthesis stage results and the analysis stage results in the Maxwell 3D Design machine with claw-poles was 9-12%.

The resulting calculation accuracy for such a complex magnetic system as the machine with claw-poles should be considered a good result. This approach can be recommended for the development of other types of electric machines that do not have flat symmetry.

7 Conclusion

Morally and physically obsolete equipment of metallurgical production requires replacement with drives that contain innovative solutions. The money invested in modernization takes a long time to pay off, and not every production can afford to replace it frequently. This raises the challenge of introducing advanced innovative solutions that will work for a long time.

The article offers a technical solution that has not yet been used in the world. As a drive motor for the pilger mill drive for the seamless pipes manufacturing is proposed to use a low-speed direct-drive brushless DC motor with a large diameter and the required inertial mass with claw-poles and a powerful permanent magnet as the drive motor for pilger mill. This solution allows you to remove the existing 120-ton flywheel by reducing the dimensions, since the necessary inertial mass is embedded in the rotor, increase efficiency by eliminating excitation losses, and reduce operating costs due to the absence of a collector unit. This offer contains a large number of technical risks. For their reduction have been developed the design system for the creation of electric machines of this class. The system contains two main parts: a synthesis subsystem and an analysis subsystem. The synthesis subsystem determines the optimal geometry of the machine based on the implementation of the multi-level single-criteria optimization problem based on the Gauss-Seidel method, the Fibonacci method, and the punitive function method. The analysis subsystem is a multi-step process using the well-established CAE complex Ansys Electronics Desktop. At the same time, minimal computer resources and estimated time are used, since at the intermediate stage a flat problem is analyzed, and only at the final stage a three-dimensional problem is analyzed. Also, the three-dimensional model thermal analysis of the machine by the finite element method in the program Ansys Icepak confirmed the possibility of using CPBM as a pilger mill drive.

The design system was actively used in the pilger mill drive development for the seamless pipes manufacturing at the Chelyabinsk pipe rolling plant.

The work was performed at SUSU with the financial support of the Russian Science Foundation (project № 14-19-00327). The author is grateful to the University leadership for the opportunity to conduct this research

Acknowledgment

The work was carried out at the SUSU with the financial support of the Russian Science Foundation (project № 14-19-00327). The authors thank the officials for the opportunity to carry out this study.

References

1. A.F. Desanti, I. Sidharta, H. Erwantono and others, “Design of Performance and Parameter Measurement System for Brushless Direct Current (BLDC) Motor”, Proceeding - 2018 International Seminar on Intelligent Technology and Its Application, ISITIA 2018, pp. 175-179.
2. H.-S. Park, S.-W. Park, D.-Y. Kim and J.-M. Kim, “Hybrid phase excitation method for improving efficiency of 7-phase BLDC motors for ship propulsion systems”, Journal of Power Electronics, 2019, 19(3), pp. 761-770.

3. S. Gandzha, D. Aminov and B. Kosimov, “Design of Brushless Electric Machine with Axial Magnetic Flux Based on the Use of Nomograms”, 2018, Proceedings - 2018 International Ural Conference on Green Energy, UralCon 2018 , 8544320, pp. 282-287.
4. Gandzha, S.A. The choice of the optimal design of the electric motor of the Pilgerstan drive for the technology of manufacturing seamless pipes. Gandzha, B.I. Kosimov, D.S. Aminov // Bulletin of SUSU. Series "Energy". - 2019. - T. 19, No 1. - P. 5-17. DOI: 10.14529 / power190101.
5. Gandzha, S.A. Comparative Analysis of Pilgerstan Drive Electric Motors for Seamless Pipe Manufacturing Technology. The choice of the optimal design / S.A. Gandzha, B.I. Kosimov, D.S. Aminov, R.R. Nimatov // Bulletin of PNRPU “Electrical engineering, information technologies, control systems”. - 2019. No 30, - P. 79-101
6. 6. R. Omri, A. Ibala and A. Masmoudi, “Characterization on the no-and on-load operations of an improved claw pole machine”, 2018, 13th International Conference on Ecological Vehicles and Renewable Energies, EVER 2018, pp. 1-8.7.
7. R. Rebhi, A. Ibala and A. Masmoudi, “MEC-based sizing of a hybrid-excited claw pole alternator”, 2015, IEEE Transactions on Industry Applications, 51(1), 6837424, pp. 211-223.8.
8. S. Gandzha, D. Aminov, and B. Kosimov, “Development of engineering method for calculation of magnetic systems for brushless motors based on finite element method”, 2019, International Conference on Industrial Engineering, Applications and Manufacturing, ICIEAM 2019. 8742976.9.
9. F. G. Zhang, H.-J. Bai and Y. Liu, “Leakage magnetic calculation on claw pole machine with outer permanent magnet rotor”, 2009, Dianji yu Kongzhi Xuebao/Electric Machines and Control , 13(4), pp. 548-552.10.
10. A. Njeh, H. Trabelsi, “New design of the claw-pole transverse flux permanent magnet machine”, 2018 15th International Multi-Conference on Systems, Signals and Devices, 8570579, pp. 1311-1316.11.
11. H. Bai, S.D. Pekarek, J. Tichenor and others “Analytical derivation of a coupled-circuit model of a claw-pole alternator with concentrated stator windings” 2002, IEEE Transactions on Energy Conversion 17(1), pp. 32-38.12.
12. B. Cristian, O. Constantin, O.Chiver and others “The advantages of numerical analysis for claw pole alternator”, 2014, EPE 2014 - Proceedings of the 2014 International Conference and Exposition on Electrical and Power Engineering, 6969928, pp. 353-357.13.
13. F. Jurca, C. Martis, “Claw-Pole generator parameters and steady–state performances analysis”, 2013, International Review on Modelling and Simulations , 6(1), pp. 41-48.14.
14. Bramerdorfer, G. Tolerance Analysis for Electric Machine Design Optimization: Classification, Modeling and Evaluation, and Example // IEEE Transactions on Magnetics – 2019, – 55(8), – 8688464.15.
15. Wang, Q., Li, J., Qu, R., Lu, Y. Design and Optimization of a Permanent Magnet Synchronous Machine for Low Vibration and Noise Applications // ICEMS – 2018 21st International Conference on Electrical Machines and Systems – 8549128, – P. 280-284.16.
16. Gandzha S., Bakhtiyor K., Aminov D. Development of a system of multi-level optimization for Brushless Direct Current Electric Machines // International Ural Conference on Electrical Power Engineering (Ural Con) 2019. 1-3 Oct. 2019 Chelyabinsk,Russia. DOI: 10.1109/URALCON.2019.8877650.17.
17. Gandzha S., Aminov D., Bakhtiyor K. Development of Engineering Method for Calculation of Magnetic Systems for Brushless Motors Based on Finite Element Method // 2019 International Conference on Industrial Engineering, Applications and Manufacturing (ICIEAM), ECF Paper 25-29 March 2019 Sochi, Russia, DOI: 10.1109 / ICIEAM.2019.8742976.18.
18. S. Gandzha, D. Aminov, I. Kiessh and B. Kosimov, “Application of Digital Twins Technology for Analysis of Brushless Electric Machines with Axial Magnetic Flux”, Proceedings – 2018 Global Smart Industry Conference, GloSIC 2018, 8570132.19.
19. P. Zheng, Q. Wu, J. Zhao and others “Performance analysis and simulation of a novel brushless double rotor machine for power-split HEV Applications”, 2012, Energies, 5(1), pp. 119–137.20.
20. Gandzha S., Kosimov B, Aminov D. Application of the Ansys Electronics Desktop Software Package for Analysis of Claw-Pole SynchronousMotor // Machines,-2019 / – https://doi.org/10.3390/machines7040065

IOP Conf. Series: Materials Science and Engineering **950** (2020) 012005 doi:10.1088/1757-899X/950/1/012005

Development of an integral criterion for evaluating the degree of aging of transformer oils

I Davidenko, A Egorov

Department of electrical machines Ural Federal University named after the first President of Russia B. N. Yeltsin.
Ekaterinburg, Russia

i.v.davidenko@urfu.ru

Abstract. The technical condition of the transformer oil affects the condition of the solid insulation, and therefore the residual life of electrical equipment, so it is important to study the aging mechanisms of the oil. The mechanism of transformer oil aging is studied using factor analysis of oil quality indicators, such as moisture content, acid number and dissipation factor tangent. Factors that have the greatest influence on the rate of oil aging have been found. In accordance with the found influence factors the permissible values of oil quality parameters of 110 kV power transformers with division by equipment tightness, oil brands and operating time have been determined. Oil aging models, taking into account the mutual influence of oil quality parameters, have been developed for transformers with different oil brands, type of oil protection and service life. An integral index, characterizing different aging processes of the oil, has been proposed. The criterion for evaluating the integral index of oil aging rates is calculated.

1. Relevance

The task of precise evaluation of technical condition of power system assets and management of this condition is always relevant. As the proportion of equipment with a long service life increases, the importance of this task only grows.

It is known that the reliability of oil-filled electrical equipment is primarily determined by the condition of its solid insulation, which is in constant interaction with the transformer oil. The quality of transformer oil affects the state of its solid insulation [1,2,3], so it is important to study the mechanisms of oil aging.

The sets of parameters characterizing the quality of transformer oil in national standards vary [4,5,6]. However, the moisture content of the oil (H_2O, g/t), the acid number (KOH, mgKOH/g) and the dielectric loss tangent (Tg 90°C, %) are included in international and national standards. These oil quality parameters describe the main processes that characterize the ageing of an oil: moisturizing, oxidizing the oil as well as the formation of colloidal compounds in it.

In our opinion, the existing approach to assessing these indicators has two drawbacks. First, the system regulating these indicators is not differentiated by the factors affecting the values of the parameters. The works [7,8,9] show that such factors are tightness of the equipment, oil grade, voltage class [7,8], load and temperature mode of the equipment, as well as its service life [7,9].

Secondly, in existing international and national standards [4,5,6] regulating values of oil quality parameters, parameters are considered independently from each other. Meanwhile, the authors [10,11] noted a significant correlation between these parameters.

2. Purpose of work

The aim of the work is to study the force of mutual influence of power transformers' (PT) oil quality parameters, as well as to develop an integral index, describing the influence of several processes, characterizing the oil aging at once.

This study is based on observation of real changes in the technical condition of 110 kV transformer oil during its operation.

3. PT operation data

The results of physicochemical analysis of the oil, accumulated in the database of the Expert Diagnostic Information System (EDIS "Albatross") for 30 years of its operation at power grid enterprises of the Russian Federation, were used as the initial data.

The initial sample for 110kV class voltage transformers was 32565 results of physicochemical oil analysis. In our previous studies [7,8,11] it was found that the quality characteristics of the oil (its aging rate) are influenced by tightness, voltage class, service life of oil-filled equipment and the content of aromatic hydrocarbons (Ca) in the oil. Therefore, the initial sample was divided according to the type of oil protection (free breathing and membrane protection) and oil brands. Ten oil brands are used in the Russian power transformer fleet, which were divided into three groups depending on the aromatic hydrocarbon content:

- with a minimum content Ca *min* ≤ 3%;
- with an average content 3 < Ca *mid* ≤ 15%;
- with a maximum content Ca *max* ≥ 18%.

The fourth group was transformers with a long service life, in which a mixture of different oil brands was formed by refilling oils different from the original oil (Oil mix).

4. Preparing data for analysis

First, the subsamples of the measurement results of the oil moisture content (H_2O, g / t), acid number (KOH, mgKOH / g) and dielectric loss tangent (Tg 90 °C, %) were cleared of errors in data entry by personnel.

Then, we needed to normalize the parameters so that they are presented equally in the factor analysis.

For normalization, permissible values (PV) of parameters were determined for each subsample by the integral distribution function at the level of 0.95.

Table 1 shows the calculated values for the most representative sample of PT 110 kV with free breathing and a group of oil brands with average aromatic hydrocarbon content of Ca *mid*.

Table 1. Permissible oil quality values of Ca *mid* group for power transformers 110 kV with free breathing

Parameter	H_2O (g / t)	KOH (mgKOH /g)	Tg 90 °C (%)
PV of parameter	28	0.06	4.9

Previous studies [7,8,9] have shown an increase in the values of these parameters as the lifetime of the transformer increases. To study the mutual influence of the parameters, we used for their normalization PV without differentiating them according to the operating life, so as not to introduce distortions into the original data array.

Further, the analyses in which the value of at least one of the three parameters was absent were removed from the investigated subsamples.

Oil breakdown voltage was excluded from the analysis of mutual influence of parameters, as this parameter is an indicator of oil moisture and contamination. As a rule, the breakdown voltage duplicates the oil moisture content of the oil, since oil contamination in modern power transformers is higher than the permissible values, this phenomenon is extremely rare.

5. Analysis of mutual influence of oil quality parameters

The correlation analysis given in the articles [10, 11] showed mutual influence of the investigated oil quality parameters (H_2O, KOH, Tg90 °C).

The task of the study is to find an integral indicator of oil aging, linking the values of three studied parameters (H_2O, KOH, Tg90 °C). The factor analysis is suitable for this purpose as it is a multivariate statistical analysis that combines methods for estimating observed variables by studying the structure of covariance or correlation matrices.

Recall that the initial sample of parameter measurements (H2O, KOH, Tg90 ° C) divided by the type of oil protection and oil brands.

Then, for each of the obtained subsamples, a factor analysis was performed with a breakdown by service life into subgroups of 6 years.

As a result of the analysis, the following equations were obtained:

$$I=k_1\cdot x+k_2\cdot y+k_3\cdot z \qquad (1)$$

where k_1, k_2, k_3 are the force coefficients of the parameters' influence on the oil aging rate;
x, y, z - normalized values of investigated parameters (H_2O, KOH, Tg90°C).

The graphs shown in Figure 1 have been used to analyze changes in the coefficients of parameters' influence on the oil aging rate depending on the operating time of transformers and the aromatic hydrocarbons content in their oil.

IOP Conf. Series: Materials Science and Engineering **950** (2020) 012005 doi:10.1088/1757-899X/950/1/012005

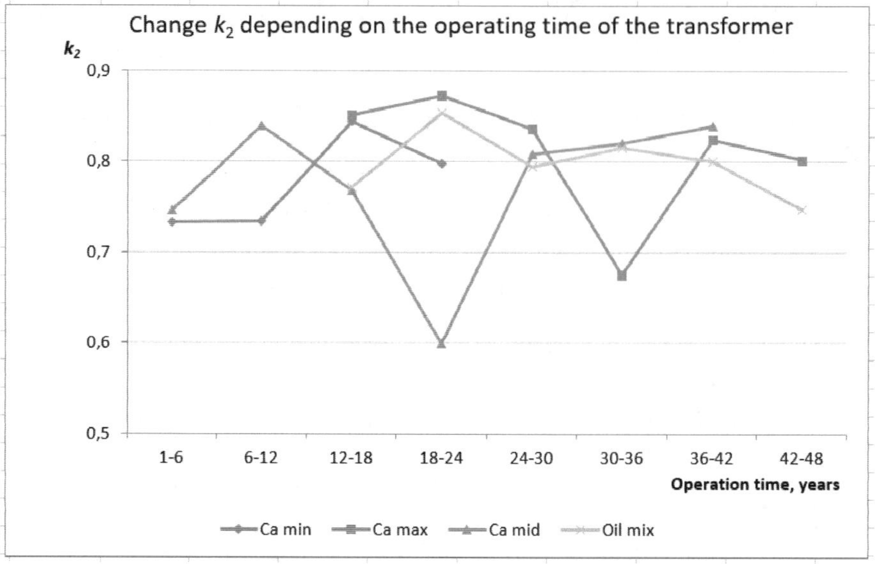

Figure 1. Change of force of influence of KOH value on oil aging rate depending on transformer operation time and aromatic hydrocarbons content in oil.

A minimal variation in the values was observed for the coefficient k2 of the acid number: 0.6-0.88. The maximum variation in the values was for the coefficient k1 of moisture content of the oil: 0.32-0.82. The variation in the values of the coefficient k3 of the tangent of the angle of dielectric loss of oil was: 0.56-0.86.

Based on the analysis of the variability k1, k2, k3 depending on the operating time PT, the following conclusions were made. With age, the strength of the mutual influence of oil aging products increases. After the operational measures (drying of the oil, change of silica gel in the thermosiphon and air-drying filters, oil regeneration, addition of antioxidant additives to the oil), the aging process slows down. During this period, the synergy of the processes of oxidation, moistening of the oil, and the formation of colloidal forms in it weakens. Then, the aging rate of the oil increases until the next operational measures. The second period of oil deterioration is shorter than the first.

Analysis of the variability k_1, k_2, k_3 depending on the content of aromatic hydrocarbons in the oil showed that the oils of the Ca *min* group are more slowly oxidized but moisturized faster than the oils of the Ca *mid*, Ca *max* groups. In addition, with longer service lives, more colloidal forms are formed in them than in Ca *mid*, Ca *max* oils. Mixtures of different brands of oils have the highest rates of moisture and oxidation, as well as the formation of colloidal forms.

6. Criteria for evaluating oil quality by integral criterion

The task of the next stage of the study was to obtain the value of integral criterion for oil quality assessment.

Recall that at the previous stage of the study for each of the data sub-samples were determined values of force coefficients of the influence of parameters on the aging rate of oil (k_1, k_2, k_3).

For example, for a 110 kV power transformer with free breathing with oil brands Ca *mid* with an operating life of 12 to 18 years, the integral value was calculated using a formula:

$$I = 0.58 \cdot x + 0.8 \cdot y + 0.73 \cdot z \tag{2}$$

x, y, z - normalized values of investigated parameters (H_2O, KOH, Tg90°C) in operation.

After calculating all the values of the integral index of the subsample I, its permissible value I_{pv} was calculated from the integral distribution function at the level of 0.95.

The value of $I_{pv} = 2.2$ is a criterion that separates cases of accelerated deterioration in the quality of the oil from its normal aging in operation.

The following figures show graphs of oil quality parameters of two transformers operating in the same substation with the same load and operating time. And the value of integral index at transformer T1 is greater than that of transformer T2. As we can see, the trends of deterioration of oil quality parameters (H_2O, KOH, Tg90°C) of transformer T1 are higher than those of transformer T2.

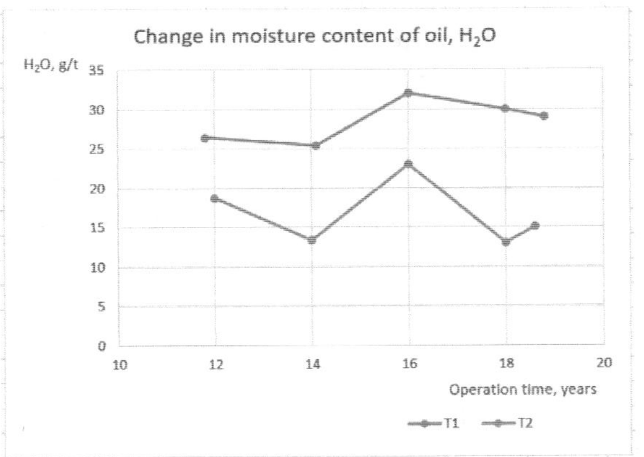

Figure 2. Change in moisture content of oil, H_2O.

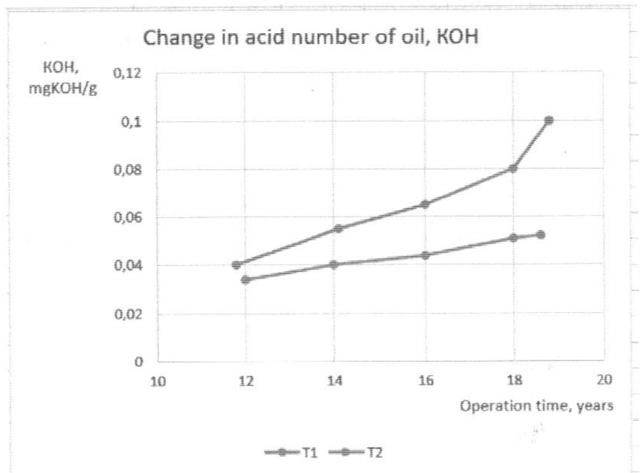

Figure 3. Change in acid number of oil, KOH.

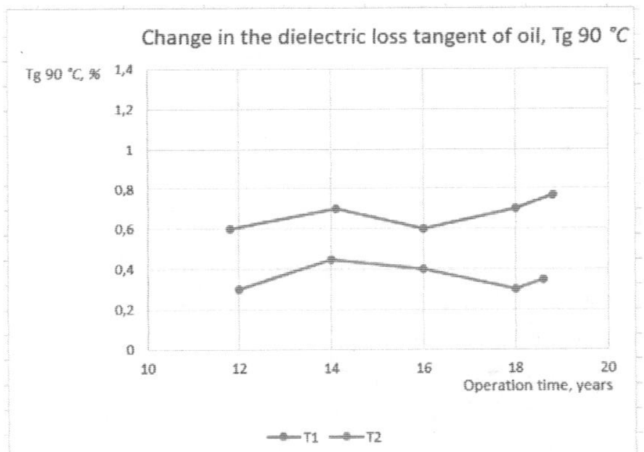

Figure 4. Change in the dielectric loss tangent of oil, Tg 90 °C.

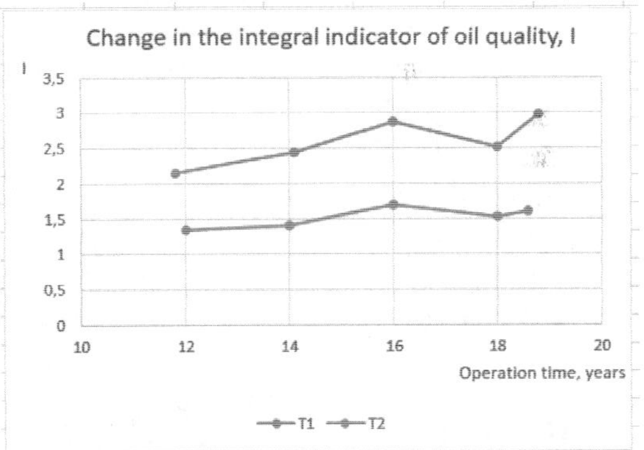

Figure 5. Change in the integral indicator of oil quality, I.

Accordingly, the assessment of oil quality by an integrated indicator allows us to identify transformers with high rates of oil aging and put them in a timely manner on frequent control. Evaluation of the parameters H_2O, KOH, Tg90 ° C individually can lead to the fact that the transformer with rapidly degrading oil will be monitored at regular intervals (1 time in 2 years).

Detection of deep oil degradation will only be possible after 2 years, when it will already have a devastating effect on the solid insulation of the transformer.

7. Conclusion

The research confirmed the necessity of differentiation of oil quality assessment criteria (H_2O, KOH, Tg90°C) by equipment tightness, aromatic hydrocarbons content in oil, operating time. Permissible values for assessment of parameters H_2O, KOH, Tg90°C with necessary differentiation have been determined.

Oil aging models including H_2O, KOH, Tg90°C parameters and their mutual influence were found for transformers with different oil grades, type of oil protection and operating life.

An integral index characterizing the oil aging processes such as moistening, oxidation, colloidal formation is proposed. The criterion for evaluating the integral index of oil aging rates has been calculated.

The obtained dependencies (formula 1) describing the oil aging process as well as the criterion of integral index evaluation can be used with the help of predictive analytics for medium-term planning of oil regeneration (replacement), determination of optimal frequency of oil quality control.

References

[1] Lipstein, R.A ., Shakhnovich, M.I.Transformer oil, Moscow, Enegry Publ., 1968. 352 p.

[2] Sokolov, V.V. Problems of estimation and reliability assurance of the power transformers [Text] / Sokolov, V.V.; Lukashchuk, V.A. // Sokolov, V.V.; Selected works / Ovsiannikov, A.G.; Osotov, V.N.; Berezhnoy, V.N. - Ekaterinburg: Autograph Publishing House, 2010.- S. More references

[3] CIGR brochure AGEING OF IMPREGNATED CELLULOSE FOR TRANSFORMERS WG D1.53

[4] IEC 60422 Mineral insulating oils in electrical equipment - Supervision and maintenance guidance. Third edition 2005-10

[5] Volume and test standards for electrical equipment. RD 34.45-51.300-97, Moscow, "Publishing house NTs ENAS", 2001g.

[6] IEEE Std C57.106-2006 IEEE Guide for Acceptance and Maintenance of Insulating Oil in Equipment

[7] Davidenko I.V., Kolushev D.N., Shyrokov A.V. Accounting of Physicochemical Characteristics Interference in Multiple-Aspect Assessment of Oil Condition. 2012 IEEE International Conference on Condition Monitoring and Diagnosis – Bali, Indonesia, September 23-27, 2012, ISBN 978-1-4673-1020-8, IEEE Catalog Number: CFP1230D-ART, pp. 361-364.

[8] Irina V. Davidenko, Alekscandr A. Egorov Determination of criteria for assessing the oil quality of current transformers type TFZM 110 kV 2020 IEEE Conference of Russian young researchers in electrical and electronic engineering (2020 ElConRus) S-Peterburg 2020.

[9] Improvement of the decision-making procedures for the transformer oils ageing estimation (in Russian) / Bondarenko V.E., Shutenko O. V. (in Russian) // Elektro. Electrotechnics. Elektroenergetika. Electrotechnical industry. - 2009. - N 1. - C. 17-21 : silt, table. - Bibliogr. : p. 21 (8 pers.) .

[10] Shutenko O.V. COMPLEX CORRELATION ANALYSIS OF QUALITY POSITIONS TRANSFORMATORY OIL Complex analysis of quality parameters in the transformer oil National Technical University "Kharkiv Polytechnic Institute". 2008 http://repository.kpi.kharkov.ua/handle/KhPI-Press/34551 Shutenko O. V. Complex correlation

analysis of the transformer oil quality indexes (in Russian) / O.V. Shutenko // Vestnik, National Technical University "KPI". : Collection of scientific papers Theme. ep. Problems of improving electrical machines and devices. - Kharkov : NTU "KPI", 2008. - – № 45. - – C. 161-172.

[11] Almukhametov I.I., Davidenko I.V. Study of the mutual influence of transformer oil parameters. Proceedings of the first scientific and technical conference of young scientists of the Ural Power Institute. May 16-20, 2016. pp. 167-170. RSCI.

IOP Conf. Series: Materials Science and Engineering **950** (2020) 012006 doi:10.1088/1757-899X/950/1/012006

Direct measurement of the current and power in the load loop of high -frequency electrotechnological devices using the Rogovsky coil

E.D. Shabaldin, V.E. Frizen, D.N. Tomashevskiy
Ural Federal University, Ekaterinburg, Russian Federation

vfrizen@yandex.ru
Abstract. The article describes the power supply system of induction furnace based on a semiconductor converter, in which it is possible to work simultaneously in the exciting mode of low-frequency currents and frequencies three times higher when switching the compensation circuit of the furnace inductor at different stages of metal melting. Calculations of electromagnetic processes have shown the advisability of operating the power supply system in the mode of tripling the current frequency at the stage of heating and melting of the metal, but in the mode of low frequency - at the stage of technological processing of the melt. The ratio between the parameters of the furnace power supply in various operating modes are established.
Keywords: Rogovsky coil, high–frequency converters, inductors, magneto-hydrodynamic (MHD) devices, induction heaters, melting furnaces.

1. Introduction

Inductors of MHD installations, induction heaters and melting furnaces are often powered by frequency converters. The frequency range of currents flowing through the inductors is in the range from fractions of Hz to several MHz. Sensors and devices used in industry for measuring electric current and power usually operate in a fairly narrow frequency range. High-frequency sensors are very expensive and often their design does not allow them to be integrated into the power supply system of induction installations, if this was not provided for during its manufacture. This problem often occurs when conducting experiments and simulating processes in research and training laboratory facilities, when verifying numerical models, and in physical modeling. The use of mass-produced sensors for these purposes is often difficult due to the fact that the measuring system during the experiment must be mounted on rigid non-disassembled busbars. An important condition is also the low price and availability of components of the measuring device.

2. Statement of the problem

The problem of direct measurement of current and power arose during the research of a laboratory induction melting plant, the scheme of which is shown in Fig. 1.

Traditionally, in such installations, power is measured in the direct current link of the frequency converter (FC), but the measured power includes losses in the semiconductor devices, the busbars, and the matching transformer. The value of the inductor current can be determined by calculating the value of the voltage drop on the compensating capacitor or by recalculating the measured current of the inverter through the transformation coefficient of the matching transformer. However, both methods are indirect and their accuracy depends on the accuracy of determining the parameters of the elements used in the measurements.

When measuring a large current directly, a method is usually used to measure the magnetic fields that occur near the conductor through which it flows.

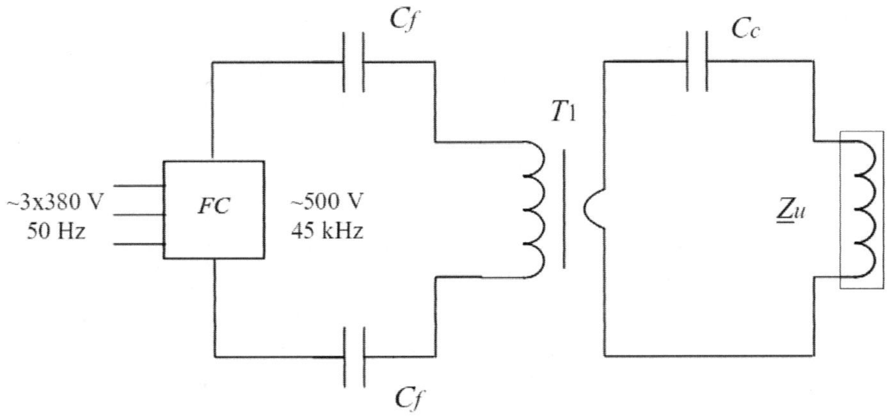

Figure 1. Electrical diagram of the device under study: *FC* - frequency converter; C_f - filter capacitor; C_c - compensating capacitor; T_1 – high frequency matching transformer; $\underline{Z}_u$ – the impedance of the inductor/

Magnetic fields occupy an important place in physics and engineering. In table 1 the main methods and sensors for measuring these fields are given [1].

Table 1. Comparative characteristics of magnetic field sensors

Sensor type	Advantages	Disadvantages
Induction probe	• simple design;	• variable magnetic fields are registered only; • a large errors at high frequencies and in strong pulse fields;
Rogovsky coil	• simple design; • the EMF value does not depend on the current distribution across the contour section;	• the sensor winding must be sectioned at high frequencies; • difficulties in registering short pulses;
Ferromagnetic probe	• relative ease of manufacturing; • high stability of operation;	• can be used for weak magnetic fields only; • signal amplification is required;
Hall effect	• records constant and variable magnetic fields in a wide frequency range; • compactness;	• a stable power supply is required; • low-noise amplifier is required.

One approach to measuring hight currents is to estimate their magnitude by the magnetic field that occurs around a conductor with a current. These measurements are non-contact, which is extremely important in the case of high voltages and hight currents. The magnetic field of a current that changes over time can be measured using an induction field sensor. However, it is only possible to associate a magnetic field measured at a single point with a current in rare cases when the connection of the magnetic field with the current in a conductor is well known, for example, in a long solenoid. If you measure the current flowing through a conductor of a complex cross-section, it is extremely difficult to establish this relationship. For pulsed and high-frequency currents, the problem is compounded by the inhomogeneous distribution of current density across the cross section due to the skin effect.

For these reasons, an integral inductive sensor, called the Rogovsky coil, is widely used [3, 4, 6-10]. According to the second Maxwell equation in integral form, the total current I_t through a pad bounded by a closed loop *L*, is associated with the circulation of a magnetic field along the boundary of this loop:

$$I_t = I_C + I_S = \oint_L \mathbf{H} dl \tag{1}$$

where **H** - is the magnetic field strength vector, $I_c u I_s$ - are the conduction and displacement currents through the closed loop *L* [1].

Measurements of the alternating magnetic field are usually performed at relatively low frequencies, when the characteristic dimensions of the conductor and circuit are much smaller than the wavelength of electromagnetic radiation. Under these conditions

$$I_t \approx I_C . \tag{2}$$

Measuring the tension integral (1) allows you to determine the value of the conduction current. It is possible to replace the contour integral with a sum along the contour:

$$I_C \approx \sum_i H_i \cdot \Delta l_i \tag{3}$$

where the vectors $\Delta l_i \Delta l_i$ form a closed polyline approximating the contour *L*. To determine the conduction current, it is possible to measure the magnetic field at a large number of points located evenly along the contour, and then use the expression (3).

Since the circulation of the magnetic field in the right part of expression (1) is determined only by the total current through the cross section , the EMF taken from the Rogovsky coil does not depend on the current distribution over the cross section of the contour *L*.

Consider the case when the Rogovsky belt is made on a frame of a dielectric tube with radius *r*, which forms a ring with radius *R*. The winding is performed evenly, and the total number of turns is equal to *N*. Using (3), it is possible to calculate the value of the current derivative over time

$$\frac{di}{dt} = \frac{2R \cdot \varepsilon(t)}{r^2 \mu_0 N} \tag{4}$$

as a function of the EMF ε, that occurs on the sensor.

The disadvantage of the Rogovsky coil for direct measurement of currents in high-frequency induction installations is the frequency dependence of the EMF obtained on its conclusions. Therefore frequency correction of the received signal to use this sensor is required [5, 9].

The correction circuits used with the sensor have a non-linear frequency response. Thus, when correcting, it is possible to obtain the signal amplitude and shape with sufficient accuracy, but the initial phase of the received signal may differ from the initial phase of the current that creates a magnetic field near the conductor. The use of phase-correcting circuits complicates the overall scheme and reduces its accuracy. This makes it impossible to use only the Rogovsky coil with a corrective amplifier as the primary current sensor in a direct power measurement system in HF circuits. To obtain an accurate signal of the current phase, it is proposed to use a Hall sensor.

3. The description of the design and measurement algorithm

The block diagram of the measuring device is shown in Fig. 2.

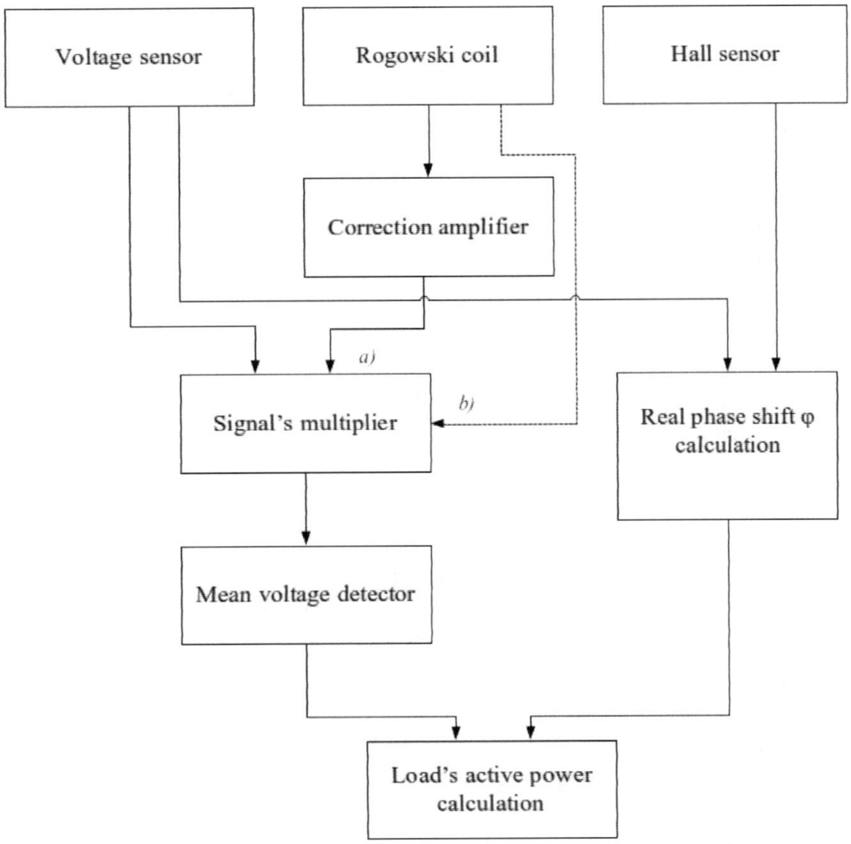

Figure 2. Block diagram of a sensor for measuring current and power
a – with the use of a corrective amplifier ; b - without the use of a corrective amplifier

The design of the Rogovsky coil is made by a removable, single-layer winding on a dielectric frame, filled with a compound (Fig. 3). Next to the winding is a Hall sensor that duplicates the current measurement.

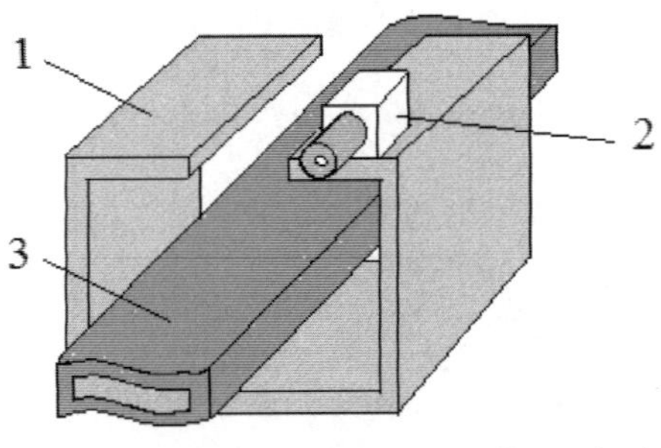

Figure 3. Placing the Rogovsky coil on the busbar:
1 - the coil; *2* - the Hall sensor; *3* - the water-cooled busbar

IOP Conf. Series: Materials Science and Engineering **950** (2020) 012006 doi:10.1088/1757-899X/950/1/012006

To calculate the active power of the inductor in the frequency range from 30 to 50 kHz it is assumed to use the following algorithm:

1. The received signal from the Rogovsky coil goes to the correction amplifier. The correction amplifier is designed to bring the output characteristic of the sensor to a form close to linear.
2. The signal from the correction amplifier is compared in phase with the signal of the Hall sensor, which is a reference, because the Hall EMF repeats the form of the magnetic induction signal.
3. Based on the data obtained, in order to compensate for the error of phase shift of the signal at the output of the correction amplifier, a table of correction coefficients is constructed for calculating the power-frequency dependence.
4. Signals from the outputs of the correction amplifier and the load voltage divider are sent to an analog signal multiplier (i.e., AD834).
5. The signal from the multiplier is fed to a digital oscilloscope, where the average power value for the period is calculated (the instantaneous power signal is processed).
6. The received power signal is corrected according to the developed algorithm using previously obtained correction coefficients for the variable frequency.
7. The correction coefficient equation is calculated analytically depending on the frequency.

In a particular case, current and power measurements in a high-frequency power oscillating circuit, the current and voltage waveforms are almost sinusoidal. In these conditions, if the frequency range is limited, it is possible to refuse to use a corrective amplifier. Then the correction factor when calculating power will take into account both the actual current amplitude and its initial phase (Fig. 2, b).

4. Numerical model of the system

Using the reference sensor in the described system is difficult , so it was decided to calibrate the measuring system using a computer model of the measurement system. When solving this problem, it was necessary to determine the value of the EMF induced on the coil of the Rogovsky coil at different frequencies, as well as the initial phase of induction near the current-conducting bus on which the sensor is installed.

The solution of the problem of determining the parameters of the magnetic field near the current-conducting bus was carried out in a two-dimensional setting in the frequency domain (all differential field parameters change over time according to the harmonic law with a fixed frequency). Solution of the field differential equations in partial derivatives

$$\begin{cases} \frac{\partial}{\partial x}\left(\frac{1}{\mu_y}\frac{\partial A_z}{\partial x}\right)+\frac{\partial}{\partial y}\left(\frac{1}{\mu_x}\frac{\partial A_z}{\partial y}\right)-i\omega\gamma A_z=-j_0; \\ \mathbf{B}=\nabla\times\mathbf{A}; \\ \mathbf{E}=-j\omega\mathbf{A}; \\ \mathbf{J}=\sigma\mathbf{E}+j\omega\mathbf{D} \end{cases} \tag{5}$$

was performed using the finite element method in the Comsol program.

The result of calculating the field for one of the frequencies is shown on Fig. 4.

When conducting numerical experiments in the range from idling to the minimum permissible load resistance of the Rogovsky coil the following values of the sensor voltages at different frequencies were obtained.

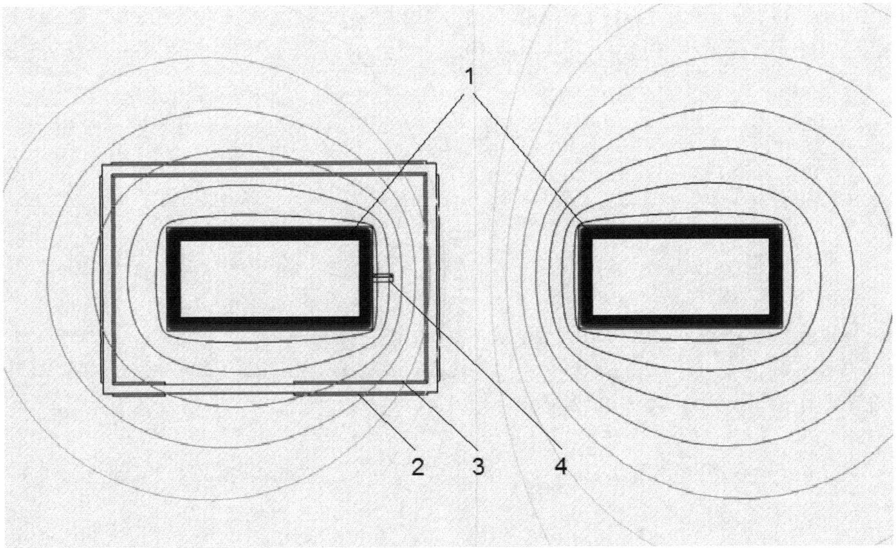

Figure 4. The distribution of the magnetic field near the current-conducting bus and installed sensors: 1-current busbars; 2, 3- the outer and inner sides of the Rogovsky coil coil, 4- Hall sensor.

Table 2. Values of measured parameters for I_m = 500 A

f, kHz	at $R_н$ = 0.1 Ohms (short circuit)		at $R_н$ = 1 MOhms (idling)	
	B_m, mT	$U_{m\ нар.}$, V	B_m, T	ε, V
30000	6.96-0.036i	0.18049+0.010i	6.93-0.038i	-0.012+16.777i
35000	6.96-0.034i	0.18066+0.0089i	6.92-0.036i	-0.013+19.575i
40000	6.96-0.032i	0.18077+0.0078i	6.92-0.033i	-0.014+22.372i
45000	6.96-0.030i	0.18085+0.0069i	6.92-0.032i	-0.015+25.170i

The results of calculations show that when the Rogovsky coil is operating in a mode close to the short-circuit mode, the load voltage practically does not depend on the frequency, and the initial phase of the voltage does not differ from the initial phase of the current. However, in this mode, the heat output in the measuring coil becomes unacceptably high. To reduce the level of interference in the measured EMF signal from the Rogovsky coil, the load resistance should be as low as possible. For the selected conductor, the permissible current is 100 mA, and the load resistance that corresponds to this current is 256 Ohms. In table. 3. the obtained values of the voltage at a load of 256 Ohms are given, taking into account the intrinsic resistance of the coil wire.

Table 3. Values of measured parameters for I_m = 500 A; $R_н$ = 256 Ohms

f, kHz	B_m, mT	$U_{m\ нар.}$, V	$I_{m\ нар.}$, mA
30000	6.93-0.037i	0.59+16.728i	65

35000	6.92-0.034i	0.81+19.508i	76
40000	6.92-0.032i	1.06+22.284i	87
45000	6.92-0.029i	1.34+25.055i	98

According to the results of the computational experiment, it can be seen that the load voltage practically does not differ in magnitude from the EMF of the coil. The maximum error is 0.3%.

the power value for sinusoidal currents and voltages can be determined by the equation

$$P = 0{,}5\, U_m I_m \cos\varphi_2, \tag{6}$$

where φ_2 – the difference between the initial phases of the voltage and current of the inductor.

The measured average value (the current signal is received from a corrective amplifier) of the product of the received signals (the initial power value calculated with an error when the initial phases do not match)

$$P' = k_{du} k_{de} f \cdot \int_0^{f^{-1}} u(t) u^{\text{im}\,Rw}(t)dt = 0{,}5 k_{du} U_m \cdot k_{de} U_m^{\text{int}\,Rw} \cos\varphi_1, \tag{7}$$

where φ_1 – difference in the initial phases of signals received from the correction amplifier $u^{\text{im}\,Rw}(t)$ and voltage on the inductor $u(t)$;

$k_{de} U_m^{\text{im}\,Rw}$ – the voltage amplitude obtained from the correction amplifier (a signal proportional to the current amplitude of the inductor);

U_m – voltage amplitude on the inductor;

k_{du} и k_{de} – the coefficients of the voltage dividers for the signal of the inductor voltage and the voltage at the load resistance of the Rogovsky coil respectively.

The power value can be obtained using the correction factor *k*, defined using the expression

$$P = P' \cdot \frac{k(f)}{k_{de} k_{du}}. \tag{8}$$

Then we get

$$k(f) = \frac{P}{P'} = \frac{0{,}5\, U_m I_m \cos\varphi_2}{0{,}5 k_{du} U_m \cdot k_{de} U_m^{\text{im}\,Rw} \cos\varphi_1} = \frac{I_m \cos\varphi_2}{k_{du} k_{de} U_m^{\text{im}\,Rw} \cos\varphi_1}. \tag{9}$$

Relation $k_{E\,\text{im}-I} = \dfrac{I_m}{U_m^{\text{im}\,Rw}}$ it can be obtained by analyzing the results of numerical modeling and the amplitude - frequency response of the correcting amplifier, and the cosine of the angle φ_1 using the measured values and the equation (7)

$$\cos\varphi_1 = \frac{P'}{0{,}5 k_{du} U_m \cdot k_{de} U_m^{\text{im}\,Rw}}. \tag{10}$$

In the absence of a correction amplifier in the current measurement system, the expression for determining the correction factor must take into account the changes in the

EMF of the Rogovsky coil when the frequency changes, since the signal from the coil (with the amplitude U_m^{Rw}) enters the signal multiplier from the linear voltage divider.

$$k(f) = \frac{k_{E-1}(f) \cdot U_m^{Rw}(f) \cdot \cos\varphi_2 \cdot U_m}{P'}, \tag{11}$$

where $k_{E-1}(f) = \frac{I_m}{U_m^{Rw}(f)}$ – coefficient that connects the voltage amplitude obtained from the Rogovsky coil with the current amplitude in the measured circuit at the frequency *f*.

To obtain the dependence of the coefficient *k* on the frequency, we will conduct a computational experiment and enter the obtained values in the table 4.

Table 4. Values of the measured parameters for $I_m = 500$ A; $R_н = 257$ Ohms

f, kHz	r_u	L_u	I_m	U_m	φ_2	$\cos\varphi_2$	$P = 0{,}5 \cdot U_m I_m \cos\varphi_2$
30	$2.953 \cdot 10^{-3}$	$1.365 \cdot 10^{-6}$	500	1.4765+129.12i	89.345	0.011	369.1
35	$3.584 \cdot 10^{-3}$	$1.366 \cdot 10^{-6}$	500	1.7920+150.64i	89.318	0.012	448.0
40	$4.207 \cdot 10^{-3}$	$1.366 \cdot 10^{-6}$	500	2.1035+172.16i	89.300	0.012	525.9
45	$4.824 \cdot 10^{-3}$	$1.367 \cdot 10^{-6}$	500	2.4120+193.68i	89.287	0.012	603.0

f, kHz	U_m^{Rw}	$k_{E-1} = \frac{I_m}{U_m^{Rw}}$	P'	$k(f) = \frac{k_{E-1} \cdot U_m^{Rw} \cdot \cos\varphi_2 \cdot U_m}{P'}$	$P = \frac{k(f) \cdot P'}{k_{de} \cdot k_{du}}$
30	0.59208+16.815i	29.717	0.022	0.17	369.1
35	0.80922+19.610i	25.476	0.03	0.152	448.0
40	1.0598+22.400i	22.296	0.039	0.136	525.8
45	1.3436+25.186i	19.824	0.049	0.124	603.0

In the future, it is unnecessary to need to measure the true value of the angle.

5. Conclusions

The usage of the Rogovsky coil as a current sensor for measuring the active power of an induction installation can be justified in cases where the installation design does not provide for the installation of other sensors (based on the Hall effect and current transformers) that have a ferromagnetic core in their design . This solution can be rational if it is necessary to measure current and power in high frequency power oscillating circuits of induction heating installations.

References

[1] Y. B. Kudasov Electrophysical measurements. - Moscow: Fizmatlit, 2010. - 184 p.

[2] A. J. Peyton, V. Walsh. Analog electronics on operational amplifiers/ - Moscow: BINOM, 1994. - 352 p.
[3] N. V. Golovinov. Use of the Rogovsky coil in electrohydroimpulse processing of materials. Bulletin of the Voronezh state technical University. 2009. Vol. 5. № 11. Pp. 103-105.
[4] Bocharov Yu. N. Technique of high stresses[Electronic resource]: textbook for students of higher educational institutions studying in the direction of bachelor's training "Technical physics" / Yu. n. Bocharov, S. M. Dudkin, V. V. Titkov; Ministry of education and science of the Russian Federation, Saint Petersburg state Polytechnic University/ Saint Petersburg: Publishing house of the Polytechnic University, 2013. - 265 p. Pp.225-229/
[5] Intelligent electrical engineering [Text] : [journal] / Ministry of science and higher education of the Russian Federation, Federal state budgetary educational institution of higher education " Nizhny Novgorod state technical University named after R. E. Alekseeva" ; 2019, № 3 (7). - 2019. - 127 p. : ill.; Pp.9-11.
[6] Ward D.A., La J. And Exon T. Measurement Engineering Science and Educational Journal 2, 3. 1993. 105 p.
[7] Sanda P., Zeljko A. Application of Rogowski coil in fast pulsed-current measurements of capillary discharge. FIZIKA A (Zagreb) 16 (2007) 4, pp. 233–242.
[8] Ray W.F. Wide bandwidth Rogowski Current Transducers Part II: The Integrator. 1993. «EPE Journal Vol 3 no. 2», pp. 116-122.
[9] Ramboz J. D. «Machinable Rogowski Coil, Design and Calibration», IEEE Transactions on Instrumentation and Measurement. 1995. vol. 45, Issue: 2, pp. 511-515.
[10] Modern tools for research of induction melting in a cold crucible / A.V. Vavilov, D. B. Lopukh, A. P. Martynov et al. // Automation and modern technologies. – 2010. – № 11. – Pp. 24–31.

Electrochemical deposition of a radionuclide nickel-63 on betavoltaic cells for a nuclear battery based on silicon p-i-n junctions

N A Ershova[1], A A Krasnov[2], S A Legotin[2], B I Rogozev[1] and V N Murashev[2]
[1] RITVERC JSC, 10, Kurchatov St., 194223, St. Petersburg, Russia
[2] NUST "MISiS", 4, Leninskiy Prosp., 119049 Moscow, Russia
e-mail: nershova@ritverc.com

Abstract. This paper highlights the manufacturing process of two types of betavoltaic cells with a source coated with ^{63}Ni: the flat one and one with a three-dimensional (3D) microchannel structure, obtained by anodizing silicon (Si). Thin-layer deposition of a nickel-63 radionuclide was carried out from the ammonium-citrate based electrolyte solution, designed for production until exhaustion.

Keywords: betavoltaic effect, anode etching, microchannel silicon, plating on silicon, radionuclide nickel-63, until-exhaustion electrolyte.

The first proposal betavoltaic power element dates back to 1951 [1], however, studies and research are still relevant nowadays [2-4]. A betavoltaic element consist of a β-radioactive source, a non-conductive layer, which electrons pass through, and collecting layer, gathering electrons. Such source is a basic unit for building no-break power cells. The efficiency of the betavoltaic element and estimated power of cell grows continuously, as well as a range of application [5-7].

The simplest way to produce a betavoltaic element seems to be a unit of a silicon wafer layer with a planar diode structure and an appressed layer of planar nickel sealed radioactive source (*SRS*). This approach is attractive as linking two or more standard items is assumed, but wasn't supported insofar as using SRS means a loss in beta-stream and additional Bremsstrahlung due to substrate absorption as a side-effect.

The second attractive approach, from the production simplicity point of view, is an element of a silicon planar or micro-channel diode structure plated by nickel-63. The downsides of the former are proved to be mitigated with the latter [8]. However, there is a production obstacle, of choosing the optimal plating process here to be solved.

Nickel-63 is used to be plated electrochemically due to avoiding significant waste of high-costed source material. Taking to account the relation between the ion flow density and massic activity of deposing radionuclide (Figure 1), using of nickel volume-maintainable electrolyte seems to be economically unviable, as far as source material cost grows exponentially as massic activity increases, and sensitivity to impurities disallows longstanding and/or multiple electroplating. The purpose of this research is the development and comparison analysis of optimal technologies for plating on silicon diode structure, planar one and microchannel one. In addition, to test the hypnotical applicability of ammonium-citrate based electrolyte solution, designed for production until exhaustion by RITVERC JSC.

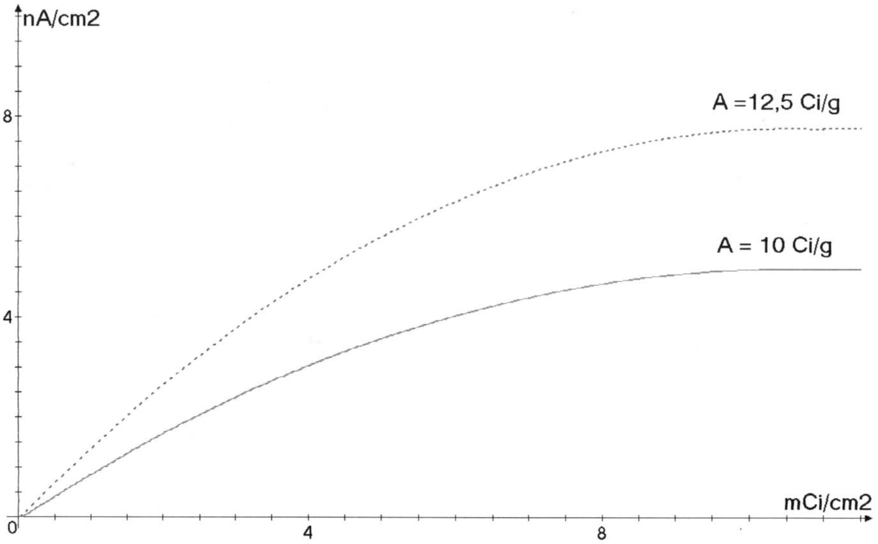

Figure 1. The relation of flow density to the massic activity of radioactive sources

Electroplating of nickel was performed from ammonium-citrate based electrolyte solution, designed for production until exhaustion by RITVERC JSC with following speciation: $(NH_4)_2C_6H_6O_7$ + $(NH_4)_2SO_4$ + $NH_3 \cdot H_2O$ (exc.) + $N_2H_4 \cdot H_2O$ + $^{63}NiCl_2$. Electrolysis was carried out in the galvanodynamic mode with a linear increase in current density from 1.5 to 30 mA / cm^2 at a temperature of 20 ± 5 ° C. An excess concentration of ammonia in the solution provided buffering capacity of the electrolyte and maintained pH value in the range of 8.5–9.0.

Silicon p-i-n structures with overall dimensions of 11x11 mm and a thickness of 470 µm were used as substrates. The square electrode was adapted to electroplating the 63Ni layer. A metal layer of Ti/Ni (20/20 nm) was evaporated to form the p+ ohmic contact. Titanium was used as an adhesive layer. The samples were annealed at 300 ° C for 60 s in N_2 ambient..

The plating area was 1 cm2. The preparation of the substrate included degreasing with dimethylformamide and activation with a 10% sulfuric acid solution. The optimal thickness of the ^{63}Ni radionuclide layer is determined by its self-shielding [9] and is approximately 1.5 µm for galvanic nickel (density 8.9 g / cm3), which is equivalent to applying 10 mCi for an agent with a massic activity of 10 Ci / g.

The activity of the layer was determined as the exhaustion ratio of electrolyte from the nickel-63 radionuclide and was monitored using the SKS-07P-B11 liquid scintillation spectrometric complex. Under these conditions, the exhaustion of the electrolyte in nickel to a concentration of 4.2–4.6 Bq / µl over a period of 490–510 minutes was archived, which corresponds to plating of 99.999% initial activity. Moreover, according to the empirical deposition curve, the deposition of 90% of the material is achieved in 210-220 minutes, and 99% of the material in 350 minutes.

Ten samples with close short-circuit currents (≈ 65 nA) were selected from the obtained betavoltaic elements (Figure 2), which were assembled into a betavoltaic cell with the following characteristics: the value of the short-circuit current was 64 nA, the value the open circuit voltage was 1 V, the maximum output power was 32 nW, the overall dimensions of the assembly without a casing were 11x11x6 mm (more in [10]).

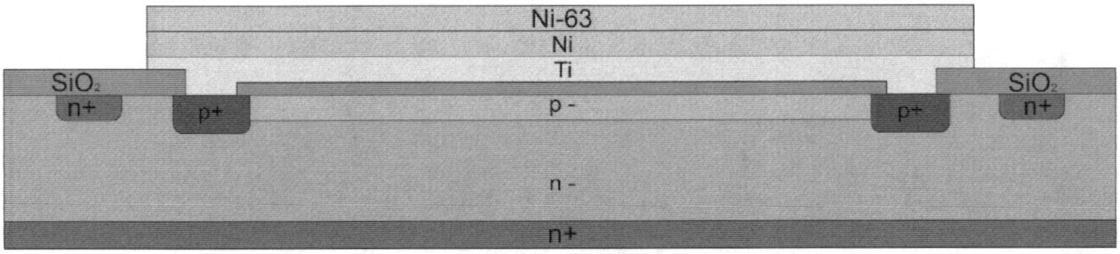

Figure 2. The schematic structure of Si betavoltaic cell

Further improvement of the characteristics of the betavoltaic power cell implies the use of a radionuclide preparation with a higher specific activity of the material (according to Figure 1) and the use of three-dimensional microchannel silicon structures as substrates [11-14]. This is due to the fact that the average geometric characteristics of the pores obtained by the deep anode fluoride etching method and the quasi-hexagonal arrangement of microchannels provide an increase in surface area by 1-2 orders of magnitude [15]. At the same time, the fact that the depth of the microchannels is, on average, 160-163 microns with a diameter of 5-6 microns and a distance between one center to another of 10 microns, significantly limits the set of available plating process applicable for such structures.

The most popular and cited method was the chemical displacement deposition proposed in 2007 [16]. It implies immersion of the substrate for 4 hours in a solution containing 1 M NiSO4, 2.5 M NH4F, 0.6 mm coumarin, 0.7 M sodium dodecyl sulfate with continuous stirring and a temperature of 60 ° C at pH 7-8 to deposition of the natural nickel sublayer of 1-micron thickness. The deposition of radionuclide nickel-63 is proposed to be carried out under the same conditions on a nickel sublayer, replacing stable nickel sulfate in the electrolyte with radionuclide sulfate nickel-63 with a massic activity of 7.5 Ci / g.

To reduce the loss of electron flow in the stable nickel sublayer and to minimize the consumption of the radionuclide agent, it was decided to modify the method. The first change was a multifold reduction in the time of applying the sublayer to reduce its thickness. The second change involved the replacement of a highly concentrated nickel fluoride electrolyte with a low-nickel variable-concentration electrolyte for the deposition of nickel-63 radionuclide. The testing of the modes of deposition of nickel in micropores was carried out on stable nickel. The efficiency of the process for filling microchannels with nickel was studied on side sections using scanning electron microscopy and X-ray microanalysis using a TESCAN VEGA LMH electron scanning microscope.

The preparation of microporous silicon included: degreasing with dimethylformamide (allowing to reduce the next stage of preparation by a factor of 3 in time), removing other organic pollutants by washing in an aqueous solution of ammonia and hydrogen peroxide (H2O / NH4OH / H2O2 = 5/1/1) at 75 ° C and activation with a 5% solution of hydrofluoric acid. The next change in the nickel plating mode was an increase in the operating temperature, forced due to the low intensity of the process led to the buildup of the metal layer from the surface deep into the material and often led to clogging of pores in the near-surface region (presumably as a result of the slow growth of SiF2 and SiF4 gas bubbles) (Figure 3).

Carrying out the process at a temperature of 70-75 ° C allowed to exclude gas clogging of the pores, however, did not provide satisfactory plating. As one can see (Figure 4), the displacement process is not uniform, which in some areas ensures the growth of the nickel layer at a speed of the order of 1.5 μm / day and leads to the complete displacement of silicon atoms by nickel atoms in the space between the pores.

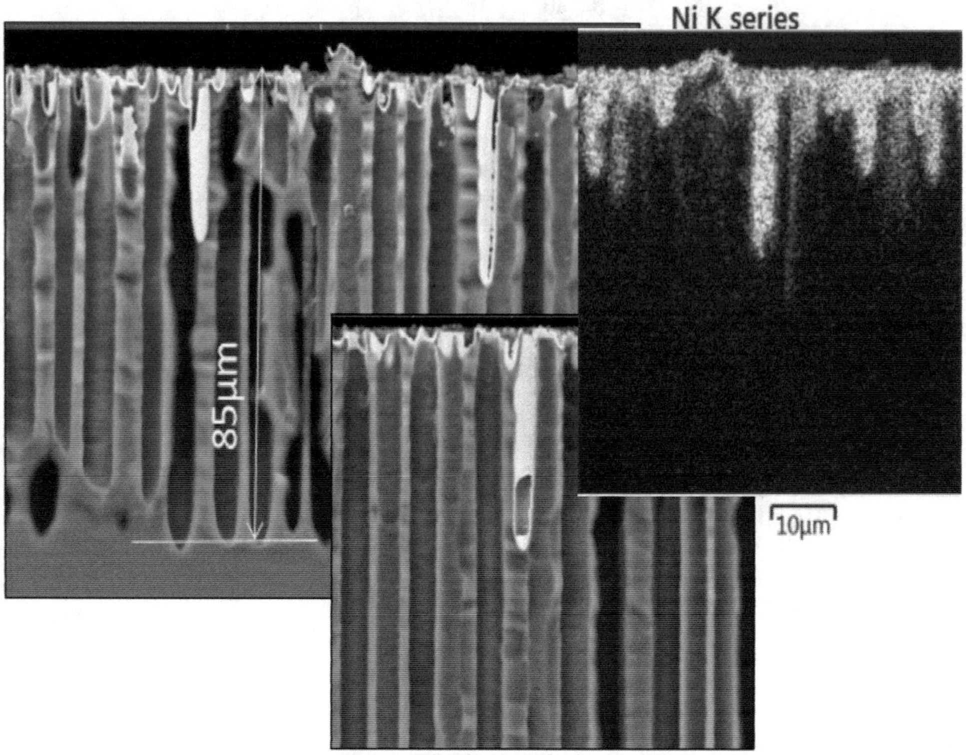

Figure 3. Surface plating on unstructured porous silicon

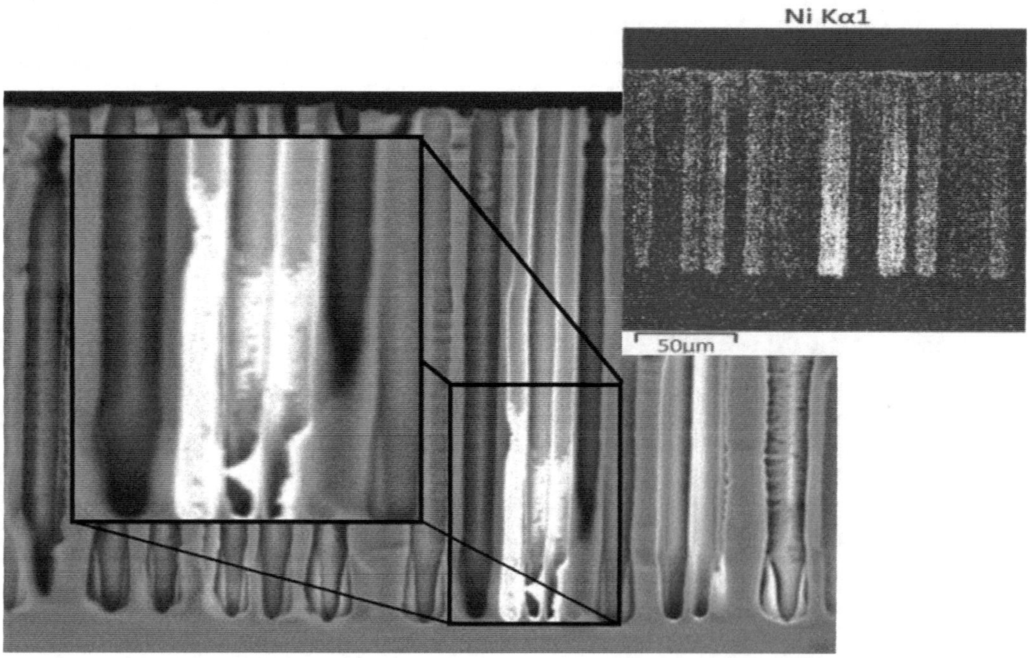

Figure 4. Plot of a microchannel structure where nickel atoms completely displace silicon.

Since the decrease in current density did not give positive results, further work was carried out with a change in the qualitative and quantitative composition of the electrolyte. The optimum plating was achieved by nickel deposition from an aqueous solution containing 154.76 g / l of nickel sulfate, 92.55 g / l of ammonium fluoride, 0.88 g / l of coumarin with the addition of 0.03 g / l of sodium lauryl sulfate as a wetting agent at 60 ° C (Figure 5). The pH of the solution was 5.6, deposition time 120 minutes.

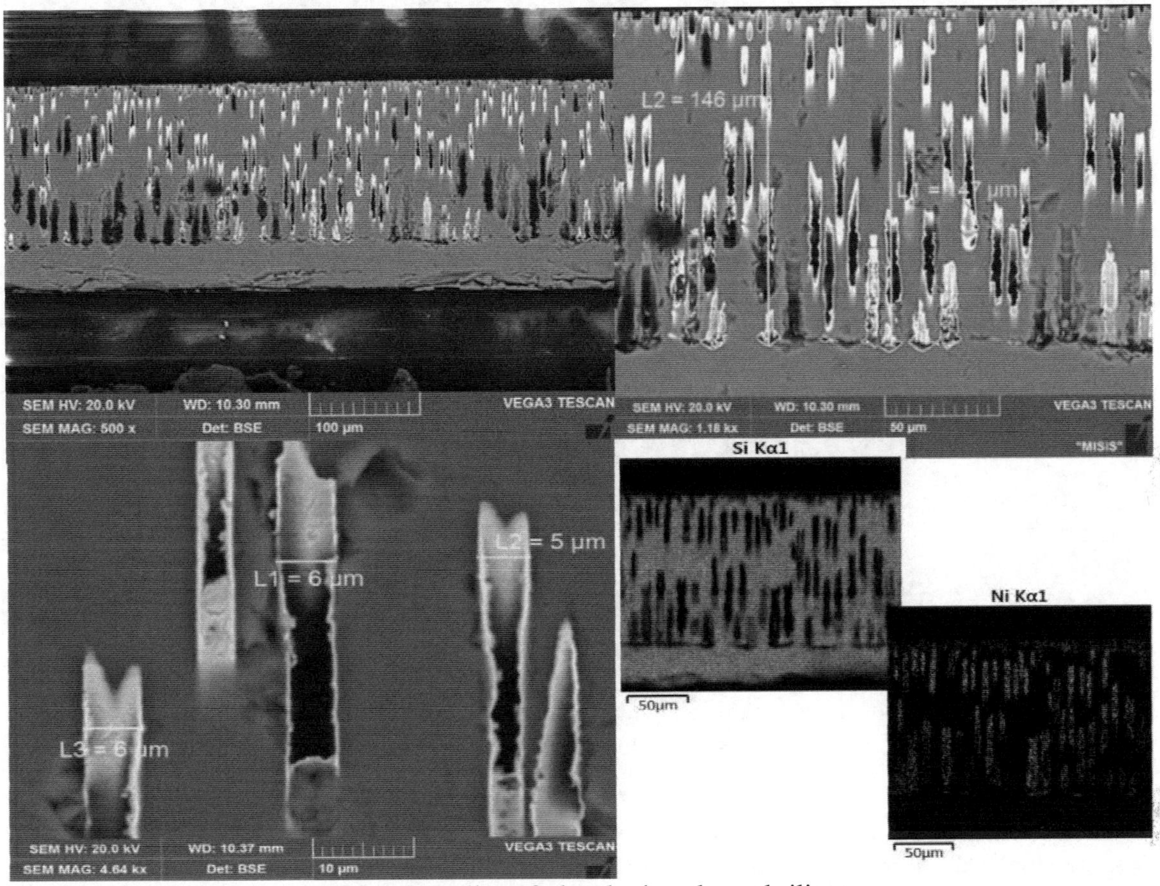

Figure 5. A slice of plated microchannel silicon

The second stage of nickel plating (deposition of a radionuclide) was carried out by the reduction reaction of citrate-ammonium nickel electrolyte. However, without any leveler in the hydrogen fluoride electrolyte, the surface of the natural nickel sublayer got a loose structure, which led to the impossibility of using a low concentrated electrolyte until exhaustion. Satisfactory results of planting without gas plugs were achieved from an aqueous solution containing 26.31 g / l sodium citrate, 14.35 g / l nickel sulfate heptahydrate, 2.14 g / l ammonium chloride, and 2.36 g / l hydrazine hydrate. The pH of the solution was maintained by ammonia at a concentration of 9.1 g / L. The process was accompanied by a slight occlusion of the solution into the coating, as evidenced by a decrease in the content of oxygen and sulfur impurities during the galvanodynamic electrolysis mode with a linear increase in current density from 2 to 30 mA per 1 cm2 of silicon wafer area for 120 minutes (Figure 6).

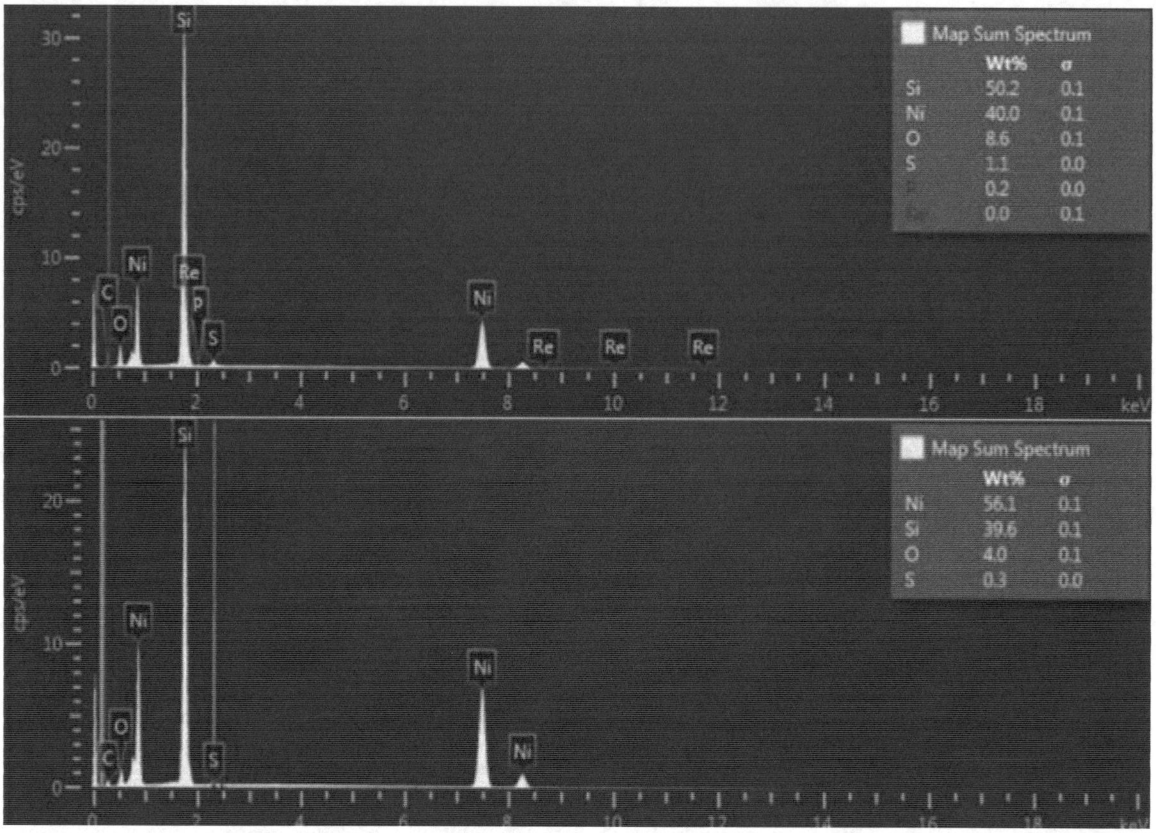

Figure 6. Analysis of the elemental composition of the structure with plated pores (the top is the elemental composition of the coating obtained in the galvanostatic electrolysis mode with a current density of 20 mA / cm2, below is the elemental composition of the plating obtained in the galvanodynamic mode)

Thus, the archived result confirms the applicability of ammonium-citrate based electrolyte solution, designed for production until exhaustion by RITVERC JSC in order to plate nickel-63 radionuclide on planar diode silicon structures. For microporous silicon, the hypothesis was not confirmed; However, the authors proposed a plating technology for such structures using a citrate-ammonia electrolyte with a relatively low concentration of nickel-63 radionuclide.

In a further, the possibility and viability of using a radionuclide agent (raw material) with a higher and / or maximum possible massic activity of a nickel-63 radionuclide should be carefully considered. The authors also find it promising to look for a way to reduce the thickness of the natural nickel sublayer, plated to the microporous structure, without loss in the uniformity of deposition.

References

[1] Ehrenberg W et al 1951 *Proc. Phys. Soc. A* **64** 424
[2] Reznev A A at al 2009 *Nano- and microsyst. technol.* **3** 14–16
[3] Kim J J, Uhm Y R, Choi B G and Son K J 2015 *J. of Nanomaterials* **2015** DOI: 10.1155/2015/283291
[4] Bormashov V S at al 2018 *Diamond and Related Materials* **84** 41-47
[5] Rahmani F and Khosravinia H 2016 *Radiation Phys. and Chem.* **125** 205–212
[6] Nejad G, Rahmani F 2014 *App. Radiation and Isotopes* **86** 46–51
[7] Nejad G, Rahmani F 2016 *App. Radiation and Isotopes* **107** 346–352

[8] Krasnov A A et al 2015 *J. of Nano-and Electronic Phys.* **7** 4 04004
[9] Alam T R, Piersona M A and Prelas M A 2017 *App. Radiation and Isotopes* **130** 80–89
[10] Krasnov A et al 2019 *Nuclear Engineering and Technology* **51** 8 1978-82
[11] Krasnov A A et al 2016 *IOP Conf. Series: Materials Sci. and Engineering* **110** 012029
[12] Maximenko S I et al 2019 *Sci Rep* **9** 10892
[13] Kharitonov I D et al 2018 *Radiochemistry* **60** 1 24-26
[14] Bondarenko V P et al 1997 *Emis datareview series* **18** 356
[15] Emel'yanov V I and Starkov V V Nonlinear 2006 *J. of surf. investigation/X-ray, synchrotron and neutron tech* **6** 116
[16] Xu C et al 2007 *J. of the Electrochem. Society* **154** 170-4

IOP Conf. Series: Materials Science and Engineering **950** (2020) 012008 doi:10.1088/1757-899X/950/1/012008

Electromagnetic separation of impurities in a conductive medium

Vladimir S. Ozernykh, Gennady L. Losev, Ilya V. Kolesnichenko
Institute of Continuous Media Mechanics, Koroleva St. 1, Perm
E-mail: ozernykh.v@icmm.ru

Abstract. A numerical study of the flow of an electrically conductive fluid is carried out in a cell under the action of a constant magnetic field. Distribution of electromotive force is investigated as well as the topology of electric current in the cell. The evolutions of velocity field are studied for different parameters. Two large scale unstable eddies are generated in the cell. The characteristic of the flow oscillations are determined using the wavelet analysis. It was found that an increase of current does not lead to the appearance of additional oscillation modes. Asymmetric pressure distribution may appear in the cell. Particle movement are possible due to the asymmetric pressure field.

1. Introduction

In our days, metals and alloys purified from impurities have a high demand. Materials and structures which made of ultrapure metals or alloys have an increased yield strength, better corrosion resistance compared to products from contaminated alloys. Metals and alloys of a high degree of purity are used in construction, in aerospace and high-tech equipment, and many other industries. Also liquid metals (for example, liquid sodium) are used as a coolant at fast neutron nuclear power plants. The degree of purity of the coolant essentially determines the lifetime of the equipment of the nuclear power plant, which is extremely important for the safe operation of the plant. In this regard, the task of detecting impurities in multiphase electrically conductive media, as well as the cleaning of melts from impurities, is one of the most important tasks for research.

There are many different methods for cleaning metals from impurities, for example, flotation methods [1, 2], sedimentation of the ore melt in a liquid [3], etc. Many existing metal cleaning methods require significant energy consumption. In this regard, the electromagnetic (em) method of mixture components separation is attractive in production. The em-method is based on the difference in the electrical conductivity of the metal and impurities. The impurities will either drown in the metal or float to the surface [4, 5] under the action of electromagnetic forces, where they can be easy removed. It depends on the magnitude of the electrical conductivity of the impurities (more or less the electrical conductivity of the metal). The method of applying on the metal the em-force is the least labor-intensive, easily implemented. Additionally it allows not only to separate impurity particles from the metal, but also to detect the impurities in the metal. It may be used in non-destructive testing technique of finished products and structures.

One of the methods for generating the em-force is to apply the external magnetic field co-directionally to the electric current. Such a configuration can be created by placing the studied cell in the solenoid. This method is the easiest to implement. The superimposed external magnetic field is significantly more intensive than the own magnetic field of the electric current.

In the cell the em-force is generated due to distortion of electric current streamlines near the impurities. The distortion appear due to difference in the electrical conductivity between the impurities and liquid metal (current distortion cases are shown in Fig. 1. Obviously, if there are no impurities in the metal, then there is no reason for the appearance of em-force.

Thus, the common aim of the investigation is to study numerically the behavior of an ensemble of impurity particles in a liquid metal under the influence of em-forces. It is necessary in the frame of mathematical model to formulate a constitutive relation for parameters describing the electrical and mechanical properties of the phases and interphase interactions. In the first stage, described in the paper, the study of em-force, current distribution and velocity field behavior is carried out for single inclusion. In the second stage, the calculations to an ensemble of impurity particles will be provide in our future work.

2. Mathematical statement

The mathematical model is based on the equation of magnetohydrodynamics. It is used the non-induction approximation, which is valid for small magnetic Reynolds numbers, this allows you to explore the electrodynamic and hydrodynamic parts of the problem separately [6]. We consider a cell of liquid metal, which is a cube, with side L. Inside the cell an spherical electrically conductive particle having radius R is placed in the center (fig. 1).

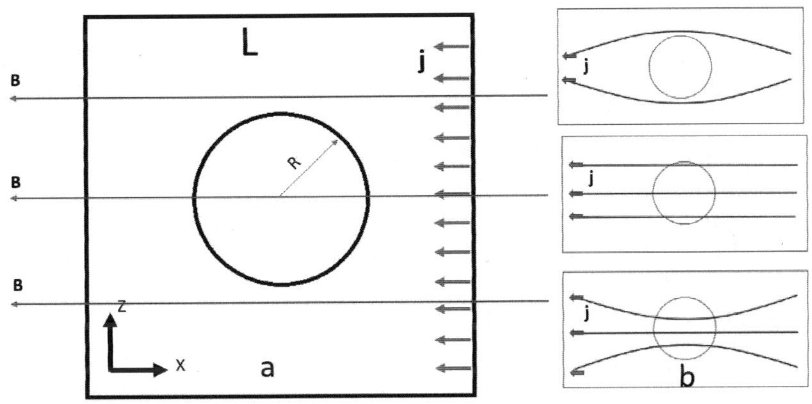

Figure 1. Scheme of the problem in the central section (a), electric current streamlines distortion schemes (b)

A potential difference is supplied to the system. The electric current of density $\mathbf{j}$ flows through the cell along the X axis. In order to suppress the own magnetic field of electric current ($\mathbf{j} = \mu\mu_0 rot\mathbf{B}$), an external magnetic field B_0 is imposed to the cell along the X axis. In this case $\mathbf{f}^{em} = \mathbf{j} \times \mathbf{B_0}$.

There are various options for the distortion of electric current streamlines in the cell, depending on the magnitude of the conductivity of the particle (less, equal to or greater than the conductivity of the metal) (fig. 1). A series of numerical experiments for the two most interesting configurations are carried out, the distinguishing feature of which are:

a) metal conductivity is greater than particle conductivity($\sigma_{Part} = \sigma_0$; $\sigma_{Me} = k \cdot \sigma_0$),

b) metal conductivity is less than particle conductivity($\sigma_{Me} = \sigma_0$; $\sigma_{Part} = k \cdot \sigma_0$).

To describe the motion of a viscous conductive fluid, it is used the Novier-Stokes equation taking into account the electromagnetic forces (1) and continuity equation (2). The (3-6) describe the electromagnetic part of the task.

$$\rho(\frac{\partial \mathbf{V}}{\partial t} + (\mathbf{V} \cdot \nabla)\mathbf{V}) = -\nabla \mathbf{P} + \eta \triangle \mathbf{V} + \mathbf{f}^{em} \tag{1}$$

$$\nabla \cdot \mathbf{V} = 0 \tag{2}$$

$$\triangle \phi = 0 \tag{3}$$

$$\mathbf{f}^{em} = \mathbf{j} \times \mathbf{B} \tag{4}$$

$$\mathbf{j} = \sigma \mathbf{E} = -\sigma \nabla \phi \tag{5}$$

$$\mathbf{B} = (B_0; 0; 0) \tag{6}$$

Electrodynamic parameters: σ_{Part} – particle conductivity, σ_{Me} – metal conductivity, I – current strength, B_0 – external magnetic field.

Hydrodynamic parameters: ρ – density, η – dynamic viscosity.

Boundary conditions: $V|_{bound} = 0$, $V|_{t=0} = 0$, $\frac{\partial \phi}{\partial n} = 0$. For potential: $\phi = \phi_1$, $\phi = \phi_2$ – on opposite walls $(x = 0, x = L)$. At the solid boundaries of the cell the non-slip condition is used.

To calculate the viscous parameters, the $k - \omega$ turbulence model is used. The problem is numerically solved in the Ansys Emag (electrodynamics) and Ansys Fluent (fluid flow) applications.

3. Results and discussion

Calculations were carried out with the following parameters: $\sigma_0 = 10^6$ Sm/m – conductivity, $\mathbf{B} = (10^5; 0; 0)$ – external magnetic field, $k = 100$ – coupling coefficient of the conductivity of a particle and a liquid, $R = 20$mm – particle radius, $L = 100$mm – cell side length.

The most interesting and indicative result is the field of the velocity profile in various sections. On fig.2 the velocity field in cross section x=0.07, y=0.05 is shown.

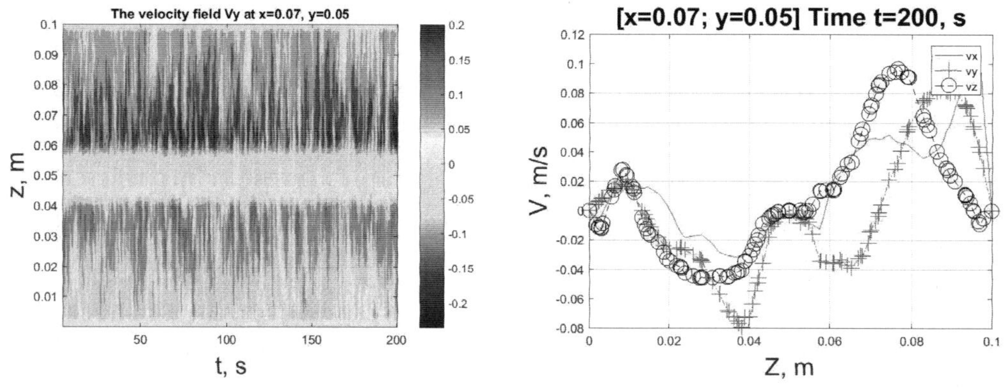

Figure 2. The velocity field in the cross section x=0.07, y=0.05; Velocity profiles

In fig.2 the oscillations (ripples) of the velocity profile are visible at a current strength of I=200A. In order to visually determine the frequencies obtained, a signal analysis was performed with a length of 60s: interval 140-200s.

Using wavelet analysis, two characteristic frequencies of velocity fluctuations can be detected: ν_1=0.15Hz, ν_2=0.07Hz.

This suggests the coexistence of large-wave oscillations in the system, as well as small-scale pulsations. Thus, the fluid flow has a complex character, but at the same time it has allocated oscillation frequencies.

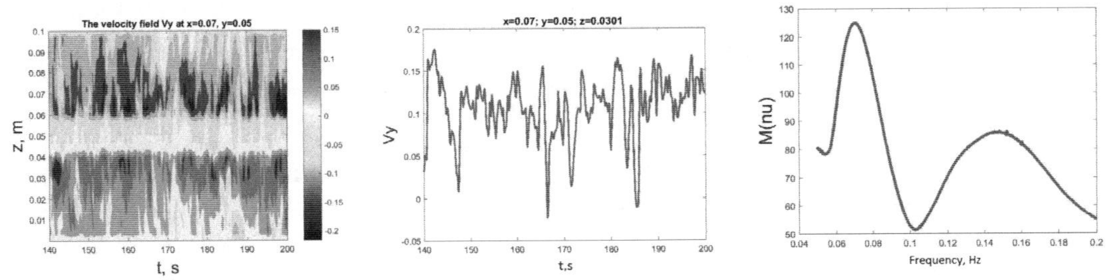

Figure 3. Velocity field, velocity profile and oscillation frequency diagram

In this case, it is important to determine the average velocity both at one point under consideration over time and the average velocity in the cross section at a fixed point in time.

One of the most indicative criteria for the intensity of the current is the average speed (in time) at a fixed point. The dependence of the average speed at the point (0.07; 0.05; 0.03) on the magnitude of the current is estimated (fig.4). A characteristic increase in the average velocity is visible as the current strength increases. The maximum speed is achieved the value V=0.22m/s for current strength of I=1000A.

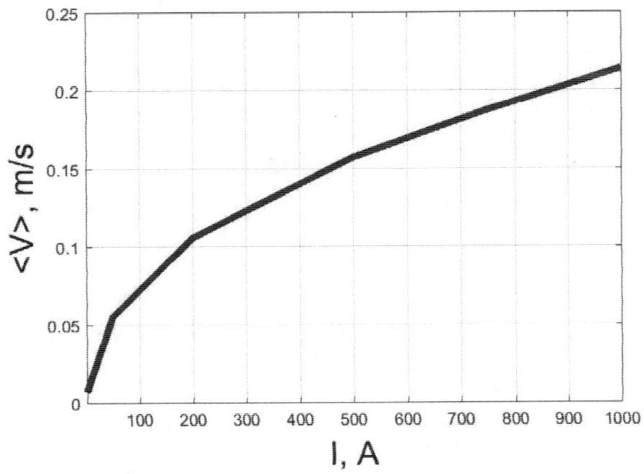

Figure 4. The dependence of the average current velocity on the magnitude of the current

The calculations were performed for currents: 1A, 50A, 200A, 500A, 750A, 1000A.

For wavelet analysis of oscillation frequencies, we choose a time interval in the range of 100-200s. This allows us to discard the process of establishing the flow and more accurately identify the frequency.

The fig.5 shows that three characteristic frequencies can be detected in the system. The values of the frequencies change slightly with the change in the current strength. Since there are several oscillation frequencies for all currents in the system, there is no need to significantly increase the current strength in order to increase the number of oscillation modes. Thus, the intensity of the flow depends on the magnitude of the current, but does not depends on the number of oscillation modes of the system.

The analysis of the pressure field in the cell around the particle is carried out. The fig.6 presents the density distribution in the plane (x,0.05,z):

It can be seen that the pressure field is not the same in different sides of the particle. It may generate the hydrostatic force which may move the particle. A pressure distribution diagram is

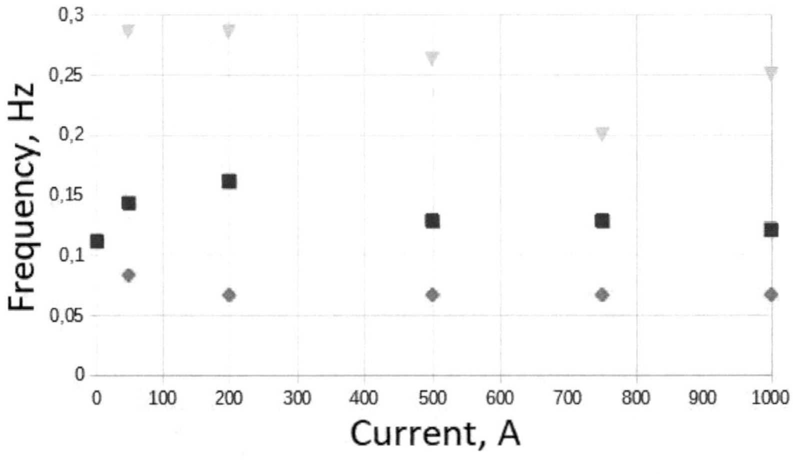

Figure 5. The dependence of the oscillation frequency on the current

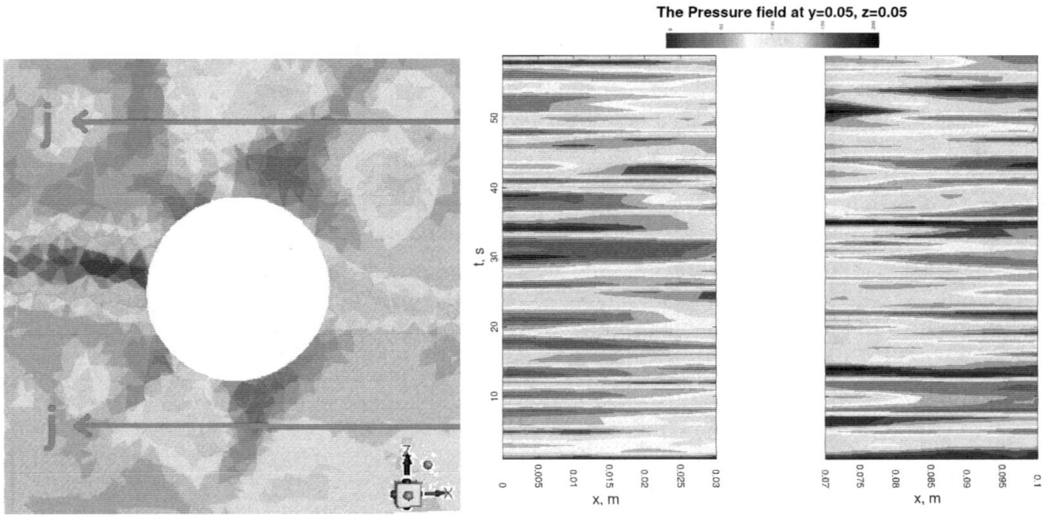

Figure 6. Pressure field: on the left - in section, on the right - along the center line depending on time

prepared along the line passing through the center of the particle along the direction of current flow (fig.7).

The fig.7 is presented their pressure distribution curves near the particle at various time moments. The figure shows that the pressure in the system varies significantly during the process having an oscillatory nature. The pressure gradient at different time moments has different (opposite) directions.

An integral estimate of the mean pressure near the particle is carried out. It was found that on one side of the particle, the time-average pressure is higher than on the other. Thus, it can be assumed that the particle under the action of em-force may shift towards a lower pressure – along the direction of the current strength.

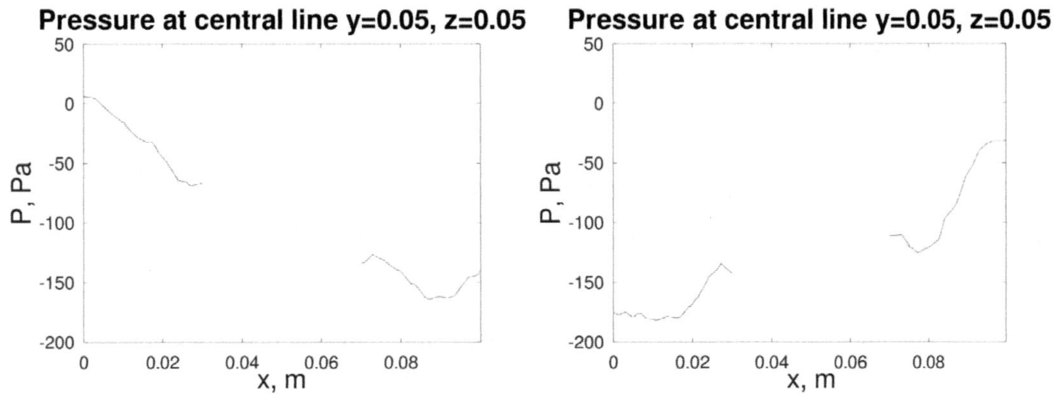

Figure 7. Pressure field along the center line

4. Conclusion

The study has shown that the vortex flow with two large-scale eddies is generated near the particle. For small value of em-force the flow is almost stationary. For moderate and large values of em-force the flow is unstable. The average flow the velocity is estimated as a function of current strength. It was found that the velocity increases with increasing current strength by square quadratic law.

A wavelet analysis of the velocity pulsations yields the characteristic oscillation frequencies. It was found that an increase of current does not lead to the appearance of additional oscillation modes. It was found that the pressure in the system is unsteady, however, it was possible to determine the integral pressure gradient. Asymmetric pressure distribution may appear in the cell. Particle movement are possible due to the asymmetric pressure field.

5. References

[1] Mamykin A, Losev G, Kolesnichenko I 2018 *Bulletin of the Perm University. Physics* **1(39)** 46-53 DOI: 10.17072/1994-3598-2018-146-53.
[2] Han J, Xiao J, Qin W, Chen D, Liu W 2017 *JOM* **69** 1563-1569.
[3] Xu Z, Li T, Zhou Y 2007 *Metallurgical and Materials Transactions A* **38** 1104-1110.
[4] Leenov D, Kolin A 1954 *Journal of Chemical Physics* **4** 683-688.
[5] Povh I 1974 *Magnetic hydrodynamics in metallurgy* (Moscow: Metallurgy).
[6] Kolesnichenko I 2013 *Magnetohydrodynamics* **1** 217-222.

IOP Conf. Series: Materials Science and Engineering **950** (2020) 012009 doi:10.1088/1757-899X/950/1/012009

Energy-efficient installations of multi-zone multi-frequency induction heating of steel billets for large deformation

V I Lusgin[1], I E Lopatin[1,3], E P Shalina[1], A Y Petrov[2], S M Fatkhullin[2], I G Alekseev[2]

[1] High-Voltage Engineering Department, Ural Power Engineering Institute, Ural Federal University, 620002, building 5, Sofia Kovalevskoy str., Ekaterinburg, Sverdlovsk region, Russia
[2] OOO NPP «RELTEC», 620078, building 51A, Studencheskaya str., Ekaterinburg, Sverdlovsk region, Russia

Abstract. This article discusses the basic construction principles of the device for multi-zone induction heating. The mathematical model of a cascade inverter is described. Various modes of heating cylindrical billets are considered depending on the ratio of powers in the heating zones and current frequencies in the inductors. The indicators of heating uniformity over the cross section and energy efficiency are studied. Based on the data obtained, a conclusion is drawn about the optimal choice of the number of heating zones, frequencies of inductors and heating modes.

Machine-building enterprises widely use metals induction heating devices in forge-and-press and rolling production. The physical processes of heating steel blanks by high-frequency currents are accompanied by significant heterogeneity of heat generation in the blank volume and, as a result, lead to a significant non-uniformity in the distribution of the heating temperature over the blank cross section and its length. The mechanical resource of high-performance forging, stamping and rolling equipment largely depends on the temperature difference of the metal heating from optimal for large deformation during rolling and pressing of blanks, which determines high demands on the quality of metal heating at a given performance of heating installations.

Current trends in the development of induction heating installations consist in increasing the controllability of the heating process, increasing energy efficiency and reducing specific energy consumption when heating a metal with a given accuracy and performance, in improving the technological parameters of installations associated with a decrease in specific mass and dimensions, and the ability to quickly reconfigure to different blanks assortment, high repair performance and ease of maintenance [1, 2].

These requirements are achieved by creating multi-zone heating systems for steel billets with independent control of the active power and current frequency of the inductors in each heating zone. Autonomous control of the power parameters of the inductors is provided by semiconductor converters generating currents of a given frequency for each inductor. The authors propose using a multi-energy channel frequency converter as a power source, consisting of a common DC converter and a cascade multi-energy channel inverter, each cell of which is assembled according to the scheme

[3] i.e.lopatin@urfu.ru

of an asymmetric current inverter [3]. Compared with individual converters for powering the inductors of each heater zone, the proposed converter can reduce the mass-dimensional characteristics of the power source by about 30% and place it in the heating installation directly under the inductors. Structurally, the induction heating installation is constructed according to the required number of inductor-inverter modules that form controlled heating zones with independent excitation at the resonant frequency of the load oscillating circuit.

Figure 1 shows a functional diagram of a multi-zone heater.

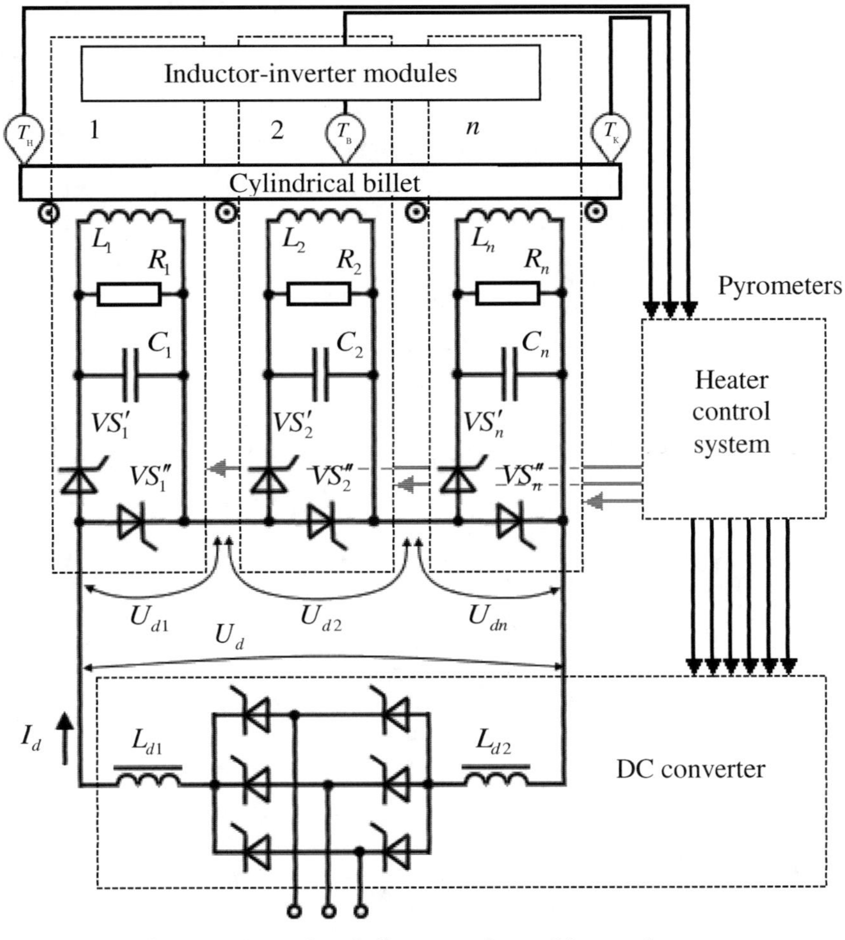

Figure 1. Functional diagram of a multi-zone heater.

Dividing the induction unit into several heating zones with an individual setting of the power and frequency of the current allows you to create the necessary power distribution profile for the heating zones and set the desired profile for the heating temperature distribution of the billets. Studies show that the process of heating blanks can be optimized according to the criterion of the heating greatest accuracy and the minimum temperature difference over the blank cross section for a given installation performance. In addition, the profile of the frequency distribution of the generated current in the heating zones can be optimized by the criterion of the highest electrical efficiency and the energy-efficient mode of operation of the installation can be determined. In multi-zone heating installations, it is possible to control the heating of the blanks under possible disturbing influences, such as a change in the speed of movement of the billets, deviation of the initial temperature of the billets, switching to another diameter of the billets, etc. By controlling the power distribution profile using temperature feedback in a certain zone of the heater (sighting point), automatic stabilization of heating and flexible

IOP Conf. Series: Materials Science and Engineering **950** (2020) 012009 doi:10.1088/1757-899X/950/1/012009

adjustment to the requirements of the technological process of rolling and stamping of products is provided.

The total active power of the cascade current inverter at the input can be determined as the sum of the active powers at the input of the inverter cascades

$$P_d = I_d U_d = \sum_{i=1}^{n} P_{di}; \quad P_{di} = \frac{U_{di}^2}{R_{di}}; \quad U_{di} = \sum_{i=1}^{n} U_{di},$$

where I_d, U_d – average values of current and voltage at the output of the DC converter;

U_{di} – average voltage at the inverter inputs;

P_{di} – active power at the inverter inputs;

R_{di} – active resistance of inductors reduced to the input terminals of DC inverters.

Assuming the voltage at the inverter outputs to be sinusoidal, their value is determined depending on the setting of the oscillatory load circuits

$$U_{\text{н}i} = \frac{U_{di}}{0{,}45\cos\varphi_{\text{н}i}},$$

where $\varphi_{\text{н}i}$ – the phase shift angels of the main harmonics of the current and voltage at the inverter output.

The output current of the inverters has the shape of rectangular unipolar pulses with an amplitude equal to I_d. The fundamental harmonic of the output current of inverters can be determined as a result of rectangular current expansion in a Fourier series with a frequency f_i

$$I_i = 0{,}45 I_d.$$

The active power at the inverter output is determined by the main harmonics of the output currents and voltages and their phase shifts

$$P_{\text{н}i} = I_i U_{\text{н}i} \cos\varphi_{\text{н}i}.$$

Based on the equalities of power at the output and input of the inverters without taking into account losses in their elements, we determine the magnitude of the reduced load resistance to the DC terminals of the inverters

$$R_{di} = 0{,}2 R_i \cos^2\varphi_{\text{н}i}.$$

The total resistance in the converter DC circuit is determined by the sum of the inverters reduced resistances

$$R_d = \sum_{i=1}^{n} R_{di}.$$

For a given average voltage U_d at the output of the DC converter, we determine the total value of the average current I_d and the input parameters of the inverters:

$$I_d = \frac{U_d}{R_d}, \quad U_{di} = I_d R_{di} = \frac{U_d}{R_d} \cdot 0{,}2 R_i \cos^2\varphi_{\text{н}i}, \quad P_{di} = P_{\text{н}i} = I_{di} U_{di} = 0{,}2 R_i \left(\frac{U_d \cos\varphi_{\text{н}i}}{R_d} \right)^2.$$

Thus, the profile of the active power distribution over the heater inductors can be adjusted by regulating the angles of the phase shift of the main harmonics of the current and voltage at the output of the inverters

$$\varphi_{\text{н}i} = \operatorname{arctg}\left(Q_i \left(q_i - \frac{1}{q_i} \right) \right),$$

where $Q_i = \frac{R_i}{\sqrt{L_i / C_i}}$ – quality factor of load circuits;

L_i, C_i, R_i – inductance, capacitance and resistance of load circuits;

$\omega_{0i} = 1/\sqrt{L_i C_i}$ – natural circular frequency of load circuits;

$\omega_i = 2\pi f_i$ – working circular frequency of inverters;

$q_i = \omega_i/\omega_{0i}$ – the coefficients of the detuning of the operating frequency of the inverter from the resonant frequency of the load circuits.

Inverters are controlled by supplying control pulses to thyristors with a certain frequency f_i, which is determined by the resonant frequency of the corresponding load circuit. To ensure the switching stability of the thyristor inverters, a prerequisite is a mode of operation with a capacitive response of the load circuits with a ratio of the operating frequency and the resonant frequency of the load circuit $f_i/f_{0i} > 1\,(q_i > 1)$. This condition corresponds to the mode of control pulses supply to the thyristors of inverters with lead angles β_i with respect to the voltage of the load circuits $U_{\text{н}i}$, at which the time intervals for applying the reverse voltage to the thyristors is longer than their turn-off time t_q

$$\beta_i \geq \omega_i t_q.$$

If fully controlled semiconductor devices (transistors, lockable thyristors) are used as electronic keys in inverters, the switching conditions are satisfied at $\beta_i \geq 0$, and their range of variation is significantly expanded, therefore, the range of power control in the heating zones increases because $\beta_i \approx \varphi_{\text{н}i}$.

The mathematical model of a cascade current inverter described above can be used to develop a general model for calculating the heating processes of billets in a multi-zone multi-frequency induction heating installation.

By the requirements for heating installations of forging blanks for the Kama Automobile Plant the parameters that they must provide are determined: uniform heating of cylindrical billets with diameter of 90÷130 mm; maximum outlet temperature $T_{\text{к}}$ = 1250°C; permissible range of temperature deviation over the cross section and along the length $\Delta T = \pm 25$°C; rated performance P_r = 3,8 t/h; maximum performance P_m= 4,2 t/h.

When dividing the induction heater into heating zones, it is advisable to limit their number to three, which corresponds to different physical conditions of steel billets, namely: in the first inductor, the billets have ferromagnetic properties; in the second inductor, the heating temperature reaches and exceeds the temperature of magnetic transformations, and in the third inductor, the billets are in a non-magnetic state. The electrical parameters of the inductors of each heating zone are also significantly different, therefore, to control the heating modes, independent control of the inverters is required to control the power parameters of the inductors in terms of power, voltage, frequency of the supply current.

Calculation of the heating process of the workpieces is advisable to carry out by methods based on the calculation of electromagnetic and temperature fields, providing a sufficiently high accuracy. Universal2D software package allows you to perform these calculations by the finite element method [4]. Joint calculations using mathematical models of a heater and a continuous converter at each time discretization step in the heat problem after calculating the equivalent inductances L_i and resistances R_i for a known value of C_i, using the specially developed search algorithm, the eigenfrequencies ω_{0i} of the load circuits and the operating frequencies of the inverters ω_i.

For three-zone heating of cylindrical billets with a diameter of 130 mm with a rated performance P_r=3.8 t/h, the calculation of heating modes at a constant current frequency $f_1 = f_2 = f_3$ = 1000 Hz was performed, with different profiles of relative powers in the heater zones P_1-P_2-P_3%. Heating modes can be conditionally divided into three groups:

- quick heat A-60-30-10%;
- medium heat AB1-50-40-10%, AB2-45-45-10%;
- smooth heating B1-40-40-20%, B2-40-30-30%, B3-33-33-33%.

Each profile can be characterized by a power profile coefficient

$$K_p = \frac{\Delta P_1 + \Delta P_2}{P_н},$$

where $\Delta P_1 = P_1 - P_2$; $\Delta P_2 = P_2 - P_3$; $P_н = P_1 + P_2 + P_3$.

Figure 2,a shows graphs of the billets heating temperature distribution over the zones of the heater. At the outlet of the heater, the billets surface temperature reaches $T_к = 1250°C$.

The heating temperature of the billets at the output of the first inductor at different heating modes is different. In the second inductor, the heating temperature of the workpieces reaches values higher than the temperature of magnetic transformations $T_{mt} = 750$ °C; the temperature in the third inductor is much higher than T_{mt}.

Billet heating modes significantly affect the nature of the heating temperature distribution over the cross section at the outlet of the third heater inductor. Figure 2,b shows graphs of the billets heating temperature distribution over the cross section for different heating modes, by which it is possible to determine the maximum temperature difference ΔT_m. Table 1 shows these values for various modes of heating the billets.

Table 1. Characteristics of heating modes at $f_1 = f_2 = f_3 = 1000$ Hz.

Heating mode	P_1-P_2-P_3%	$K_p = \frac{\Delta P_1 + \Delta P_2}{P_н}$	ΔT_m, °C	$W, \frac{kW \cdot h}{t}$
A	60-30-10	0,5	18,95	300
AB1	50-40-10	0,4	19,05	293
AB2	45-45-10	0,35	19,42	291
B1	40-40-20	0,2	46,55	276
B2	40-30-30	0,1	90,61	269
B3	33-33-33	0	102,91	260

In addition, table 1 shows the values of specific energy consumption for different heating modes, which is determined by the ratio of the total active power of inverters and performance $W = P_i / P$.

Figure 3 shows the dependences ΔT_m and W on the power distribution coefficient K_p, from which it can be seen that the most effective should be considered the average heating modes AB at $K_p = 0.3 \div 0.4$, where a small temperature difference over the workpiece cross section ΔT_m and average W is achieved.

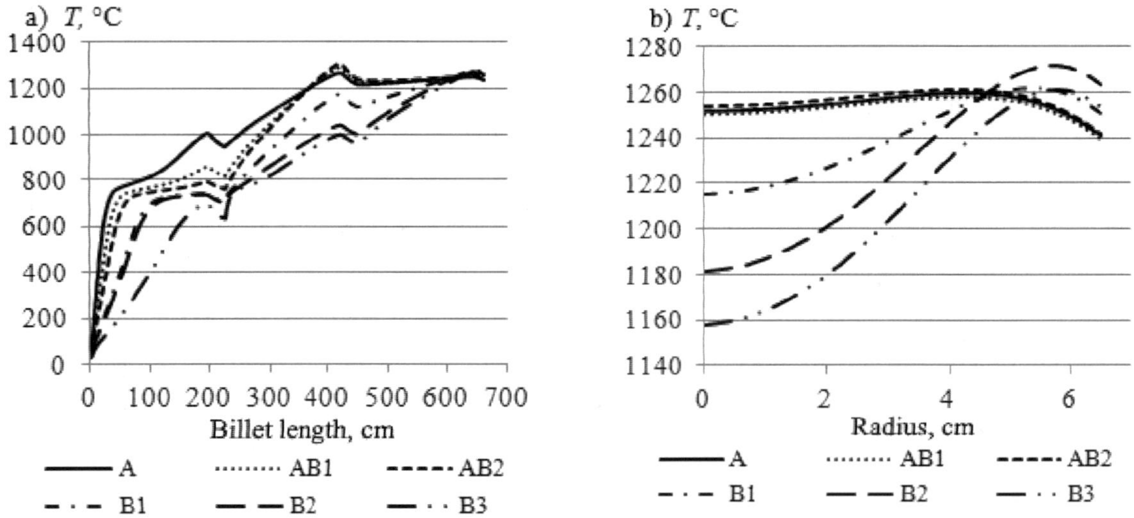

Figure 2. Distribution of the billets heating temperature along the length (a) and radius (b) for different modes at a current frequency in the inductors $f_1 = f_2 = f_3 = 1000$ Hz.

In the mode of fast heating of workpieces at $K_p = 0.4 \div 0.5$, small ΔT_m are provided, but W have increased values. In this mode, intense scale formation and welding of billets among themselves is observed.

In the mode of smooth heating at $K_p = 0 \div 0.3$, the characteristic ΔT_m has a large slope and the temperature drops along the billet cross section take unacceptably large values, while W reaches the smallest value

Based on the requirements for heating accuracy, energy efficiency, manufacturability, it is advisable to accept the modes of the average billet heating rate, where the most effective modes are AB1 and AB2.

The accuracy of heating the workpieces and energy efficiency depend on the choice of the current's operating frequency in the inductors. When heating steel billets, a rational choice of frequency can be made according to the criterion of maximum energy efficiency and efficiency of single-frequency heating of cylindrical billets [1]

$$f = \frac{6}{D^2} = 355 \div 740 \text{ Hz.}$$

According to the mathematical model of the installation, the calculations of the process of heating billets with a diameter of 130 mm at the same current frequency in the inductors in the power distribution mode AB1 are made. Figure 4 shows the dependences of the maximum temperature difference ΔT_m, the efficiency of the inductors η_1, η_2, η_3, and the specific energy consumption W as a function of the current frequency in the inductors. As the frequency increases, ΔT_m decreases and reaches its minimum value at $f = 1200 \div 1400$ Hz, the specific energy consumption W also decreases to the lowest value at a frequency $f = 1200$ Hz.

The efficiency curve of the first inductor η_1 has a weakly expressed maximum corresponding to the current frequency $f_1 = 500 \div 600$ Hz. The efficiency graphs of the second and third inducers η_2, η_3 do not have a maximum and increase monotonically with increasing frequency, while their growth rate decreases at $f > 1200$ Hz. It should be noted that η_2 and η_3 differ little in value at the same frequency from each other; therefore, it is advisable to choose the optimal current frequency of the second and third heating zones according to the energy efficiency criterion in the range $f_2 = f_3 > 1200$ Hz.

When heating billets of different frequencies, it is advisable to choose the optimal current frequency according to the criterion of the highest efficiency of the inductors in the heating zone and the lowest specific energy consumption W with a minimum temperature difference over the billet cross section ΔT_m at the heater outlet.

Figure 5 shows graphs of the temperature difference over the ΔT_m section and the specific energy consumption W as a function of the frequency coefficient in the heater zones $K_f = f_2/f_1$. As K_f increases, the temperature difference ΔT_m drops to a minimum value in the range $K_f = 2 \div 3$, which also corresponds to the lowest value of specific electric energy consumption W. If the optimal current frequency in the first inductor is taken to be $f = 500$ Hz, then the optimal frequencies in the second and third inductors are advisable to take $f_2 = f_3 = 1000 \div 1500$ Hz.

In the technological process of pressing products, it is often required to control the rate of delivery of heated billets, which corresponds to a change in productivity and, consequently, an increase or decrease in the speed of pushing billets through heater inductors. In this case, the accuracy of heating the billets must meet the requirements for heating quality.

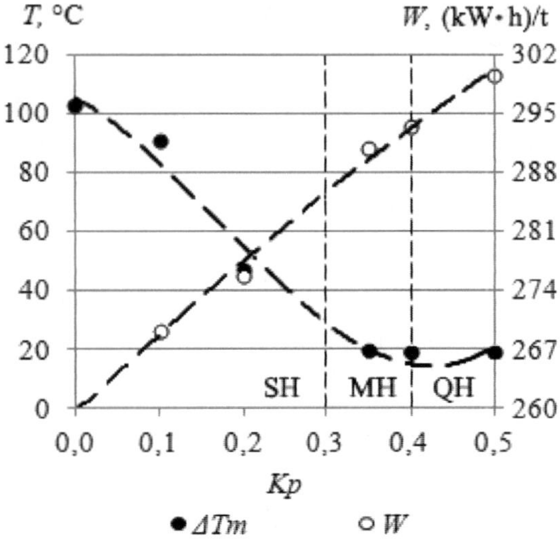

Figure 3. Dependences of the temperature difference over the cross section of the workpieces ΔT_m and specific energy consumption *W* on the power profile coefficient K_p;

SH – smooth heating, MH – medium heat, QH – quick heat.

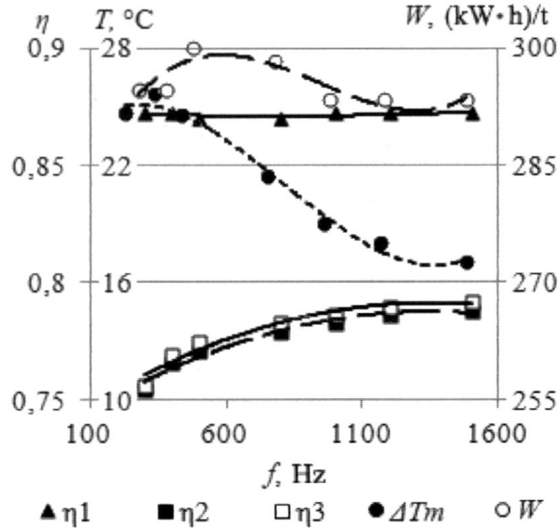

Figure 4. Frequency characteristics of the temperature difference over the cross section of the workpiece ΔT_m, the efficiency of the inductors η_1, η_2, η_3 and specific electricity consumption *W*.

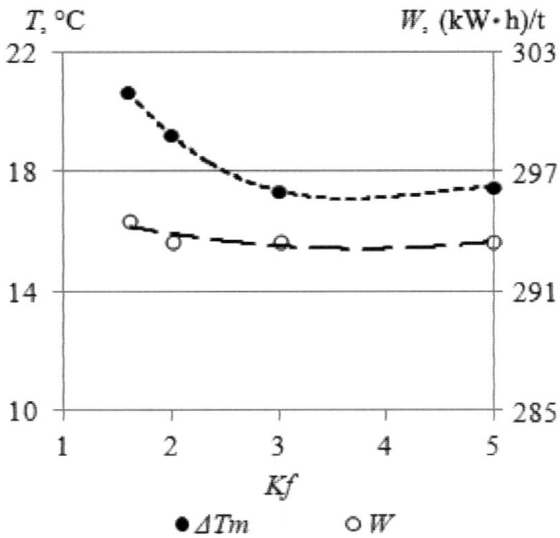

Figure 5. Characteristics ΔT_m and *W* as a function of the frequency multiplicity of the inductor current in heating zones.

In the three-zone heating installation under consideration, it is possible to control the heating by a controlled change in the relative power profile in the heater zones, which allows you to create a control system using temperature feedback in a certain heating zone. If power control is carried out in the middle zone of the heater when the movement speed of the workpieces changes, then the sighting zone will be inside the second inductor [5]. The position of the point of sight can be set when calculating the heating of billets with different capacities.

Figure 5 shows graphs of the billets heating temperature distribution for different capacities, which correspond to the profiles of relative power in the heater zones, shown in Table 2.

Table 2. Characteristics of operating modes for different capacities.

P, t/h	P_r, kW	P_1-P_2-P_3, %	K_p	ΔT_m, °C	$W, \frac{kW\cdot h}{t}$
3,4	1027	55-35-10	0,45	21,5	301
3,8	1118	50-30-10	0,40	19,4	295
4,2	1221	46-45-9	0,37	18,3	291

Since power regulation is carried out in the second heating zone when the installation performance changes, and the powers in the first and third zones remain unchanged, the relative power profile in the heater zones changes, which leads to a slight change in the temperature difference over the billet cross section ΔT_m and specific electric energy consumption W.

Figure 6 shows graphs of the billets heating temperature distribution along the length of the heater for different plant performances in the heating mode AB1 and K_f = 2, which have a point of intersection in the zone of the second inductor at a distance $2/3\ell$ from the beginning of the inductor. When a pyrometric temperature sensor is placed at this point, continuous monitoring of the billets heating mode and automatic control of the power distribution profile across the inductors can be carried out when the workpiece speed changes, which allows to stabilize the heating temperature and flexibly adapt to the conditions of the product pressing process.

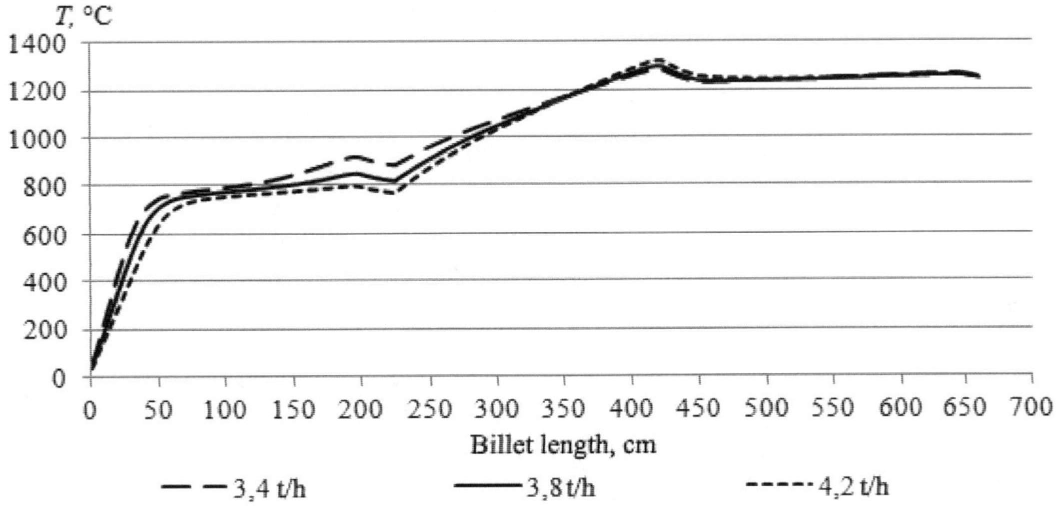

Figure 6. Billets heating temperature distribution along the length of the heater for different plant performances.

According to the results of the calculations, it is possible to determine the optimal power distribution profile for the heating zones, which corresponds to K_p = 0.34, and the optimal ratio of the current frequencies of the inductors of the first zone and subsequent heating zones, $K_f = f_2/f_1$ = 2, and the current frequency of the latter is assumed to be the same. Based on these relations, the division of the installation into three heating zones should be considered as rational, and the frequency converter can be made dual-channel with two inverters operating independently of each other and regulating the power and frequency of the current in the first inductor, as well as the power and frequency of the current in the inductors of the second and a third heating zone connected in parallel. Figure 7 shows a functional diagram of a three-zone dual-frequency heating installation with a dual-energy frequency converter.

In this case, the converter circuit is simplified, the load of the inverters is leveled, its mass-dimensional and cost indicators are reduced.

The parallel connection of inductors with the same parameters in the second and third heating zones implies their operation at the same voltage and equal current frequencies. Since when choosing AB heating mode, the workpieces in these zones are in a non-magnetic state, the active powers of the second and third inductors will be approximately the same ($P_2 = P_3$). Thus, the power distribution control in this installation is determined by the ratio of power in the first zone and the sum of the capacities in the second and third zones, and the coefficient of the power distribution profile is

$$K_p = \frac{(P_1 - P_2) - (P_2 - P_3)}{P_1 + P_2 + P_3} \approx \frac{P_1 - P_2}{P_1 + P_2 + P_3}.$$

Figure 8 shows graphs of the temperature difference over the billet cross section ΔT_m and specific electricity consumption W in the function K_p when heating billets with a diameter of 130 mm at a current frequency $f_1 = 600$ Hz and $f_2 = 1200$ Hz ($K_f = 2$), which show that the smallest ΔT_m is achieved at $K_p = 0.34$, while W depends little on K_p. Therefore, a rational choice of power distribution over heating zones corresponds to P_1-P_2-P_3% = 56-22-22%.

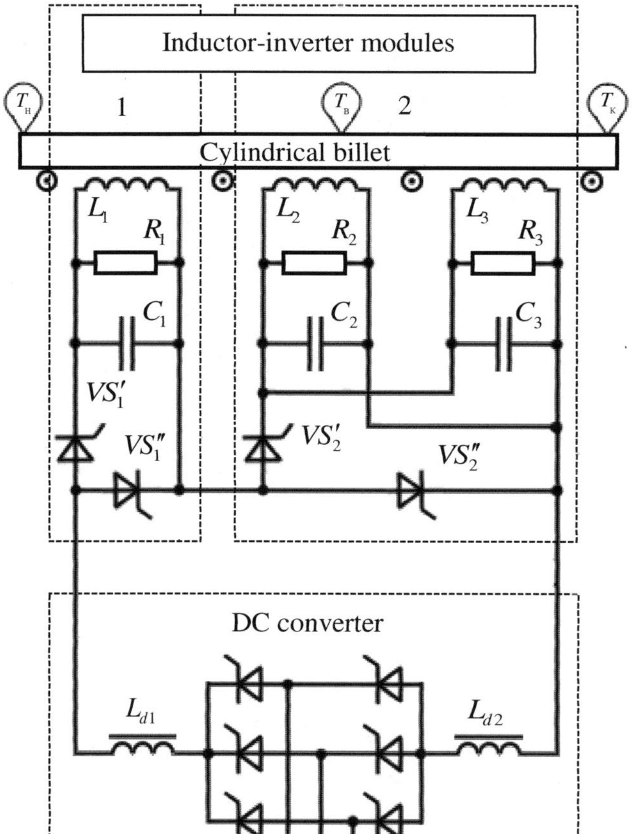

Figure 7. Functional diagram of a three-zone two-frequency heating of forging blanks.

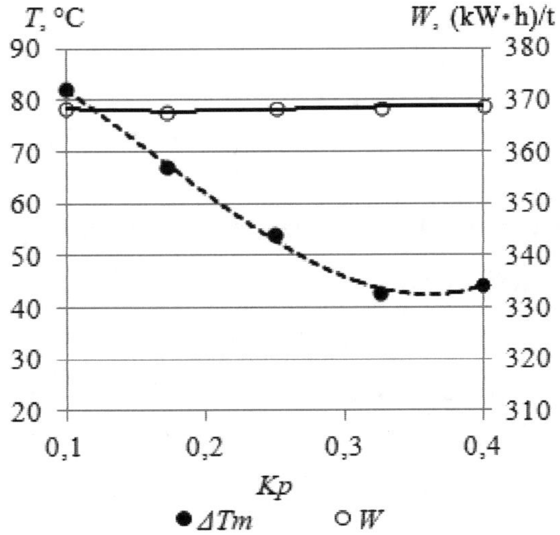

Figure 8. Characteristics ΔT_m and W from K_p for three-zone dual-frequency heating.

Thus, the method of three-zone two-frequency heating of billets makes it possible to make the optimal choice of the inductors parameters when creating induction heating plants with minimal weight and size indicators and wide functional capabilities.

The enterprises of OOO NPP «RELTEK» (Ekaterinburg) and Roboterm (Czech Republic) have developed multi-zone heating of forging blanks with a diameter of 90 ÷ 130 mm with a maximum performance of 4.2 t/h and a power 1700 kW. These plants are in pilot operation in the forging and pressing industry of the Kama Automobile Plant.

References

[1] Sluhotsky A E, Ryskin S E 1974 *Inductors for induction heating* (Leningrad: Energy) p 264

[2] Sluhotsky A E, Nemkov V S, Pavlov N A, Bamuner A V 1981 *Induction heating plants: Textbook for universities* ed A E Sluhotsky (Leningrad:. Energoatomizdat, Leningrad Department) p 328

[3] Luzgin V I, Petrov A Yu, Chernykh I V et al. 2010 *Converter device for induction heating and its variants* (Rus patent No. 2394400, Bull. Number 19)

[4] Demidovich V B, Chmilenko F V 2013 *Computer simulation of induction heating devices* (St. Petersburg: Publishing House of St. Petersburg Electrotechnical University "LETI") p 160

[5] Demidovich V B, Olenin V A, Grigoryev E A, Sitko P A 2013 Modular induction installations in forge production *Induction heating* Issue No. **4** (**26**) pp 9-12

IOP Conf. Series: Materials Science and Engineering **950** (2020) 012010 doi:10.1088/1757-899X/950/1/012010

Free metal surface resonance under alternating magnetic field low-frequency oscillations.

N Sizganov[1,2], M Khatsayuk[1]

[1]Siberian Federal University, 79 Svobodny, Krasnoyarsk, 660041, Russian Federation

[2]E-mail: vitz1234@gmial.com

Abstract. The article describes the definition of relationship between parameters of induction crucible furnace and meniscus shape using a mathematical model. The data obtained has been analyzed and the mathematical model qualitatively assessed using the open non-profit software Elmer and OpenFOAM. The models derived have been checked out to study the occurrence of free-surface oscillation resonance when subjected to pulsed action of an alternating electromagnetic field. Transients in an induction furnace crucible have been analyzed. A mathematical model with the electromagnetic field pulsed action has been created and free surface resonance occurrence and stability examined.

1. Introduction

Molten metal in a pulsating magnetic field of an induction crucible furnace (ICF) affected by the radial component of the bulk Lorentz forces $\boldsymbol{f}_L$ starts moving and shaping a meniscus on the free melt surface (Figure 1). The natural double-streamline electrodynamic circulation of the melt being formed has both positive effects, i.e. composition temperature equalization, metal-slag reaction, and negative one, i.e. the oxide film breakthrough, inefficient slag use, relatively poor mixing of circulating streamlines, etc. [1, 2, and 3].

In magnetic hydrodynamics (MHD), the issue of primary melt circulation streamline mixing or transient mass transfer intensification is resolved in different ways [4]. One of them is a liquid resonance driven by the pulsed action of a single-phase alternating electromagnetic field [5]. This paper is devoted to the study of free melt surface resonance oscillations under low-frequency pulsed action of an alternating electromagnetic field.

We emphasize the already known relationship for the meniscus in the ICF: meniscus height increases as the specific power rises; at a constant power transmitted to the melt, the force impact thereon strengthens as frequency decreases [2]. It has also been established that the meniscus intensity will depend not only on electromagnetic (EM), but also on the geometric parameters of the EM system, namely coil end position relative to the malt surface, melt to coil distance [1, 2].

Both analytical and numerical models can be constructed to solve this issue. The major restriction for the analytical method is typically rough assumption in the form of relatively simple (one-dimensional) models needed to achieve the solution. Numerical special case-oriented methods (FDM, FEM, FVM) are used for more accurate calculations. There is both commercial (Ansys, COMSOL, etc.) and free software (OpenFOAM, Elmer, etc.) with various types of partial differential equations solvers for specific problems in applied physics.

Previous meniscus studies were carried out using analytical methods [1,2], one-dimensional numerical models ignoring the hydrodynamics [6], two-dimensional axisymmetric numerical models with account for the hydrodynamics [7], as well modern software improved to link EM and hydrodynamic problems: Ansys classic + CFX [8], Ansys classic + Ansys fluent [9], Elmer + OpenFOAM [10] and Comsol + Ansys fluent [11]. The free melt surface in an alternating electromagnetic field is an MHD problem falling into a category of complex physics problems. To solve this type of problem, an open EOF-Library [12] is proposed, which allows linking software products with open source code in order to solve hydrodynamics (OpenFOAM) and electromagnetism (Elmer) problems. The method chosen for the numerical solution of the MHD problem has been repeatedly verified against a problem with a levitating melt drop in a conical coil field, which is common for this category [13].

This paper is aimed to conduct a theoretic experiment to check the occurrence of a free surface oscillation resonance under pulsed action an alternating electromagnetic field (EMF).

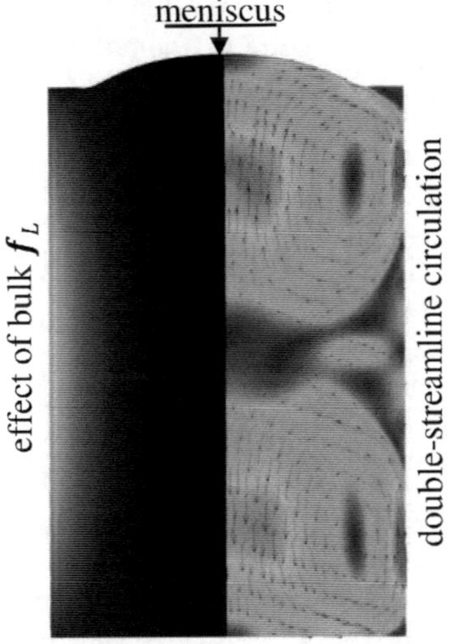

Fig. 1 - Meniscus formation in the induction furnace crucible

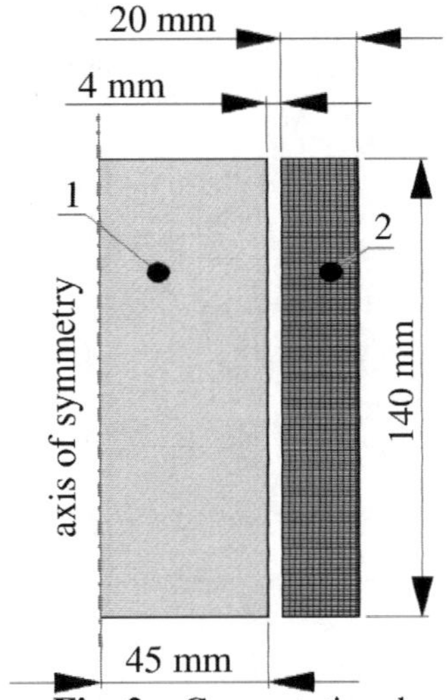

Fig. 2 – Computational region geometry

2. Materials and Methods

The ICF electromagnetic process studies consider the inductor download system. Based on the system's energy parameters, the inductor may be a natural air-cooled multi-turn alternating current coil or a massive forced water-cooled tube. The charge, if a non-conductive crucible is used or a crucible EMF-transparent at these frequencies, is only the melt.

For the purpose of numerical solution, it has been assumed that the actual inductor's turn are replaced by a single turn with the corresponding magnetomotive force (MF); non-conductive and minor geometric structural elements are neglected; the problem is solved in a two-dimensional axisymmetric arrangement; in the electromagnetic problem, the field changes harmonically; thermal processes and natural convective motion of the melt are neglected. Assumptions are standard for this class of problems and do not drive any significant increase in the result deviation, but allow for significant saving of computing resources. See Figure 1 for a scheme of computational region geometry indicating main dimensions, where 1 is the melt, 2 is the inductor. The dimensions were taken on the scale of an induction crucible furnace physical model to study MHD phenomena using a low-temperature eutectic alloy with metallic conductivity.

The hydrodynamic problem was solved in the melt region only. The expected Reynolds number in the melt is $Re \approx 38\,000$, which is much more than the critical value, so turbulent flows should be expected. These were taken into account using the k-ε turbulence model. Under alternating magnetic field conditions, the hydrodynamic problem took into account the force action through an additional source of motion. A free surface is simulated using the volume of fluid method (VOF), where α characterizes the volume fraction of fluid in the mesh cell.

The electromagnetic problem was solved in a quasi-stationary arrangement, indicating the equivalent current density. The computational region, in addition to the inductor and the charge, is filled with air with boundaries in the electromagnetic problem distanced from the inductor to reduce the impact of boundary conditions.

During the calculation of the MHD problem, the EM part recalculation would be initiated based on the change in the free surface position determined during the solution of the hydrodynamic part. This approach reduces the number of inner solutions of the electromagnetic part greatly declining the total amount of calculations.

The pulsed Lorentz force in the melt is created by interrupting current generation in the inductor with a frequency of $f_{pls} = 1/T_{FS}$. The ratio of the pause time when the melt moves driven by inertia to the period of time during which the Lorentz force affected the melt is called relative pulse duration and is always equal to $\psi = 1$ in this study. The created impulse action at the mathematical model level is implemented by adding the Lorentz force matrix multiplier in the EOF-library $\boldsymbol{f}_{pls}(t) = \boldsymbol{f}_L \cdot k_{10}(t)$. Pulse ratio used k_{10}:

$$k_{10} = \frac{1}{2}\left(1 + \frac{\sin(2\pi f_{pls} \cdot t)}{|\sin(2\pi f_{pls} \cdot t)|}\right)$$

where f_{pls} -is resonant frequency, t is process time.

Hereinafter in the text and in the graphs, the total current $I, \mathrm{A} \cdot turn$ is used to indicate the MF value. All study results related to the media interface will be shown for $\alpha = 0{,}5$. Properties of materials used necessary to solve the MHD problem are shown in Table 1.

Table 1. - Physical properties of materials

Item no.	Parameter	Symbol	Units of measure	Value
GaInSn melt				
1	Density	ρ	kg/m^3	6080
2	Conductivity	σ	S/m	3.85×10^6
3	Kinematic viscosity	ν	m^2/s	2.6315×10^{-7}
Copper inductor				
4	Relative magnetic permeability	μ	rel. units	1
5	Conductivity	σ	S/m	5.96×10^7
Environment – Air				
6	Density	ρ	kg/m^3	1
7	Kinematic viscosity	ν	m^2/s	1.48×10^{-5}
8	Melt-air surface tension	γ	N/m	0.72

3. Results and discussion

3.1. The study of meniscus shape as a function of energy and geometric parameters

The determination of relationships connecting ICF parameters and the meniscus shape has required the introduction of the following evaluative attributes: total meniscus height $\Delta h = h_{max} - h_{min}$, where h_{max}, h_{min} are meniscus' maximum and minimum axial position; meniscus' deviation from equilibrium $\Delta h_0 = h_{max} - h_0$, where h_0 is initial melt level (Figure 3); U_{ave}, is the average melt circulation rate.

These parameters have allowed to characterize the observed processes in the melt in terms of hydrodynamics. In course of the numerical experiment, two EM parameters of the system have been varied – inductor power frequency f_0 and full current I, as well as the fullness of the crucible with the melt.

We consider the dependence of evaluative parameters on the total current, as shown in Figure 4.

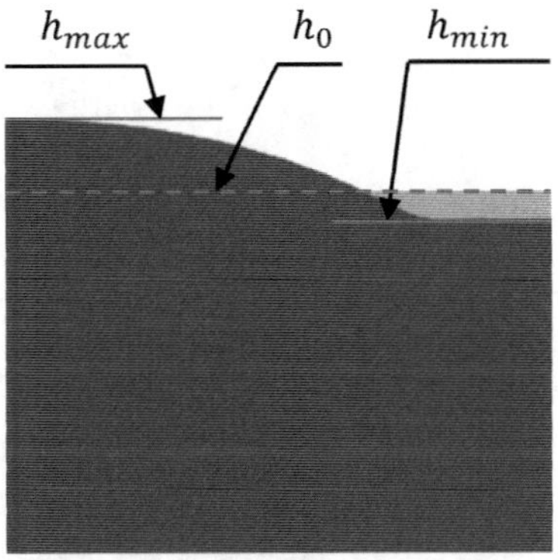

Fig. 3 - Axial position of the meniscus

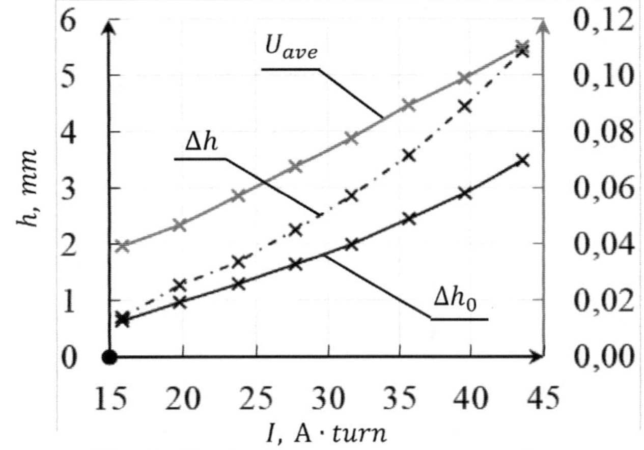

Fig. 4 - Evaluative parameters vs. total current

Figure 4 shows a quadratic dependence of the meniscus height (Δh) on the current and, accordingly, a linear dependence on the system capacity. The average speed (U_{ave}) is in turn linearly related to the current magnitude.

Figure 5 shows the relationship between evaluative parameters and the relative penetration depth $\Delta' = 1/(R\sqrt{\pi f_0 \sigma \mu_0})$, where R is the radius of the melt, μ_0 is the magnetic constant, σ is the melt conductivity, f_0 is the inductor power frequency.

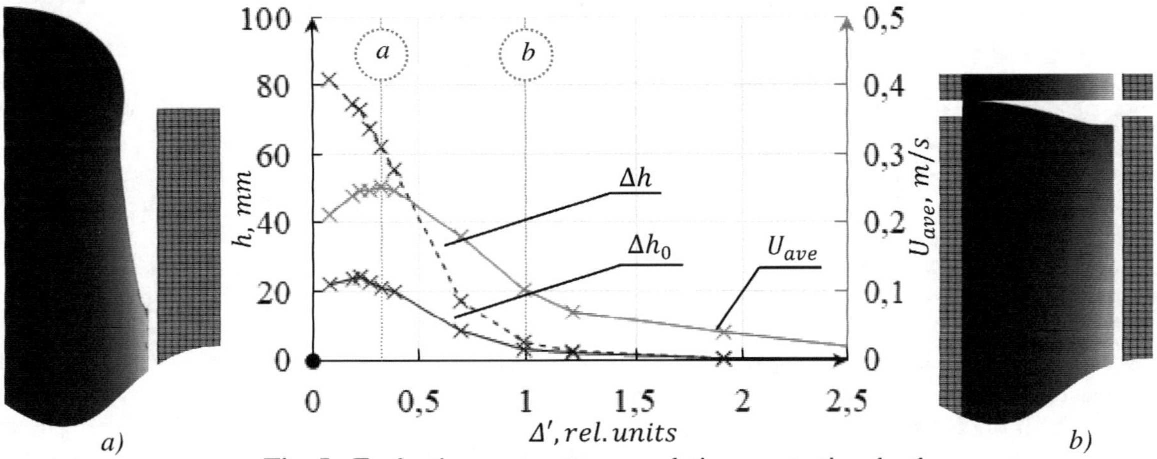

Fig. 5 - Evaluative parameters vs. relative penetration depth

Increasing inductor power frequency (decreasing Δ') results in a steady increase in the integral value of the Lorentz forces F_L, which is shown by an increase in all the evaluative parameters. When the penetration depth approaches the workpiece radius (for the sizes used in the calculation of about 500 Hz), an increase in electromagnetic pressure induces an increase in hydrostatic pressure to balance the system, as can be seen from the meniscus deviation from the equilibrium position (Δh_0). Melt speeds (U_{ave}) depend on the value F_L, while free surface parameters and Lorentz forces are of a more complex relationship. The meniscus shape depends on the size, however the local nature of F_L force distribution, namely the axial distribution in the meniscus zone due to the boundary effect has a greater effect on the free surface, which is characterized by different values Δ' of the specified parameter maxima. A further increase in the frequency results in a decreased integral value, but an increased density of Lorentz forces, which can explain differences in the behavior of parameters Δh_0 and Δh within this frequency range. This meniscus height (Δh) increasing behavior is explained by the fact that the meniscus point's h_{min} decline is due to increased EM pressure in the boundary layer exerted on the melt surface up to the complete separation of the melt from the crucible walls.

To indicate the fullness of the crucible with the melt relative to the inductor, $k_m = h_0/h_{ind}$ parameter is introduced, where h_{ind} is the inductor height; its effect on meniscus characteristics is analyzed (Figure 6).

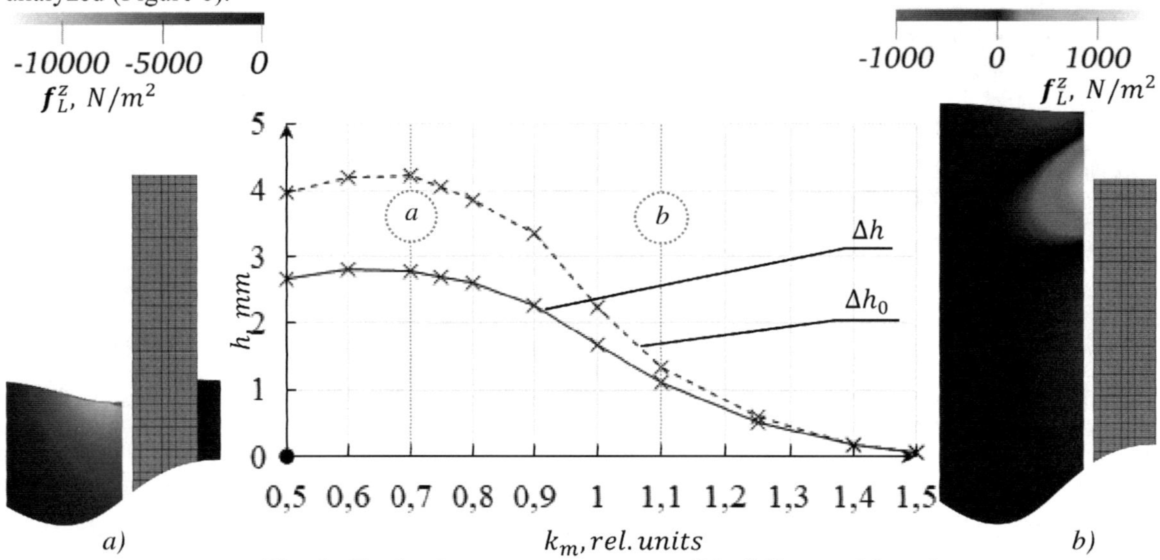

Fig. 6 - Evaluative parameters vs. crucible fullness with melt

When emptying the crucible $k_m < 1$ (Figure 6, a) the meniscus height will increase, this is explained by an increase in the axial component of the Lorentz force due to edge effects. The additional created pressure exerted on the free surface near the contact point of the three media (air-melt-crucible) intensifies the upper circulation streamline and increases the meniscus height. Further emptying of the crucible leads to a slight decrease in the meniscus height due to a limited total melt volume for closing hydrodynamic flows. When the crucible is overfilled $k_m > 1$ (Figure 6, b) the axial component of the electromagnetic forces disappears along with the meniscus. From [1] it is also known that this region can form secondary flows over the inductor.

3.2. Description of free surface-related transients.

A numerical experiment was carried out with the MHD system configured to achieve an apparent meniscus height of 3 mm. These system parameters were selected with the intention to continue the research on the physical model. To make it practicable, industrial power frequency $f = 50\ Hz$, total current allowing to construct the unit with no considerable financial costs for additional power supplies, $I = 40$ кА $\cdot$ $turn$ as well as a small volume of liquid metal ~0,02 m^3 were used.

Figure 7 shows the results of a non-steady MHD problem calculation with several states. The bottom graph is given to indicate the corresponding process control behavior, namely, the dynamics of instantaneous currents and effective electromagnetic forces.

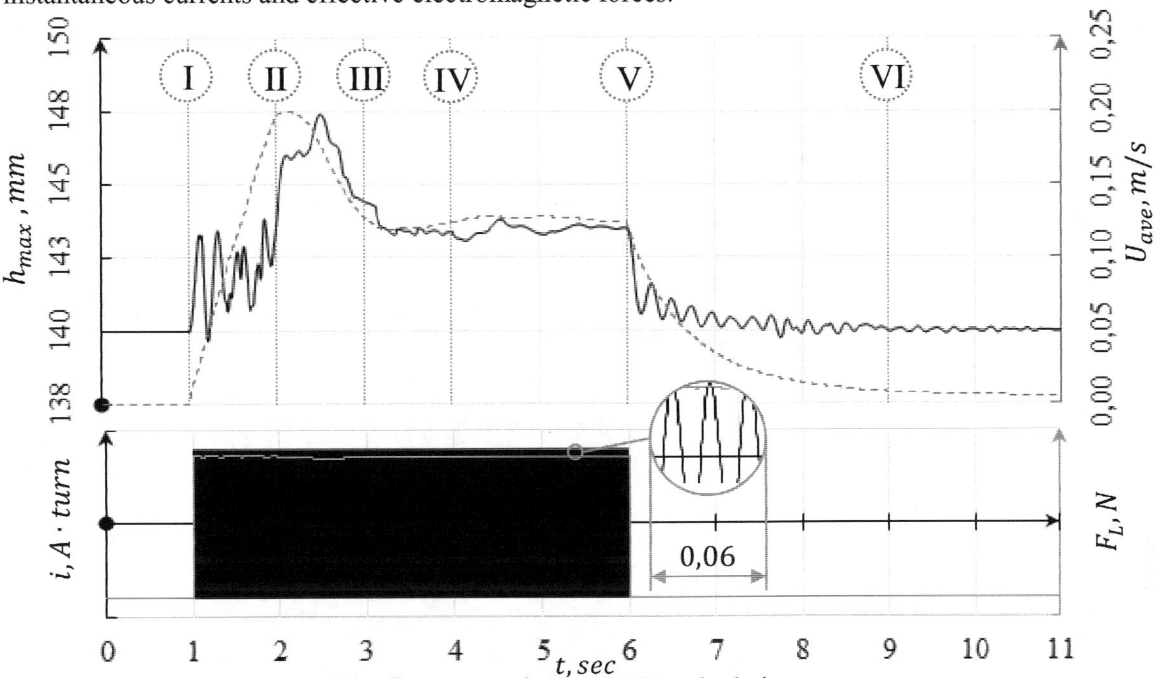

Fig. 7 - Non-stationary MHD calculation

Figure 8 shows the dynamics of the MHD processes at characteristic points of time indicated in Figure 7. The left side shows the bulk Lorentz force vector field modulus for the melt $\boldsymbol{f}_L = Re\{\dot{\boldsymbol{\delta}} \times \dot{\boldsymbol{B}}^*\}$where $\dot{\boldsymbol{\delta}}$ is the complex current density vector, $\dot{\boldsymbol{B}}^*$ is the complex conjugate magnetic induction vector. The right side shows the velocity modulus distribution.

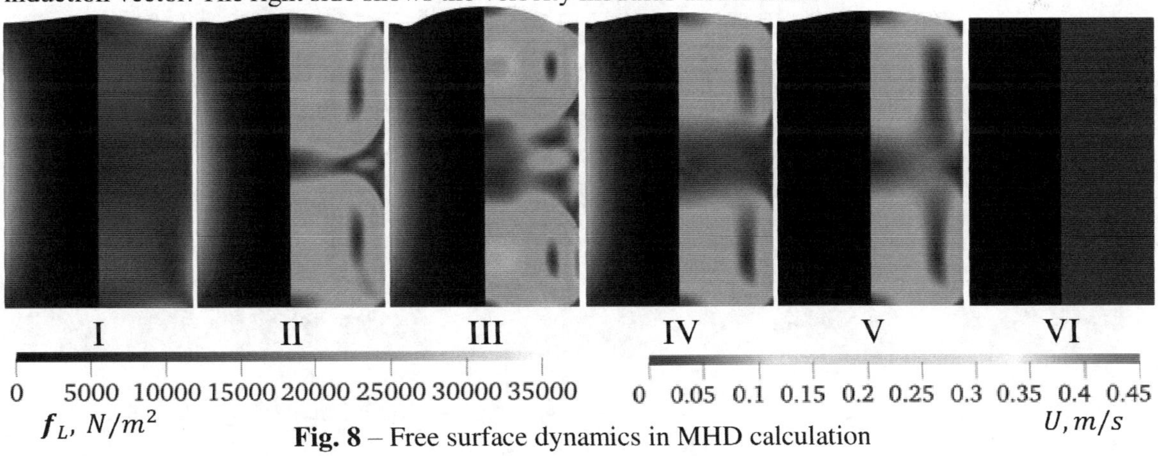

Fig. 8 – Free surface dynamics in MHD calculation

In the initial period of time (before I), the equilibrium state of the melt unaffected by Lorentz force is shown. At time point I, the coil is energized and the current is created in it. The force effect of induced fields interacts with the melt and a velocity transient begins with 2 distinguishable phases:

(I-II). Instantaneous reaction of the melt to EM pressure, characterized by meniscus extrusion with cylindrical damped waves formed on the surface, which cause a complex nature of meniscus pulses due to interaction with their part reflected from the crucible;

(I-III). The hydrodynamic effect has its transient rate up to $\sim 3\ sec$ during which the average melt velocity rises and the average vibrational position increases, higher than in phase (I-II), to a maximum value.

Then, within the interval (III-IV), the hydrodynamic and EM pressures inside the melt begin to equalize and a steady state with a "permanent" meniscus (IV-V) sets in. At time point V, the ICF turns off and an inverse transient occurs with damped cylindrical vibrations, after which the free melt surface reaches a calm state (VI) again, while velocities inside the melt tend towards zero.

Based on observations of transients in such systems, it was hypothesized that the free surface resonance is possible [14], and the possibility to maintain the resonance state by a pulsed action of EMF forces with a frequency equal to that of free surface vibrations was assumed.

3.3. Determining the natural surface oscillation period

Determining a more accurate value of the natural surface oscillation period T_{FS} for the crucible, has required the solution of an additional hydrodynamic problem with no effect of Lorentz forces. In this case, the initial conditions in the free surface problem, are represented by an arbitrary meniscus height (Figure 9, a).

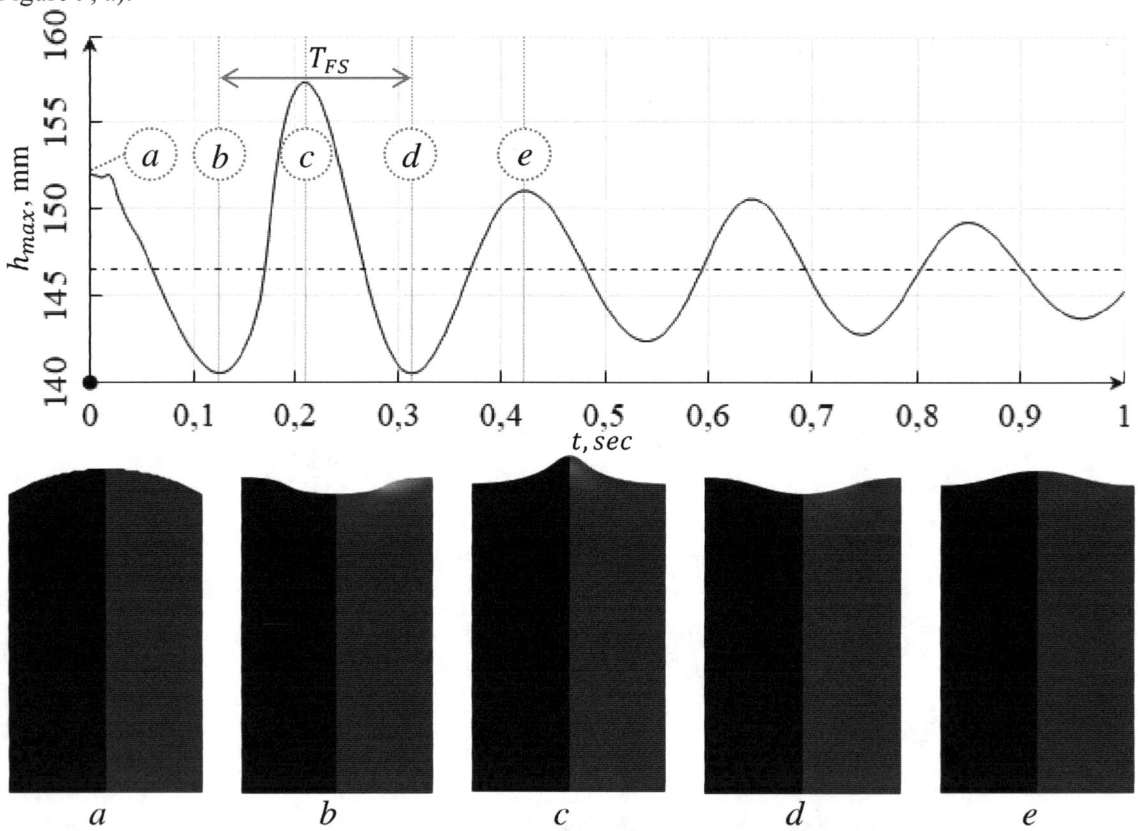

Fig. 9 – Non-stationary calculation of the hydrodynamic problem with no effect of Lorentz forces, with free melt surface dynamics

The graph shown in Figure 9 demonstrates the dynamics of damped oscillations h_{max} The bottom of Figure 9 (a - e) shows the distribution α at time points indicated on the graph. The adjusted free surface oscillation period was equal to $T_{FS} = 0{,}211\ sec$. Therefore, Lorentz force pulsed action frequency $f_{pls} = 1/T_{FS} \approx 4.7\ Hz$.

3.4. Analysis of non-steady processes with pulse behavior of EM forces.

Figure 10 shows the nature of process development within a single oscillation period.

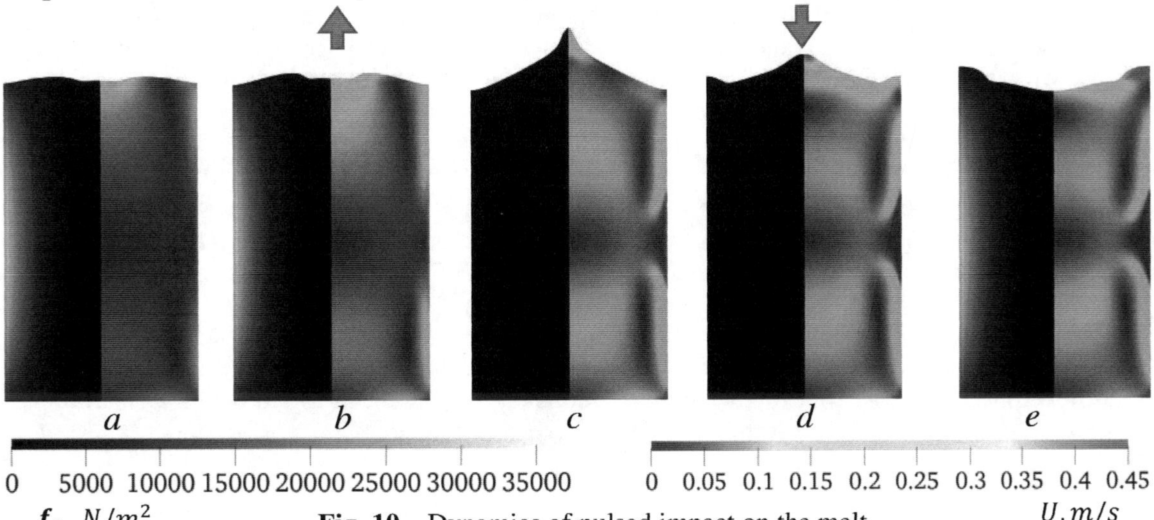

Fig. 10 – Dynamics of pulsed impact on the melt

At time point (a) the meniscus moves up driven by Lorentz forces, which coincides in phase with the free surface oscillation period, thereby "neutralizing" the attenuation of oscillation. Upon reaching a peak in the meniscus height (c) the coil is deenergized, which allows the meniscus to move freely downward under the action of the gravity field. When the minimum height is reached, the coil is energized again and the cycle described continues.

For ease of comparison, Figure 11, similarly to Figure 7, shows transient dynamics for main parameters when affected by pulsed Lorentz forces.

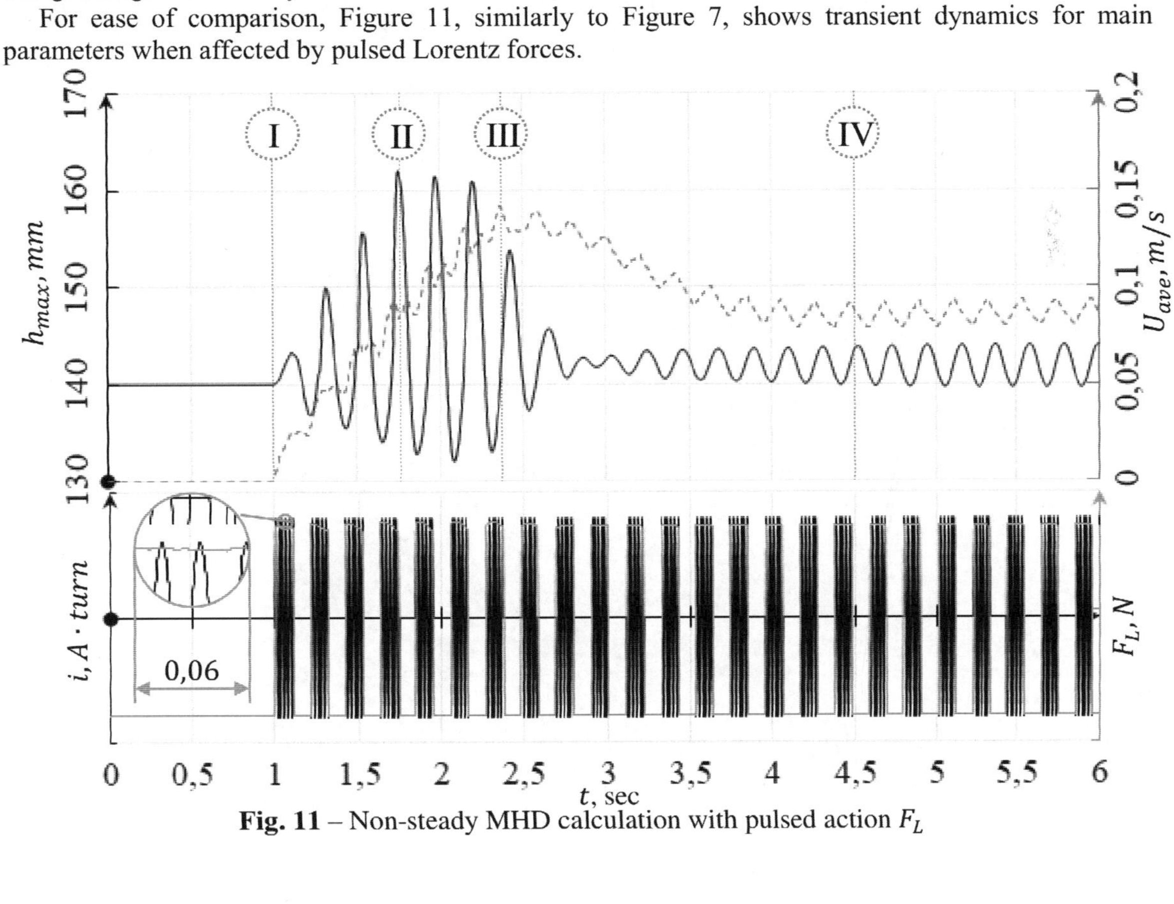

Fig. 11 – Non-steady MHD calculation with pulsed action F_L

The hydrodynamic transient begins at time point I and, as in the constant exposure case F_L (Figure 7), consists of 2 phases. Under pulsed action during the first phase of the transient, free surface oscillation amplitude intensifies sharply reaching the maximum value at point II, which indicates the onset of the free surface oscillation resonance. It is clearly seen that when the maximum melt (III) velocity is reached, the resonance starts decaying, which suggests the dependence of the melt surface resonant frequency on the melt circulation rate. In addition, given the studies of relationship of the fullness of the crucible with the melt, described in the first part, it is now interesting to study the local effect on the free surface by the pulsed method to minimize the velocity field by reducing the radial component F_L. At time point IV, a quasistationary state of a hydrodynamic system with a constant pulsation amplitude U_{ave} and h_{max} sets in. In this case, the velocity fluctuations are saw-toothed. Drastic free surface changes noticeably affect F_L value, which is indicated by uneven pulse maxima in the lower graph of Figure 11.

To study the effect of the velocity field on the resonance frequency, a series of numerical experiments was carried out to determine the dependence of the free surface oscillation amplitude in the steady state on the force pulsed action frequency. Resonance frequency f_{pls} has been varied within a small range from the onset of steady-state condition under the resonant pulsed action of Lorentz force (Figure 11, IV).

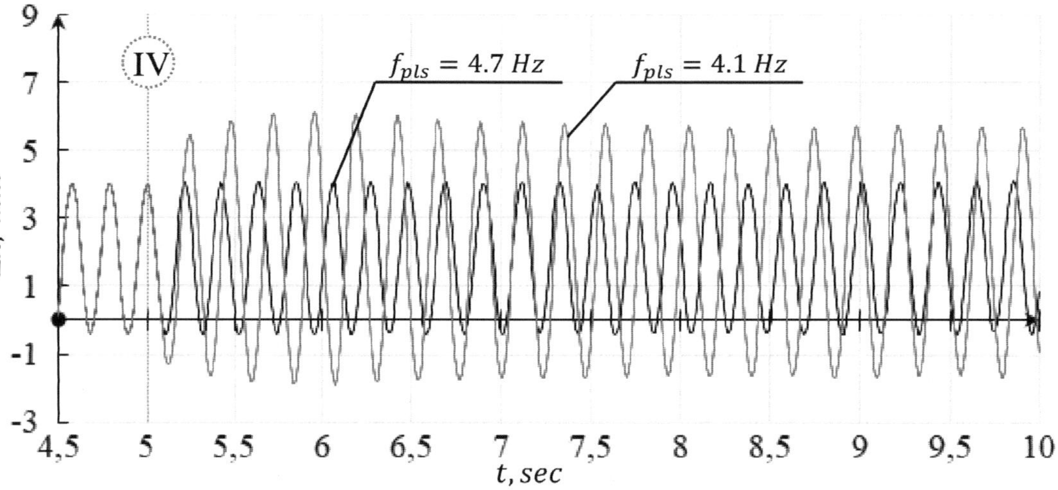

Fig. 12 – Comparison of calculations at different frequencies f_{pls}

Figure 12 shows an example with a pronounced increase in the free surface oscillation amplitude after a slight decrease f_{pls}, which confirms the assumed velocity field effect. Obviously, when using an automated electromagnetic force fluctuation meniscus position feedback control system for an ICF, the system becomes self-oscillating i.e. resonates automatically.

Figure 13 shows the dependence of the meniscus height and average melt circulation rate on frequency f_{pls}.

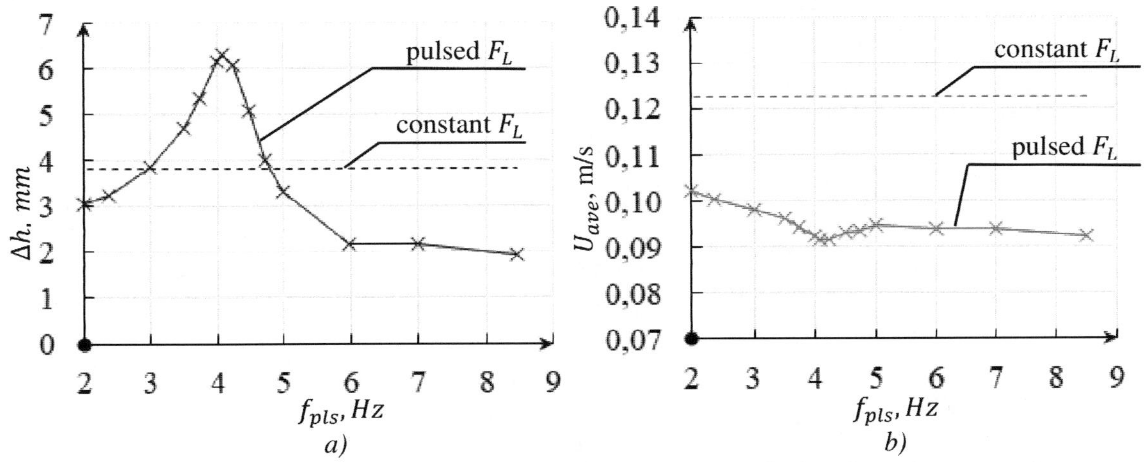

Fig. 13 – Dependencies of meniscus height *(a)* and average velocity *(b)* on frequency f_{pls}

As can be seen on the graphs, the velocity field shifts the resonant frequency towards the decrease. It is also seen that with the free surface resonance, the meniscus' equilibrium state deviation (Δh_0) is almost twice as high as with constant exposure to EMF, despite the fact that the average value F_L, and hence the power is actually half as great. On the other hand, apart form the obvious general decrease in the average velocity, the resonant frequency demonstrates its additional deviation indicating an increase in the kinetic energy dissipation under these conditions.

4. Conclusions

1. The study of the basic laws of meniscus formation in a steady-state electromagnetic field has allowed to assess the conformity of the models created to the verified results of authoritative papers in these areas of study and to checkout them. The results obtained have allowed the selection of basic values of main MHD system parameters for further solution of the unsteady problem aimed to study oscillatory processes.
2. The presence of the free surface resonance with negative and positive properties has been confirmed.
3. The creation of self-oscillations using an automated control system is proposed, which can be verified by reworking the numerical model or creating a physical one.
4. A relationship has been found between the melt flow velocity and the frequency of the Lorentz force's pulsed action on the melt, which is of interest for further studies in various conditions and system configurations.
5. Also, this method has been assumed as a separate exciter of local EM pressure to create a mixing streamline due to free surface oscillations in an inert medium.

Acknowledgments

The study has been carried out with funding from the Russian Federal Property Fund and the Government of the Krasnoyarsk Territory as part of a scientific project No. 18-48-242013 The Research of the Influence of Spatial and Time-And-Frequency Electromagnetic Force Distributions in the Melt on MHD Process Behavior in Liquid Metal.

References

[1] A.E. Sukhotsky, V.S. Nemkov, N.A. Pavlov, A.V. Bamuner, Induction heating plants, Energy Publishing House, Saint Petersburg, 1981.

[2] A.M. Weinberg, Induction melting furnaces, State Energy Publishing House, Moscow, 1960.

[3] E. Baake, M. Langejuergen, M. Kirpo, A. Jakovics, Analysis of transient heat and mass transfer processes in the melt of induction channel furnaces using LES, Magnetohydrodynamics 45 (3) (2009) 385-392. http://doi.org/10.22364/mhd (http://mhd.sal.lv/contents/2009/3/MG.45.3.11.R.html)

[4] V.I. Luzgin, A.Yu. Petrov, F.N. Sarapulov, V.E. Frizen, Turbo-induction complex for melting and casting of metal composites, Induction heating 4 (2011) 32-36. https://elibrary.ru/item.asp?id=17104867

[5] D. A. Musaeva, V. K. Ilyin, V. Geža, E. Baake, Experimental investigation of low-frequency pulsed Lorentz force influence on the motion of Galinstan melt, St. Petersburg Polytechnical University Journal: Physics and Mathematics 000 (2016) 1-8. http://dx.doi.org/10.1016/j.spjpm.2016.08.002

[6] H. Nakata, J. Etay, Meniscus Shape of Molten Steel under Alternating Magnetic Field, ISIJ International 32 (4) (1992) 521-528. https://doi.org/10.2355/isijinternational.32.521

[7] J. Sakane, B.Q. Li, J.W. Evans, Mathematical Modeling of Meniscus Profile and Melt Flow in Electromagnetic Casters. / Metallurgical and Materials Transactions B 19 (3) (1988) 397-408. https://doi.org/10.1007/BF02657737

[8] S. Spitans, A. Jakovičs, E. Baake, B. Nacke, Numerical modelling of free surface dynamics of conductive melt in induction crucible furnace, Magnetohydrodynamics 46 (4) (2010) 425–436. https://dx.doi.org/10.22364/mhd.46.4.12

[9] V. B. Demidovich, M. Y. Khatsayuk, I. I. Rastvorova, V. N. Timofeev, A. A. Maksimov, Numerical simulation of non-crucible melting of titanium alloy in the alternating electromagnetic field, Magnetohydrodynamics 51 (4) (2015) 785-794. https://dx.doi.org/10.22364/mhd.51.4.11

[10] J. Vencels, A. Jakovics, V. Geza, M. Scepanskis, J. EOF Library: open-source Elmer and OpenFOAM coupler for simulation of MHD with free surface. XVIII International UIE-Congress Electrotechnologies for Material Processing (2017). https://www.researchgate.net/publication/318418395_EOF_Library_Open-Source_Elmer_and_OpenFOAM_Coupler_for_Simulation_of_MHD_With_Free_Surface

[11] R. Bourrou, O. Budenkova, P. Charvin, C. Lafon, F. Lemont, A. Gagnoud, Numerical study of the hydrodynamics in a two-phase induction melter for nuclear waste treatment under various operating parameters, European Journal of Mechanics/B Fluids 79 (2020) 181–189. https://doi.org/10.1016/j.euromechflu.2019.09.005

[12] J. Vencels, P. Råback, V. Geža, EOF-Library: Open-source Elmer FEM and OpenFOAM coupler for electromagnetics and fluid dynamics, SoftwareX 9 (2019) 68-72. https://doi.org/10.1016/j.softx.2019.01.007

[13] J. Venčels, A. Jakovičs, V. Geža, Simulation of 3D MHD with free surface using open-source eof-library: levitating liquid metal in an alternating electromagnetic field, Magnetohydrodynamics 53 (4) (2017) 643–652. https://dx.doi.org/10.22364/mhd.53.4.6

[14] M. Khatsayuk, Theory and modeling of magnetohydrodynamic processes in electrotechnological complexes of metallurgical purposes (Doctor of technical sciences thesis), Saint Petersburg Electrotechnical University "LETI", Saint Petersburg, 2020. https://etu.ru/assets/files/nauka/dissertacii/2019/hacayuk/dissertaciya_hacayuk-myu.pdf

IOP Conf. Series: Materials Science and Engineering **950** (2020) 012011 doi:10.1088/1757-899X/950/1/012011

Induction heating system for die tooling of press for isothermal stamping of large-sized parts

V. Frizen, F Tarasov, S. Bychkov, S. Fatkullin, S. Sarapulov
Ural Federal University, Yekaterinburg, Russia
E-mail: F.E.Tarasov@urfu.ru

Abstract.
The paper describes the installation for stamping large-sized parts. The use of induction heating for heating the working tool of the stamp in the operating mode is considered. A system with 4 flat inductors is described. This solution allows for the most efficient hot stamping operation compared to other known solutions. An example of the result of numerical simulation of the temperature field distribution is demonstrated.

1. Introduction

In the production of parts from wrought aluminum alloys by stamping, the costs of their machining are significantly reduced. Therefore, this method is in demand for mass production. To obtain a stamped part by this method, it is required to provide certain conditions that ensure the plasticity of the workpiece during pressing. In particular, the billet must be heated to a certain temperature when loaded into the press. This temperature must be maintained during the deformation of the billet in direct contact with the die tooling. The pressing of relatively small workpiece is fast enough. During pressing, the workpiece does not have time to transfer the accumulated heat to the die tooling. A certain amount of heat is released in the workpiece itself during the deformation process. Therefore, additional heating is not required to ensure the plasticity of the workpiece during pressing. However with an increase in the pressed part dimensions, the die tool dimensions, as well as the pressing time, grow. If you do not take additional steps to heat the die equipment in this case, the load on the die tool increases and the structure of the resulting part is disturbed. Therefore, in the production of large-sized stamped parts from aluminum alloys, preliminary heating of the stamping equipment is required.

The Fig. 1 shows a diagram of a die tool used for the production of large-sized parts from wrought aluminum alloys. The die and punch of the die is fixed to the parts of the hydraulic press using a series of traverses (intermediate steel plates). The area of these plates is gradually reduced to evenly distribute and transfer the force developed by the press directly to the die tool. These elements of the die equipment, as well as the tool itself, have a significant mass and are capable of accumulating a large amount of heat. The Fig. 2 shows a flow diagram of the production of a part using the above-described tool. The pressing section is completed with furnaces in which the stamped parts are heated, as well as the preheating of the stamping tool. At the end of a long heating process of the stamping tool, it is installed in a hot state on the press, after which the blanks are stamped. The stamping process can be continued as long as the surface of the stamping tool remains hot and is in direct contact with the blank being stamped.

Figure 1. Die tooling design 1 - punch; 2 - matrix; 3 - traverses

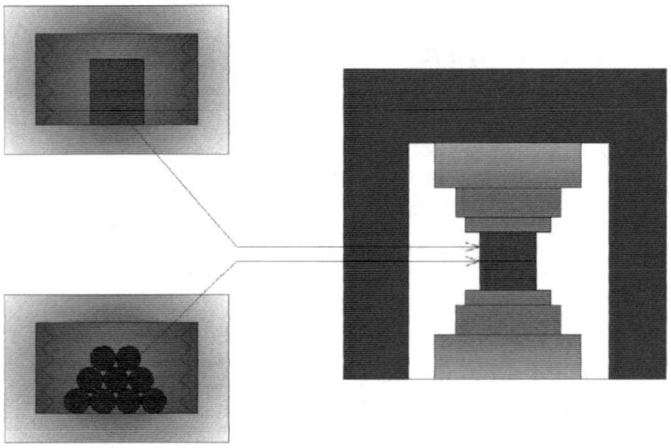

Figure 2. Technological scheme for preheating the stamping tool using 2 electric furnaces

As the tool surface temperature drops below the critical value, the die tool is removed from the press and loaded into the furnace for heating. Such a technological scheme is associated with significant labor costs for tool replacement. Since the die is mounted on cold die tooling plates, it loses heat at a high rate. Therefore, the tool replacement is necessary to make several times per day. To increase productivity and reduce energy consumption for the production of stamped blanks, tool heating is required directly during the pressing of parts. When compensating for the heat losses of a die tool, pressing of parts can be carried out without changing the tool during a work shift.

Previously, various schemes for heating the die tooling of presses were proposed (Fig. 3). For relatively small presses, variants of the design of the system for heating the outer surface of the die tool with gas burners or electric heaters [1, 2] were proposed. In this case, the die tooling is placed inside the heat-insulating chamber. This design is ineffective for large presses due to the insufficient specific surface power of the heaters to compensate for the heat loss of the preheated die tool for heating the traverse. In addition, stamping often requires the use of a lubricant that partially burns out when applied to the working surface of the stamp. The combustion products will inevitably settle on the inner surface of the heat-insulating chamber, which necessitates frequent stoppages of equipment for cleaning.

Since the main heat flux from the preheated die tool is directed towards the traverses, schemes for heating the tool with the location of the heat source directly in the traverses [3, 4] were

Figure 3. Variants of arrangement of die tooling heating systems

proposed. The possibility of placing heating elements in the traverses is limited by the conditions for ensuring the strength of the traverses, since these elements of the die equipment are under high mechanical stress. On the other hand, to accommodate heating elements of sufficient power, the number of cavities inside the traverse must be large. The use of electric heating elements in such heating systems for large die tooling did not justify itself due to insufficient efficiency in the presence of the above limitations.

Of particular interest are the systems for heating the working surface of the stamping tool [5, 6, 7, 8, 9] in the intervals between stamping operations. The use in such heating systems as heat sources of electric heaters with a relatively low specific surface power is associated with an increase in the time between stamping cycles. This leads to a change in the rate in the technological production cycle. The maintenance of stamping equipment becomes more complicated. The use of induction heaters in this arrangement is practically impossible when changing tools, since the inductive wire must follow the inner surface of the die. It is also difficult to ensure the required uniformity of heating of the working surface during its induction heating.

In view of the above, the most promising in the production of large stamped parts from aluminum alloys is the induction heating system of the outer surface of the stamping tool. In this case, the surface of the stamp will be sufficient to accommodate the inductor of the required power, in addition, such a system is quite simple to control.

The inductor of the heating system can be a rectangular coil-solenoid made of a copper water-cooled tube, put on the corresponding part of the stamping tool [10]. The induction wire is insulated from the die tool. This design is the easiest to manufacture, but such an inductor is not universal. It is required to keep several inductors for different tools used in production. In addition, heating control in this case is only possible in one zone.

More multipurpose is the design of the inductor, made of flat coils attached through a layer of thermal insulation to each surface of the stamp separately (Fig. 4). In this case, each coil can be powered from a separate power source and the power can be controlled more flexibly on either side of the die. The problem with the versatility of the heating system is also removed. This inductor system can be placed on a large number of dies without changing the coil design.

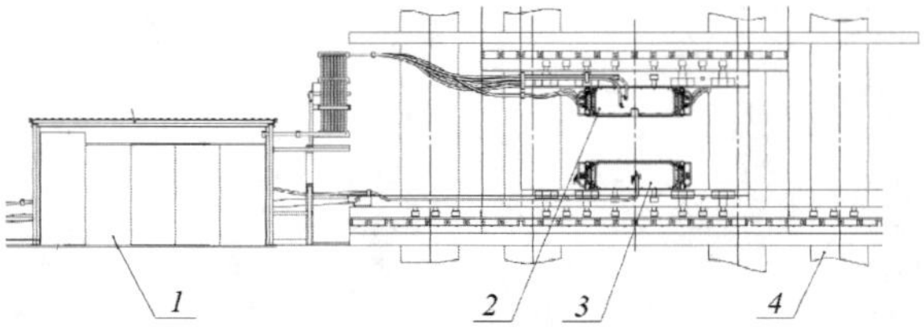

Figure 4. Induction heating system for die tooling with flat inductors 1 - power cabinet for the induction installation; 2 - a block of 4 inductors of the upper part of the tool; 3 - a block of 4 inductors of the lower part of the tool; 4 - press supports

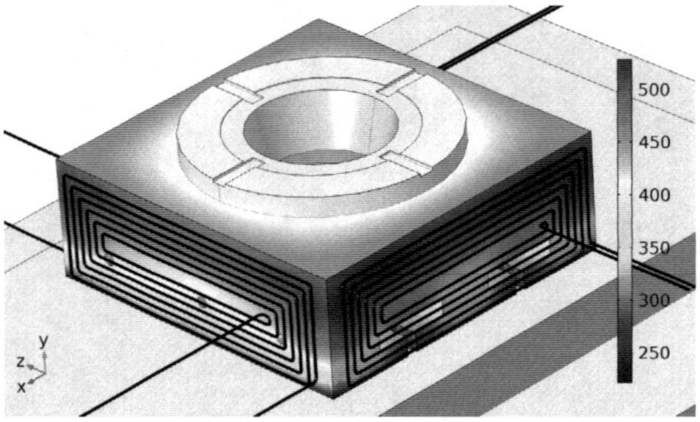

Figure 5. Temperature in a stamp working tool

Preliminary calculations have shown the high efficiency of the induction heating system for die tooling of the above-described design. The calculation results are shown in Fig. 5.

2. Conclusions

The review has shown the relevance and perspective of the development of systems for induction heating of die tooling for the production of large parts. In most cases, this heating method is the most acceptable both from the point of view of minimizing capital costs, and from the point of view of convenience and safety of operation, as well as maintenance of the press equipment.

Acknowledgments

This work was financially supported by the Ministry of Science and Higher Education of the Russian Federation, project No. 075-11-2019-028.

3. References

[1] Izakov I A 2005 *Patent RU 2 259 901 C1*
[2] Vasil'evich P N 2012 *Patent RU 2 464 172 C1*
[3] Bryan F E, Mitchell W D and Wheelon O A 1978 *U.S. Patent US3783669A* URL `https://patents.google.com/patent/US3007427?oq=US3783669A`
[4] Johns C L 2017 *U.S. Patent US10267968B2* URL `https://patents.google.com/patent/US10267968B2/en?oq=US201706`
[5] Urban J DeSouza Robin M Forbes Jones B B H J A L L R S M S A S 2018 *U.S. Patent US20190015890A1* URL `https://patents.google.com/patent/US20190015890A1/en?oq=RU2550455`
[6] Kunio Tomioka Masaaki Kita T S 1977 *U.S. Patent US4088000A* URL `https://patents.google.com/patent/US4088000A/en?oq=US4088000`
[7] 2018 *China Patent CN109909439A* URL `https://patents.google.com/patent/CN109909439A/en?oq=CN109909439`
[8] 2017 *China Patent CN207735540* URL `https://patents.google.com/patent/CN207735540U/en?oq=CN207735540`

IOP Conf. Series: Materials Science and Engineering **950** (2020) 012011 doi:10.1088/1757-899X/950/1/012011

[9] Shur I A 2013 *Patent RU 2 475 329 C1*
[10] Bychkov S A, Tarasov F E, Frizen V E and Smolyanov I A 2017 Stamp tool induction heating system *2017 18th International Conference on Computational Problems of Electrical Engineering (CPEE)* pp 1–4

IOP Conf. Series: Materials Science and Engineering **950** (2020) 012012 doi:10.1088/1757-899X/950/1/012012

Identification of working reactance of industrial arc furnace equivalent scheme based on known specific electric power consumption for melting of charge

A M Kruchinin, M Ya Pogrebisskiy, E S Ryazanova and A Yu Chursin

Department of Electric Power Supply of Industrial Enterprises and Electrotechnologies, National Research University "MPEI", 14 Krasnokazarmennaya, 111250, Moscow, Russia

PogrebisskiyMY@mpei.ru

Abstract. Method of identification of working reactance of industrial arc furnace equivalent scheme based on known specific electric power consumption for melting of charge taking into account complex (radiation and convection) nature of arc heat exchange in melting space is presented. The solution of the multi-factor problem of determining the share of arc radiation power and convective heating of charge by arcs in the furnace useful power was proposed at MPEI on the basis of analytical methods of the developed electric arc heat exchange model (EAHEM). Method of calculation of specific electric power consumption for melting of solid charge for specified type of arc furnace design is considered.

1. Introduction

Identification of parameters of industrial arc furnace equivalent electric circuit, including reactance, is necessary for obtaining power and performance characteristics of furnace, design of electric modes, synthesis and adjustment of automatic furnace control system.

In order to identify the working reactance, the electric thermal problem is solved taking into account the complex (radiation and convection) nature of the electric arc heat exchange.

2. Accounting for the complex type of charge arc heating in the furnace melting space

Experience of arc furnaces operation and methods of modern theory of working processes in arc furnaces indicate complex nature of arc heat exchange in melting space caused by radiation and convection caused by arc column plasma movement. The changing structure of charge heating during the smelting process affects the rate of charge melting and thus the specific power consumption for charge melting.

Column plasma radiation has a volumetric character. In this regard, the calculation of arc emission power P_{ε} is not possible without knowing the temperature profile of the column $T(r)$

$$P_{\varepsilon}=l_a\cdot 2\pi \int_0^{r_a} \sigma_{\varepsilon}(T(r))\cdot \text{¿} r\cdot dr\, \text{¿} \qquad (1)$$

where l_a – length of arc, r_a – radius of arc.

This calculation is illegal on the basis of the Stefan-Boltzmann law relating to body surface radiation.

This methodological problem in the calculation of arc emission power and the allocation of this power in the furnace useful power results in significant errors in the calculation of furnace performance when the furnace useful power is taken as the power of arc emission in the calculation of thermal characteristics. This is due to the fact that the share of arc radiation power in the melting process depends on many factors and primarily on:

- melting period and heat exchange conditions of the arc in each period of the working cycle;
- parameters of the furnace equivalent circuit;
- operating current of the arc.

Convective heating of the charge by arcs [1] is mainly due to the intense plasma motion in the column and depends not only on these factors, but in addition depends on the thermal regime of the graphite electrode [2], which varies from period to period in the melting process.

The solution of the multi-factor problem of determining the share of arc radiation power and convective heating of charge by arcs in the furnace useful power was proposed at MPEI on the basis of analytical methods of the developed electric arc heat exchange model (EAHEM):

- the method of universal characteristics of the arc [3] and
- the method of structural characteristics of the arc [2].

These methods are based on the provisions of the modern theory of arc heating [1] and allow us to link the projected furnace electric mode of the furnace with the conditions of its heat exchange in the melting space [2] and energy-physical processes in the plasma of the arc column of known chemical composition for a specific melting period. The mathematical apparatus of the developed methods was used to form a package of computer programs "*tomed_dsp*" for designing an arc furnace rational electric mode [4].

To illustrate the possibilities of the EAHEM method, we calculate the performance characteristics for the melting period of the charge in the well-known DSP-200 industry arc furnace (with a capacity of 200 tons) with transformer with a power S_t of 32 MVA and a diameter of 550 mm graphite electrodes [5]. Based on the furnace operation data, there are reliable experimental data on the specific electric power consumption for melting the charge 400kW·h/ton. Electric mode parameters during the melting of the charge: the secondary voltage of the transformer U_2 is 591.5 V, the operating arc current is 43.9 kA. Reduced values of furnace equivalent circuit resistance and reactance: resistance is 0.65 mOhm, reactance for the first harmonic of the short-circuit experience is 4.6 mOhm.

As a result of the calculation, without taking into account the influence of arc current distortion on the furnace equivalent circuit reactance, it is possible to identify the current parameters of the furnace electric mode. Calculation by the well-known methods of electrical characteristics (useful power, lining wear coefficient (LWC)) allow to calculate the current parameters of the furnace at the maximum points of these characteristics:

- arc current for the maximum value of useful power I_{pd} is 48.6 kA;
- arc current for the maximum value of LWC I_{rf} is 38.6 kA;
- maximum value of useful power $P_{U\,max}$ is 33 300 kW.

In the practice of designing a rational electric mode of arc furnaces [6], the operating arc current is usually taken within

$$I_{oper} = (0,87 - 0,90) \cdot I_{pd} \tag{2}$$

The conditions of electric mode rationality, which should ensure the fullest use of the installed power, the introduction of as much energy into the furnace as possible per unit time and ensure that no overheating of any part of the furnace masonry. However, the solution of the thermal problem based only on the electrical characteristics is not possible without taking into account the complex type of heating by charge arcs.

In this regard, the EAHEM method allows to calculate the structure of arc heating in the melting space of the furnace, taking into account the heat exchange conditions characteristic of a given melting period, and to link the heating characteristics with the electric mode. Calculations for DSP-200 furnace showed that the main components in the useful power of furnace going on the formation of heat flow in the melting space, the cost of electricity on the formation of the radiation power arcs and heat flux forced convection due to the plasma column motion.

Figure 1 shows the dependence on the arc current share of heat flow into useful power analyze DSP-200 in current range from 0.5 to 1.1 for the period of melting without taking into account the influence of the arc current distortion on the furnace reactivity.

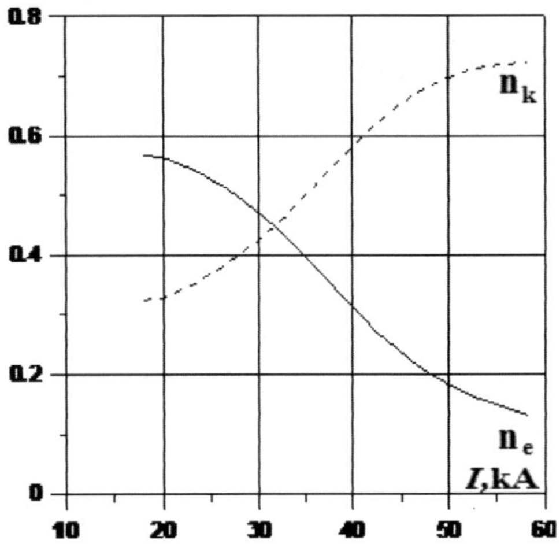

Figure 1. Change in the structure of heat flows in the melting space of DSP-200 furnace during the melting of the charge.

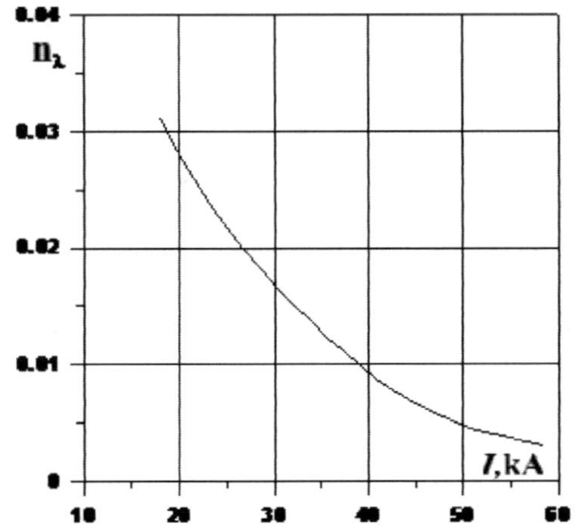

Figure 2. The share of electric power expense for the formation of heat flow in the useful power of DSP-200 furnace during the melting of the charge.

As follows from the figure 1 graph the structure of heat flows in the furnace melting space depends significantly on the furnace impedance, supported by a power controller. For DSP-200 furnace operating current is 200 kA, the share of arc radiation is 0.252, the share of forced convection due to plasma motion in the column is 0.64. As the furnace impedance decreases, the proportion of arc radiation decreases to 0.2.

The convective heat flux withdrawn from the surface of the column under conditions of natural convection is associated with the processes of thermal conductivity in the plasma and does not play a significant role in heating the charge. Figure 2 shows the dependence of the share of electric power expense for the formation of this flow in the useful power of DSP-200 furnace during the melting of the charge. The share of electric power expense for the formation of this flow in the useful power of the DSP-200 furnace does not exceed 0.02. However, the values of the heat flux are the basic physical EAHEM factor determining the temperature profile of the arc, and hence the level of other heat fluxes generated by the arc in the melting space for the conditions of heat exchange of the arc in a particular melting period [2].

The EAHEM method thus allows to calculate the temperature profile of the arc column, to identify the arc length and, as a result, to calculate the arc radiation power in the melting space of the furnace

IOP Conf. Series: Materials Science and Engineering **950** (2020) 012012 doi:10.1088/1757-899X/950/1/012012

on the basis of (1). The operating characteristic of the radiation power and the position of its maximum usually does not coincide with the maximum point of the operating characteristic of the LWC. This, in particular, is evidenced by the calculated characteristics for the DSP-200 furnace, presented in figure 3.

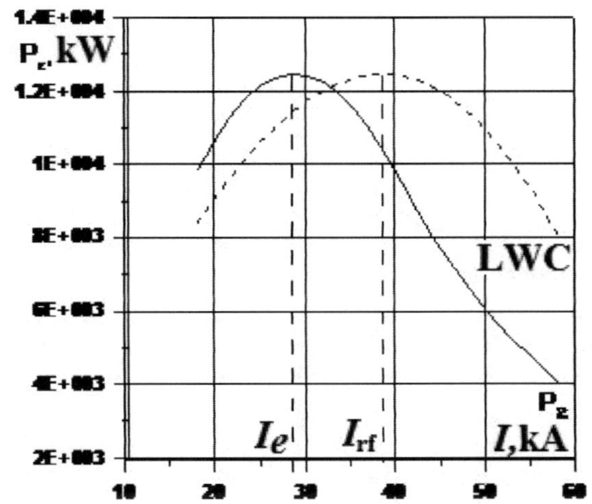

Figure 3. Characteristics of arcs radiation power and LWC for DSP-200 furnace during the melting of the charge.

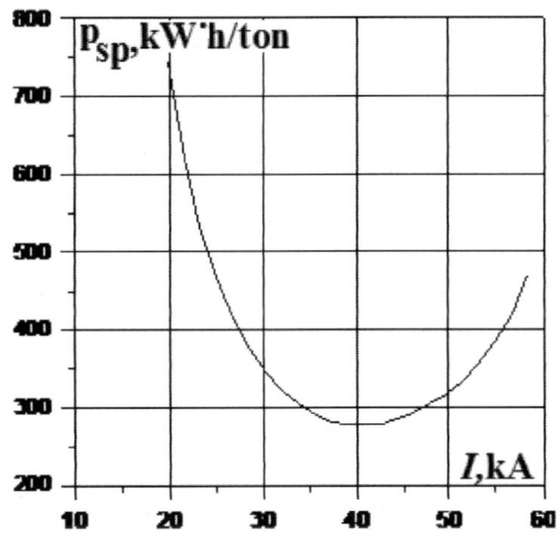

Figure 4. Operating characteristic of the specific power consumption for melting the charge in DSP-200 furnace.

The current value I_e for the maximum arc radiation ($P_{\varepsilon\,max}$=12 460 kW) is 28.9 kA. The current value I_{rf} for the maximum LWC, respectively, is 38.6 kA. In this regard, the operating characteristic of the arc furnace is an empirical electrical characteristic and does not reflect the position in the operating mode of the furnace currents of the maximum irradiation and wear of the lining.

3. Calculation of specific electric power consumption for melting solid charge for a given type of furnace construction

As a result, based on the calculation of performance characteristics, which can be performed for a specific industrial furnace of a given type of construction, to predict the electric power consumption for melting the solid charge, you can use the formula of the EAHEM method for specific power consumption

$$p_{sp}(I)=\left\{274,39+\frac{189,025}{\left[1+\left(G_f/29,65\right)^{0,6}\right]}\right\}\cdot\left(\frac{\left(K_{pl}\right)_b\cdot\left(0,85+0,15\cdot n_{pm}\right)}{\left(\frac{P_U}{G_f}\right)\cdot\left(\frac{P_\varepsilon(I)}{\left(P_\varepsilon\right)_{max}}\right)^{0,5}\cdot\left(\frac{P_U(I)}{\left(P_U\right)_{max}}\right)}\right),\ \frac{kW\cdot h}{ton} \qquad (3)$$

where G_f – furnace capacity in tons, n_{pm} – indicator of the type of furnace design, P_U – useful power.

For low power furnaces n_{pm}=1, for powerful furnaces n_{pm}=2, for high power furnaces n_{pm}=3, for high power furnaces n_{pm}=4, for ultra-high power furnaces n_{pm}=5.

In the formula (3), the base rate of melting the charge of low-power furnaces (n_{pm}=1)) can be calculated by the generalized formula

$$\left(K_{pl}\right)_b = \frac{529{,}5}{\left[1+\left(\left(G_f+144{,}74\right)/181{,}485\right)^2\right]} \tag{4}$$

The melting time of the charge under current taking into account (3) and (4) can be calculated as

$$\tau_m = \frac{G_f \cdot p_{sp}}{P_U} \tag{5}$$

As studies have shown [7], the generalized formula (3) allows with a high degree of reliability to predict the electric power consumption for the melting of a solid charge when setting up and designing an energy-saving electric mode of the active period of the charge melting in the furnace.

For figure 4 on the basis of the described EAHEM method, the operating characteristic of the specific power consumption for melting the charge for the analyzed DSP-200 furnace without taking into account the influence of arc current distortion on the reactivity of the DSP is calculated.

As follows from the calculated characteristics (figure 4), the operating arc current of 43.9 kA on the operating furnace is near the current of 40 kA, at which the electric power consumption will have the lowest values. However, the analysis of the current electrical mode on rationality in this case is difficult because of not taking into account the influence of arc current distortion on the furnace reactivity. This problem of setting up the furnace electrical mode remains unsolved to date because of the methodological uncertainty of the methods used in practice [7].

4. Method of identification of industrial furnace equivalent circuit working reactance

The EAHEM method opens a practical opportunity to identify the index of influence of arc current distortion K_X on the furnace inductance

$$X_e = K_X \cdot X_{SC} \tag{6}$$

where X_e – the furnace equivalent circuit working reactance; X_{SC} – the reactance of the first harmonic of the short-circuit experiment.

If for a particular furnace it is known, obtained from the experience of operation, the reliable value of the specific power consumption for the melting of the charge, then it is possible to identify the indicator of the influence of arc current distortion on the furnace inductance. For the case of DSP-200 furnace with an operating current of 43.9 kA, the calculated value of the specific power consumption for melting the charge without taking into account the influence of arc current distortion (K_X = 1) is 283.3 kW·h/ton, which is significantly lower than the value determined on the basis of experience – 400 kW·h/ton [5].

Using the EAHEM method for the melting period of the charge in the DSP-200 furnace, taking the values K_X>1, we find for the operating current of 43.9 kA the working point at which the specific power consumption for melting the charge will be equal to a certain accuracy to the experimental value of 400 kW·h/ton. When K_X=1.2215, the calculated value of the specific power consumption for melting the charge is 399.4 kW·h/ton. For the analyzed DSP-200 furnace, the value of the coefficient K_X during the melting period of the charge is 1.2215, which indicates an increase in the furnace working reactivity compared to the reactance for the first harmonic from the short-circuit experiment X_{SC} by 22.15%.

Table 1 shows the results of identification of the arc current distortion effect coefficient K_X for a number of industrial furnaces [5] with the values of specific power consumption for charge melting and charge melting time known from operating experience.

Table 1. Indicators of melting of the solid charge in the industrial furnaces.

G_f	tons	5	100	200
n_{pm}		1	1	1
p_{sp} – experience	kW·h/ton	510	420	400
S_t	kVA	2800	25000	45000
U_2	V	257	573	591.5
I_{oper}	kA	6.3	34.6	43.9
K_X		1.1367	1.2225	1.2215
p_{sp} – calculation	kW·h/ton	507.1	420.9	399.4
τ_m – calculation	h:min	1:12	2:19	2:55
τ_m – experience	h:min	1:05	2:00	3:06

Evaluation of the calculated and experimental melting time of the charge indicates a high degree of reliability of the proposed method of identification of the coefficient of influence of arc current distortion on the furnace reactivity during the melting of the charge.

It should be taken into account that the melting time of the charge and the specific power consumption for melting the charge, in addition to the parameters of the electric mode depend on operational technological factors, including the mass and properties of the charge loaded into the furnace.

Thus, on the basis of the experience obtained, the value of the specific power consumption for the melting of the charge for the furnace operating in the industry can be obtained close to the real value of the working reactance.

5. Summary

Thus, on the basis derived from experience, values of specific electricity consumption for the melting of the charge for serving industry arc furnaces by using the EAHEM methods, get close to the actual values of the working reactance.

The described TOMED methods open the possibility to identify the effect of arc current distortion on the furnace reactivity on the basis of calculation of the operating characteristic of the specific power consumption for melting the charge. This requires a reliable value of the specific power consumption for melting the charge, obtained on the basis of generalization of the operating experience of a particular furnace.

References

[1] Bowman B and Kruger K 2009 *Arc Furnace Physics* (Dusseldorf: Verlag Stahleisen GmbH) p 245

[2] Kruchinin A M and Juri D P 2015 *To the question of stability of operation of the electric arc furnaces at the beginning of melting. Differential and average arc voltage gradient function* (*Electrometallurgy* vol. 6) pp 20–32

[3] Kruchinin A M, Ryazanova E S and Fomenko O J 1995 *Intercollegiate collection of papers* (Tver) pp 28–32

[4] Kruchinin A M, Ryazanova E S and Chursin A Y *Program is to rationalize the electric modes of operation of the existing electric arc furnaces* (Certificate No. 2017619675 on state registration of computer programs) Date of state registration in the Register of computer programs 01 September 2017

[5] 1978 *Electrical equipment and automation of electrothermal installations (Reference)* ed. A P Altgauzen, etc. (Moscow: Energy) p 304

[6] Lutsenko V T, Pavlov V A and Dokshitskaya A I 2005 *Arc furnace* (Ekaterinburg: GOU VPO USTU-UPI) p 141

[7] Kruchinin A M, Pogrebissky M Ya, Ryazanova E S and Chursin A Yu 2018 *Fedorov readings-2018: XLVIII international scientific and practical conference with elements of the scientific school* (Moscow: publishing house of MPEI) pp 75–82

IOP Conf. Series: Materials Science and Engineering **950** (2020) 012013 doi:10.1088/1757-899X/950/1/012013

Induction heat treatment of large rolls with two independent power sources

Yu Perevalov [1], V Demidovich [2]

[1] Department of Electrotechnology and Converter Engineering, Saint-Petersburg Electrotechnical University (LETI), 197376 St.Petersburg, prof. Popov 5, Russia

[2] Russian Technologies of Induction Heating (RTIH) Ltd., 193313 St.Petersburg, Iskrovsky prospect 14/12, Russia

E-mail: yyperevalov@yandex.ru

Abstract. The two-inductor heat treatment system for rolls of rolling mills is highly flexible and can improve the quality of heat treatment while solving energy-efficiency and productivity problems. A digital model of induction heat treatment of rolls with the computation of electromagnetic, temperature and stress fields is developed taking into account the characteristics of power supplies and control systems for heat treatment modes. Particular attention is paid to the reliability of frequency converters in the presence of magnetic coupling between inductors. Recommendations on frequency selection and inverter settings are given. A universal method of digital design and control of thermal conditions in induction heat treatment systems is considered.

1. Introduction

In the metallurgical industry the induction heating devices have a power level up to several hundreds of kilowatts or megawatts. The frequency range of used power sources is from tens of Hz to MHz.

Thermal processing of pipes and rolls is an important stage in manufacturing of high quality steel products like lengthy pipes and rolls with big diameters for rolling mills. Solidification and annealing are main types of heat treatment of pipes and rolls. Modern heat treatment requires strict implementation of temperature evaluation during heating, controlled cooling and, probably, repeated heating. [1]

The developed numerical models of technological thermal processing include two-dimensional modeling of electromagnetic and thermal fields in the system. Thermal tension during processes of heating and cooling the pipes and rolls is simulated. These information is essential for defining the structure, hardness, size of a grain and other characteristics of the pipes' and rolls' metal.

The developed two-dimensional model was extremely effective not only for the induction heat treatment systems design of pipes and rolls, but also for a digital control of these complexes. However, it is necessary to use the temperature feedback under the presence of perturbations. The original way of controlling the temperature at the heater' exit is proposed by the control of the temperature at the special point in the "active zone".

The use of modular structure of induction heaters has a significant advantages in induction heating technology, since it allows optimizing the operation modes of equipment in relation to: energy consumption; scale formation; ability to maintain a constant exit temperature.

The advanced digital control allows to effectively process the possible disturbances, such as: changes of speed, diameter and initial temperature of the workpiece, and, therefore, to minimize the amount of defects during heating.

2. Problem

To Figure 1 shows the temperature distribution along the roll during quenching in 2 different modes. Each mode should provide the exit temperature of 1000 °C after the second inductor on the surface [1]. Two modes of hardening of the roll are presented in fig.1. 1 (indicated by 1) – is the hardening mode of the rolling mill's roll from two power supplies (each power source operates on its inductance coil) power of the first is 570 kW, power of the second power source is 200 kW. The operating frequency of both power supplies is 100Hz [2…5]. Tempering mode 2 (indicated by 2) – is the hardening mode from one power source, inductor connected in series, power 820 kW, frequency 65 Hz. The inductive speed in each case is 0.1 cm/s. 3 shows temperature at a depth of 100 mm for two quenching modes.

Inductors must be placed in a short distance from each other to ensure the required temperature distribution. In the 2 mode, the surface temperature has a strong unevenness. Just a small area reaches the required temperature, which requires a decrease in the speed of the inductor to maintain the necessary time for the recrystallization of the metal [6…10]. Extension of the holding time can lead to overheating of the local zones of the part.

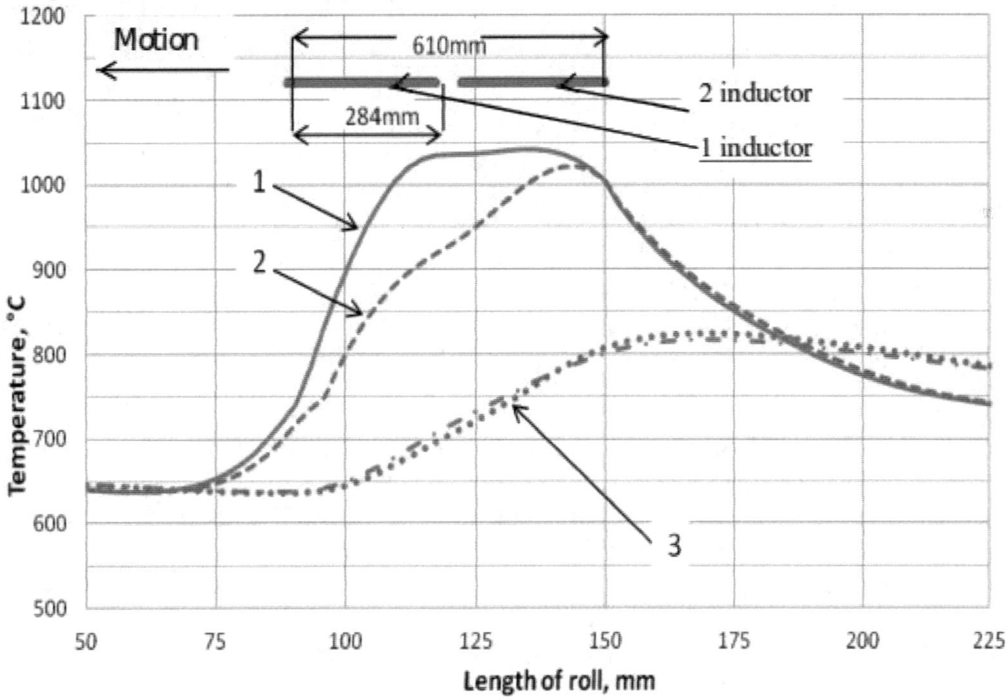

Figure 1. Hardening of the rolling mill roll

The mode 2 is much more complex, and the derived surface has a much lower quality, in comparison with the mode 1.

The 1 mode allows to achieve uniformity of the temperature almost under the entire surface of the 2nd inductor, avoiding the appearance of local maxima of the temperature distribution.

Thus, the use of the mode 2 allows obtaining a surface with improved characteristics, reducing the probability of defects, while increasing the productivity, in comparison with the mode 1.

Since the inductor is located at close range, there is instability in the operation of power sources due to the presence of magnetic coupling between the inductors.

Fig. 2 shows the hodograph of the induced reaction on the adjacent contour in conventional units of frequency, where f1, f2 are the resonant frequencies of the magnetically coupled circuits, the frequency f1 is the resonant frequency of the load circuit, and f2 is the magnetically coupled circuit.

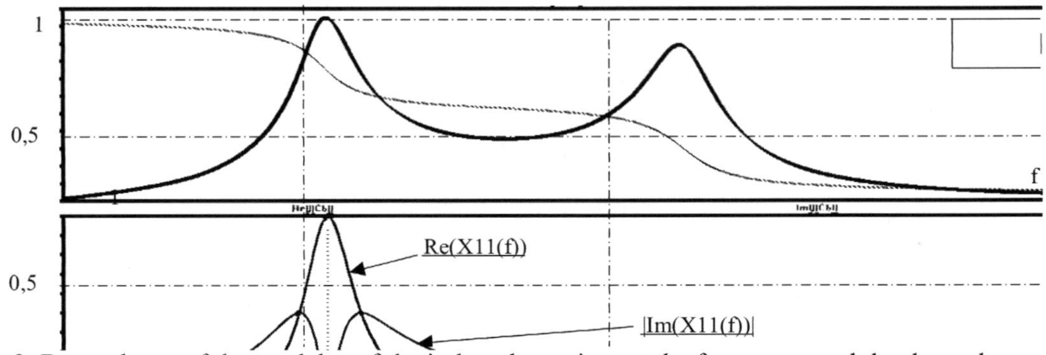

Figure 2. Dependence of the modulus of the induced reaction on the frequency and the dependence of the active and reactive components of the frequency

It can be seen that X12(f) has two local maximas: at the resonance frequency of the intrinsic circuit and at a frequency close to the resonance frequency of the neighboring contour. Also there are the modules of the imaginary and real parts of the response of the loading circuit in the graph. Thus, it can be seen that with a deviation from the frequency of the resonance, the amplitude of the induced interference sharply decreases, and the active component of the reaction also decreases, but the reactive component increases (within certain limits).

From the foregoing, we can determine the conclusion that the constructed model allows you to accurately determine the area of stable operation of inductors and power supplies for hardening rolls of rolling mills [11…15].

3. Conclusions

New technologies for induction heat treatment of large diameter rolls of rolling mills with a barrel diameter of up to 65" are developed. It was found that the digital model of the induction system must necessarily include the calculation of electromagnetic and thermal fields, the calculation of stress and strain fields, as well as the calculation of power source parameters. This digital model can be used to create an intelligent control system for the installation of induction heat treatment of large-sized rolls of rolling mills [16…20].

References

[1] A. Muehlbauer: History of Induction Heating and Melting. (2008), Vulkan Verlag.

[2] V. Demidovich , V. Andrushkevich , Yu. Perevalov, Induction heat treatment of large diameter pipes and coils // 9th International Symposium on Electromagnetic Processing of Materials (EPM2018): IOP Conf. Series, 2018. - p.1-4. doi:10.1088/1757-899X/424/1/012064

[3] Perevalov, Yu.Yu. Modeling and development of an automated induction hardening plant for large rolls of rolling mills: author. dis. Cand. tech. Sciences: 05.09.10 / SPb: SPbGETU, 2017.- 16 p.
[4] Influence of Magnetic Steel Induction Heating Power Density on Inductor Resistance Range / S. Dzliev, K. Pishchalev // Proceedings of the 2019 IEEE Conference of Russian Young Researchers in Electrical and Electronic Engineering (ElConRus), Spb, 2019.- pp ... 956-959.
[5] Demidovich V.B., Perevalov, Y.Y. Andrushkevich V.V. Induction heat treatment of large diameter pipes and coils. 9th International Symposium on Electromagnetic Processing of Materials (EPM2018)14–18 October 2018, Hyogo, Japan, IOP Conf. Series: Materials Science and Engineering 424 (2018) https://doi.org/10.1088/1757-899X/424/1/012064
[6] Baldan M., Nikanorov A., Nacke B. Hierarchical optimization approaches in designing surface hardening induction systems. 9th International Symposium on Electromagnetic Processing of Materials (EPM2018)14–18 October 2018, Hyogo, Japan, IOP Conf. Series: Materials Science and Engineering 424 (2018) https://doi:10.1088/1757-899X/424/1/012067
[7] Barglik J. Induction contour hardening of gear wheels made of steel 300M. 9th International Symposium on Electromagnetic Processing of Materials (EPM2018)14–18 October 2018, Hyogo, Japan, IOP Conf. Series: Materials Science and Engineering 424 (2018) https://doi:10.1088/1757-899X/424/1/012062.
[8] Nacke B., Dietrich A. Potentials of single stage induction heating for press hardening of steel blanks. 9th International Symposium on Electromagnetic Processing of Materials (EPM2018)14–18 October 2018, Hyogo, Japan, IOP Conf. Series: Materials Science and Engineering 424 (2018). doi: 012058. 10.1088/1757-899X/424/1/012058
[9] Huu Kien Bui, Abdoulaye B., Berthiau G. Trichet D. (2018). Fast 3D simulation of various applications of induction heating. 9th International Symposium on Electromagnetic Processing of Materials (EPM2018)14–18 October 2018, Hyogo, Japan, IOP Conf. Series: Materials Science and Engineering 424 (2018). https://doi: 012048. 10.1088/1757-899X/424/1/012048
[10] Gagnoud A. Du Terrail Yves, Budenkova O. 3D Numerical Modeling for Inductive Processes. 9th International Symposium on Electromagnetic Processing of Materials (EPM2018)14–18 October 2018, Hyogo, Japan, IOP Conf. Series: Materials Science and Engineering 424 (2018). https://doi: 012045. 10.1088/1757-899X/424/1/012045
[11] Ando, T., Katayama, D., Hirota, N., Koike, O., Tatsumi, R., Yamato, M.. Structure Formation of Magnetic Particles under Magnetic Fields toward Anisotropic Materials. IOP Conference Series: Materials Science and Engineering. https://doi: 424. 012076. 10.1088/1757-899X/424/1/012076
[12] Demidovich, V.B., Tchmilenko, F.V., Perevalov, Y.Yu., Rastvorova, I.I. Electromagnetic processing of metal as coupling of multi-physics phenomena., Proceedings of the 14th International Conference on Computational Plasticity - Fundamentals and Applications, Barcelona, Spain, COMPLAS 2017
[13] Rudnev V. Trends in Induction Hardening. Gear Technology. March/April. 22-23. (2019).
[14] Hájek Jiří, Rot D., Jirinec J. Distortion in Induction-Hardened Cylindrical Part. Defect and Diffusion Forum. 395. 30-44. (2019). DOI: 10.4028/www.scientific.net/DDF.395.30
[15] Bukanin V., Ivanov A., Zenkov A., Vologdin V. Induction Hardening of External Gear. IOP Conference Series: Materials Science and Engineering. 327. 022016. (2018). DOI: 10.1088/1757-899X/327/2/022016
[16] Aliferov A., Lupi S., Forzan M., Milliseconds pulse induction hardening. International Journal of Microstructure and Materials Properties. 13. 73. (2018). DOI: 10.1504 / IJMMP.2018.10014735

[17] Hirao M., Ogi H. Measurement of Induction Hardening Depth. In: Electromagnetic Acoustic Transducers. Springer Series in Measurement Science and Technology. Springer, Tokyo. 2017. https://doi.org/10.1007/978-4-431-56036-4_13

[18] Zheng C., Wu Z., Wang S.. Induction hardening process control of wind turbine gear ring. 38. 100-103. (2013).

[19] M. Yermekova and S. Galunin, "Numerical simulation of induction through heating for forging," 2018 IEEE Conference of Russian Young Researchers in Electrical and Electronic Engineering (EIConRus), Moscow, 2018, pp. 524-527, doi: 10.1109/EIConRus.2018.8317149.

[20] A. Bondar, O. Streltsova, F. Chmilenko and D. Bondarenko, "Prognostification of Specific Steel Resistivity for Modeling Electrothermal Processes," 2019 XXI International Conference Complex Systems: Control and Modeling Problems (CSCMP), Samara, Russia, 2019, pp. 370-373, doi: 10.1109/CSCMP45713.2019.8976724.

IOP Conf. Series: Materials Science and Engineering **950** (2020) 012014 doi:10.1088/1757-899X/950/1/012014

Inductive level sensor: experiment and calculation

V Eltishchev, S Mandrykin, I Kolesnichenko

Institute of Continuous Media Mechanics, Perm, Russia

E-mail: eltishchev.v@icmm.ru

Abstract. A non-contact induction technique for registering the free sufrace position of liquid metal is studied numerically and experimentally. A localized inductive level sensor (ILS) generates an alternating magnetic field and measures the resulting magnetic field. If an electrically conductive medium appears in the sensitive area of the sensor, this field changes. In the experiment, signals from the ILS were measured for a rotating cylinder with a beveled upper boundary, for different frequencies of alternating current, supplying the sensor generating coil. A set of calibration curves was obtained, that can be used to determine the cylinder rotation angle from the signal. In the numerical simulations, the three-dimensional fields of magnetic induction were obtained. The influence of the alternating current frequency, as well as the cylinder rotation angle on the ILS measurement, was studied. Numerical simulations show that increasing current frequency results in higher ILS sensitivity. The numerical and experimental results are in agreement. The developed technique is applicable for measuring the oscillation frequency of the liquid metal free surface.

1. Introduction

An integral part of many technological industries is the constant monitoring of certain process parameters. It can be a variety of physical parameters that are typical to a specific task. However, of particular interest is the industry associated with electrically conductive media, namely, with liquid metals. In such problems, an additional parameter is the position of the free surface of the melt. It is important to determine the position and shape of its free surface, for example, during continuous casting of metal. In the development of liquid-metal batteries [1], it is also important to monitor the shape and position of the metal-electrolyte interface in order to evaluate the effect of various factors on the stability of the multilayer structure. Classical ultrasonic and optical methods in this case are not applicable, due to high temperatures, optical opacity and, as a rule, high aggressiveness of the medium.

The most obvious way to determine the level of liquid metal in a tank is a conductive method. However, in application to metallurgical problems it may have insufficient accuracy, owing to the formation of oxide caps on the melt surface. Another method is the inductive one, based on the excitation of eddy currents in a metal and the subsequent measurement of the resulting magnetic field. Already there are industrial devices based on this method for measuring, for example, the level of aluminum [2] and steel [3] melts. At the same time, the disadvantage of such systems is the need to place the sensor directly in a container with liquid metal, above its surface. As a result, the sensors are exposed to high temperatures and metals, which complicates their design and use. There are alternative configurations of such devices [4], however, they are rather narrow-profile and are developed for a specific installation. In addition, such systems also cannot guarantee adequate measurement accuracy.

It should be noted that there are measuring devices of yet another configuration [5], allowing non-contact measurement of the position of the free surface of liquid metal, while the same sensor can be used on different installations. A set of such inductive level sensors (ILSs) provides a picture of the two-dimensional structure of the melt free surface, which is similar to the idea of magnetic tomography. However, at the moment there is only information about the prototype of this sensor. Also, in [5], studies were performed on a gallium alloy in a liquid state at room temperature and a transparent non-conductive wall, while in real industrial applications, the material of containers is usually opaque and electrically conductive.

The possibility of detecting vibrations of the free surface of a liquid metal using an inductive level sensor (ILS) is studied numerically and experimentally in this paper. In numerical and laboratory experiments, the distorted free surface of a conducting medium is modeled using a beveled cylindrical body made of duralumin.

2. Methods

2.1. Experimental setup

A solid duralumin cylinder with a beveled upper surface acts as a model for the rotation of the free surface of a liquid metal (see Fig. 1). The bevel angle is 45 degrees. The cylinder is mounted on a platform that is connected to an electric motor. Opposite the cylinder, an inductive level sensor (ILS) is installed coaxially with the radial axis at the height of the middle of the bevel.

The sensor is a system of three coils connected in series and coaxially. With the central coil magnetic field is generated, and the two outer coils are measuring. The generating coil contains 50 wicks of enameled wire with a diameter of 1.12 mm. Each of the measuring coils consists of $\sim$ 1300 wicks of enamel wire with a diameter of 0.3 mm. The external dimensions of the sensor are $72 \times 46 \times 42$ mm^3. A more detailed description of the inductive level sensor is presented in [6].

The ILS is placed on three mutually perpendicular micrometric platforms, which allow moving the sensor up to 50 mm in each direction. Using a sinusoidal signal generator (GZ-109), alternating current is supplied to the sensor generating coil. Two measuring and one generating coils of ILS are connected to the National Instruments data acquisition boards (NI9225 and NI9227), through which experimental data is recorded using the LabView system.

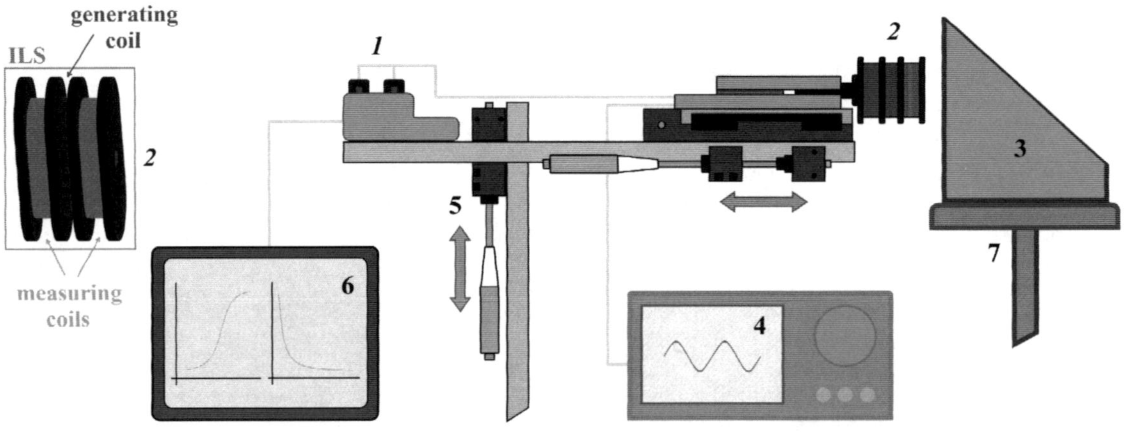

Figure 1. Experimental setup: 1 – National Instruments data acquisition boards; 2 – inductive level sensor (ILS); 3 – duralumin cylinder with a beveled upper surface; 4 – generator of a sinusoidal signal; 5 – micrometric platforms; 6 – computer with a LabView system; 7 – rotating platform

The distance between the side surface of the cylinder and the ILS does not exceed 1 *mm*. The "peak" of the beveled cylinder corresponds to a polar angle equal to 180°.In the experiment, the ILS output signal from the angle of rotation of the beveled cylinder is studied at various values of the frequency of the alternating current on the generating coil. The effective value of the current is 1.5 A. During the experiment, a specific angle of rotation of the beveled cylinder is set. The value of the alternating current (AC) frequency on the generating coil is set and the output signal of the ILS is recorded. Then the AC frequency value changes and the measurement is repeated ($\nu \in [20, 600]$ Hz). Next, the beveled cylinder is rotated at a certain fixed angle and the series is repeated. The angle of rotation of the beveled cylinder occurs in the range from 0 to 350 degrees in increments of 10.

2.2. Numerical simulations

Numerical simulations are carried out using the Ansys Emag software. The computational domain consists of the cylinder 1, beveled at an angle of 45° (see. Fig. 2) and a generating coil 2, surrounded by a non-ferromagnetic non-conductive space. The ILS consists of a generating coil, sandwiched between two measuring coils, is 1 mm away from the cylinder and is positioned so that its axis of symmetry passes through the center of the beveled surface of the cylinder. An unstructured mesh, consisting of triangular elements with maximum linear size of 5 mm is used. The rotation of the cylinder is determined by angle α, as pictured in Fig. 2b.

The dimensions and material parameters of the model are selected in accordance with the experimental study. Namely, D16 duralumin with conductivity of $\sigma = 26 \cdot 10^6$ S m^{-1} is used as the material of the cylinder. The cylinder diameter is $D = 0.172$ m, the height is $H = 0.222$ m. The generating coil has 50 wicks of wire, and the effective current value is 1.5 A. The coil size is determined by three parameters: the inner and outer radii $R_1 = 0.006$ m and $R_2 = 0.015$ m, respectively, as show Fig. 2c. The thickness of the generating coil is 0.006 m.

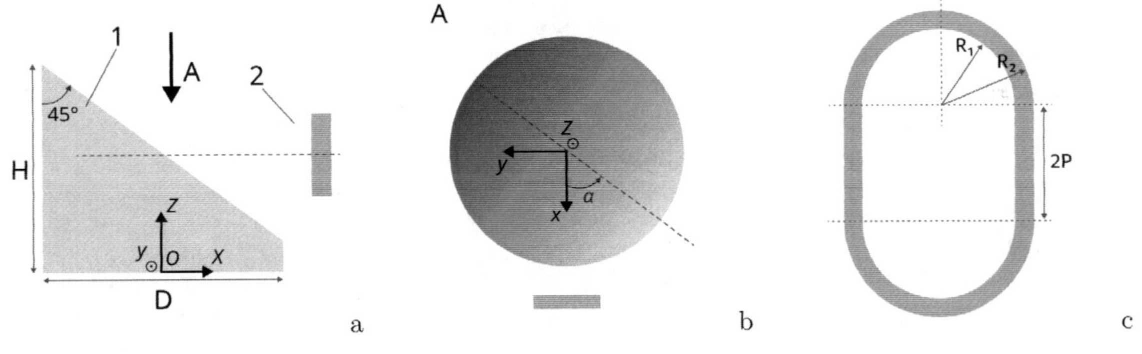

Figure 2. Computational domain scheme

The Cartesian coordinate system (see Fig. 2a), the origin of which is located in the center of the cylinder bottom end face, is used in calculations and results interpretation. The Oz axis of this system coincides with vertical symmetry axis of the cylinder, and the Ox axis is pointed to the sensor.

3. Results

3.1. Experiments

Based on the experiments, the dependencies of the ILS output signal on the angle of rotation of the beveled duralumin cylinder are constructed (see Fig.3).

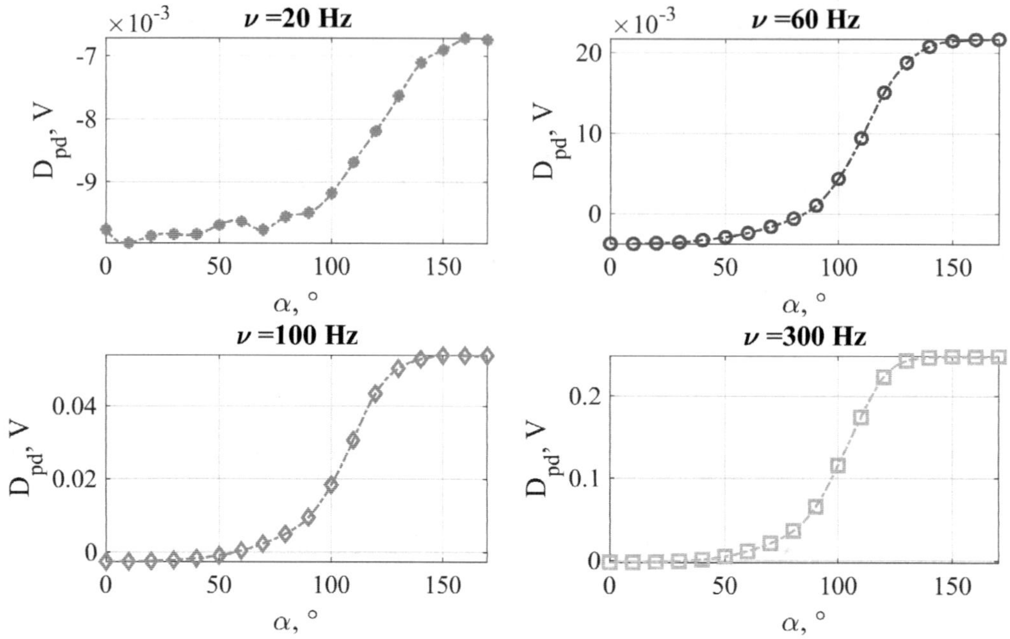

Figure 3. The output signal of the sensor for different angles of rotation of the beveled cylinder at different frequencies of alternating current on the generating coil.

The value of D_{pd} is defined as the difference of the root mean square voltage values on the measuring coils of the sensor. As can be seen from the figure 3, with increasing frequency of the alternating current on the generating coil, the sensitivity of the inductive level sensor increases. Qualitatively, the dependencies are identical and have a sigmoidal appearance. The maximum of the function $D_{pd}(\alpha)$ corresponds to the state when the peak of the beveled cylinder is at the smallest distance from the sensor, otherwise, the function takes the minimum value.

Thus, it is possible to introduce the normalized function $F(\alpha)$, which takes on a zero value when there is no metal opposite the working side of the sensor (the beveled surface of the cylinder is far) and a unit value when the peak of the beveled cylinder is opposite the sensor (see Fig.4).

From the obtained dependence $F(\alpha)$, it is possible to restore the position of the beveled surface of the cylinder at any time during its continuous rotation. In this case, the sensitivity of ILS depends both on the frequency of the alternating current on the generating coil of the sensor and on the speed of rotation of the beveled cylinder.

The measurements for the remaining angles of rotation of the beveled cylinder ($\alpha \in [180, 350]^\circ$) are mirror symmetric and coincide within the measurement error, which does not exceed 7% (for an alternating current frequency of 20 Hz). With increasing frequency of the alternating current on the generating coil, the relative measurement error decreases. The measurements were carried out for various AC frequencies up to 600 Hz. Qualitatively, the dependencies are similar to those presented above.

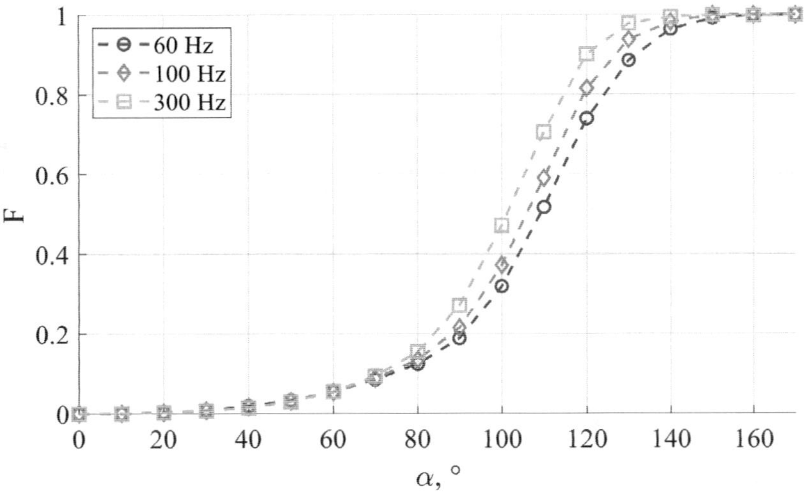

Figure 4. The normalized function $F(\alpha)$ for various frequencies of alternating current on the generating coil

3.2. Numerical simulations

The cases of different electric current frequencies on the generating coil are considered for each angle α in the range from 0 to 140 degrees. A three-dimensional fields of magnetic induction $\vec{B}$ are obtained. The imaginary part of the magnetic field x-component $\operatorname{Im} B_x(x)$ is used in results analysis. If there is no metal in the ILS proximity, then there is no induced field, thus $\operatorname{Im} B_x(x) = \text{const(x)} = 0$.

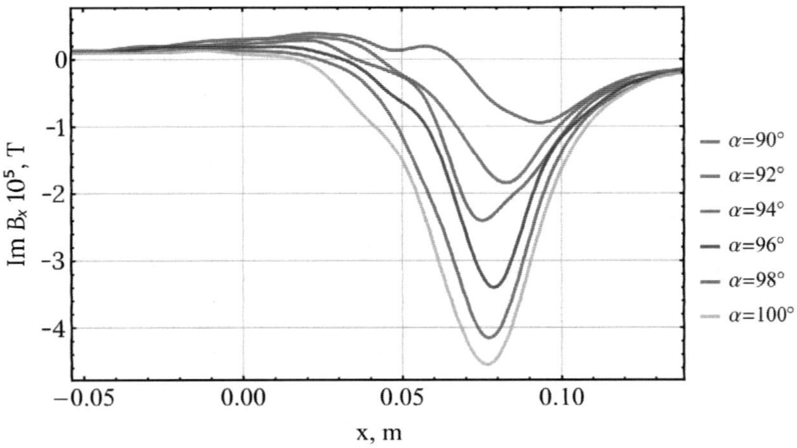

Figure 5. The $\operatorname{Im} B_x$ as a function of x, for different cylinder rotation angles α

Fig. 5 shows the example of the $\operatorname{Im} B_x(x)$ profiles, plotted for different cylinder rotation angles α. As follows from Fig. 2b, the case $\alpha = 90°$ corresponds to the metal-air interface. At this value of α there is a moderate deformation of the resulting magnetic filed profile, which intensifies with a further increase in the rotation angle. That is, the profile deformation is higher when there is more metal in the ILS sensitivity area.

To quantitatively describe this result, the function $f(\alpha)$ is introduced as follows:

$$f = \left|U\left(B_x(X-\delta)\right) - U\left(B_x(X+\delta)\right)\right|, \tag{1}$$

where $U\left(g(X-\delta)\right)$ is the average value of some arbitrary function $g(x)$ on the interval $[X-\delta; X]$ (see Fig. 6). In this study X is assumed to be the coordinate of the sensor center (generating coil), that is, $X = 0.110$ m. The parameter $\delta = 0.013$ m is selected as the distance between the centers of the generating coil and measuring coils. In the following, the value of the normalized function f will be denoted as F. That is, $F = 1$ means that the maximum amount of metal is located near the sensor. The metal-air interface corresponds to some intermediate value (between 0 and 1), which is found in the sensor calibration process.

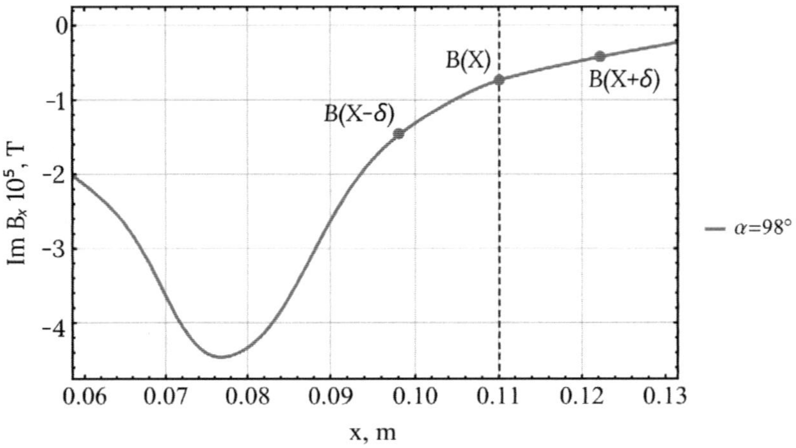

Figure 6. To the definition of the function F

In the initial position of the sensor, that is, when $\alpha = 0°$, the least amount of conductive medium is located in the sensor sensitivity area. This case corresponds to $F = 0$, as shown in Fig. 7. In this Figure, the dependencies $F(\alpha)$ for several considered frequencies $\nu = 10$, 60, 100 and 300 Hz are pictured. Here the lines correspond to numerical results, and the symbols — to experimental ones. With an increase in the angle α, a gradual increase in the value of F occurs, which sharply increases increases when approaching $\alpha = 90°$. In this position, the axis of the ILS is parallel to the beveled plane of the cylinder. This position corresponds to the metal-air interface case. With further rotation of the cylinder, the value of F tends to 1, because in this range of α the change in metal amount near the sensor is small. It can also be seen from the Figure that for small angles (below 80 degrees) the data practically do not differ. At the same time, when approaching the value $\alpha = 90°$, the function F value is larger, the higher is the electric current frequency. This means that increasing the frequency ν can result in higher ILS sensitivity. It should be noted, that in real industrial applications, where liquid metal is placed in a container with conductive walls, one has to make sure that the skin layer is greater than the thickness of the container wall.

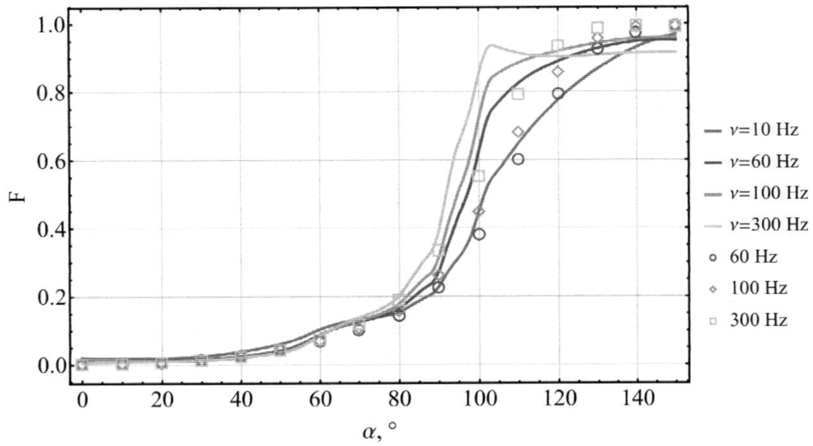

Figure 7. The dependencies of F on the angle α for the frequencies $\nu = 10$, 60, 100 and 300 Hz. Lines and symbols correspond to numerical and experimental results, respectively

4. Conclusions

The possibility of determining oscillations of the metal-air interface by an inductive level sensor (ILS) is studied numerically and experimentally. The interface was modeled by a duralumin cylinder, beveled at angle of 45°, rotating around its vertical axis. The position of the ILS was chosen so that the direction of its sensitivity passes though the center of the cylinder beveled surface. The calculation parameters are chosen in accordance with the experimental study. Various alternating current frequencies on the generating coil were considered, ranging from 10 to 600 Hz. In the experiments, the signals from the ILS are obtained, as well as the set of sensor calibration curves. The three-dimensional fields of magnetic induction are obtained in numerical simulations.

It is shown that the ILS of considered configuration allows to determine the presence of metal near the sensor. The sensitivity of the ILS depends on the electric current frequency on the generating coil. Namely, the best sensitivity was observed for the highest frequencies considered. The numerical and experimental results are in agreement.

Using an array of such sensors, after calibrating them for a given application, can determine not only the level of liquid metal in the container, but also the position of the free surface of the melt at each time point. Thus, this technique can be applied to measure the oscillation frequency of the free boundary of a liquid metal.

References

[1] Kelley D H and Weier T 2018 *Applied Mechanics Reviews* **70**
[2] Micro-Epsilon Measurement 2016 Liquid aluminum level
[3] Agellis Group AB 2016 Emli
[4] Khalilov R I, Khripchenko S Y, Frik P G and Stepanov R A 2007 *Measurement Techniques* **50** 861–866
[5] Zürner T, Ratajczak M, Wondrak T and Eckert S 2017 *Measurement Science and Technology* **28** 115301
[6] Eltishchev V, Dimov I, Pavlinov A, Khalilov R and Kolesnichenko I 2019 *IOP Conference Series: Materials Science and Engineering* **581** 012004

IOP Conf. Series: Materials Science and Engineering **950** (2020) 012015 doi:10.1088/1757-899X/950/1/012015

In-melt electromagnetic forces at MHD-stirring with the non-sine supply

E.R. Vinter[1,2] **, V.N. Timofeev**[1,2]**, A.V. Fomin**[1,2]**,**

[1] Siberian Federal University, 79 Svobodny pr., 660041 Krasnoyarsk, Russia
[2] RPC Magnetic hydrodynamics, LLC Russia, Krasnoyarsk, Akademika Kirenskogo Street., 9A, b.225

The crucial task in production of multi-component alloys is a development and study of MHD-stirring conditions which make it possible to achieve new process effects such as microinhomogeneous homogenization or MHD-resonance. One of the advanced ways to provide such stirring conditions is the supply of the non-sine periodic current of the given waveform to inductor coils. The paper presents the results of numerical simulation of a magnetohydrodynamic stirrer with non-sinusoidal power supply. The analysis of the behavior of electromagnetic forces at various power supply parameters is carried out. A comparative analysis of the obtained results with the characteristics of the MHD stirrer with a sinusoidal power supply system is carried out.

1. Introduction

At the time the area of metallurgical products' application is heavily expanded. This is the case, for example, of aluminium alloy products which are nowadays widely used in the aircraft industry, in additive technologies and in other spheres. Therewith MHD-stirrers are one of the most effective way to increase the metallurgical equipment efficiency [1]. Thus the in-furnace MHD-stirring of a molten metal presents a widely occurring technology for the modern metallurgical production. MHD-stirrers allow automation of homogenization, decreasing the preparation time and power consumed to produce the good-quality alloys [2, 3].

The crucial task in production of multi-component alloys is a development and study of MHD-stirring conditions which make it possible to achieve new process effects such as microinhomogeneous homogenization or MHD-resonance [4]. One of the advanced ways to provide such stirring conditions is the supply of the non-sine periodic current of the given waveform to inductor coils.

A number of the studies of non-sine pulse EM fields' efficiency in the molten metal stirrers have been conducted up to date. To sum up, MHD-stirrers of non-sine supply can be divided into two groups. The first group includes stirrers where the supply current waveform presents sequential pulses of sine current modulated by the frequency much less than the basic one. Stirrers with industrial mains frequency are supplied in such a way. The second group includes stirrers where the supply current waveform presents the non-sine periodic function of the given waveform. The supply current may be as the single-pole or alternating square-wave pulses and triangular pulses. Low-frequency stirrers (0.5 - 5 Hz) are supplied in such a way.

A number of computational and experimental studies have been conducted for different MHD-stirrers of the first group [4-6]. These works revealed a certain advantage of sequential modulated sine pulses supplied. The work [4] showed that such supply allows the so-called MHD resonance which leads to more intensive heat and mass transfer in melt.

As regards the MHD-stirrers with the non-sine periodic supply current of the given waveform, the electromagnetic and hydrodynamic processes underwent the theoretical researches for some simple systems [7, 8]. The work [9] showed that non-sine supply of MHD-stirrer inductor with the current waveform being nearly square at the constant amplitude makes it possible to increase the integral EM force applied to the melt. Still it is evident that in this case the real power loss in the inductor will also increase as the effective mains current will actually get bigger. The work [10] includes computational studies of EM forces' distribution in melt with the inductor being supplied with non-sine periodic rectangular voltage with various duty factor.

However, with such type of supply the current waveform may be corrected by altering only the coil parameters (turns, geometry, resistance) which is a hard-to-implement task. Therewith the up-to-date current supplies actually allow generating the output current of various waveforms [11] whereby control of the force applied to melt is possible.

This work includes computational study of spatial and time-and-frequency distribution of EM forces in MF-influenced melt such magnetic field being generated by the MHD-stirrer inductor supplied with non-sine currents of the given waveform.

2. Statement of the Problem

2.1. Electromagnetic field equation and boundary conditions

Two-phase MHD-stirrer with intersecting winding produced by RPC MAGNETIC HYDRODYNAMICS, LLC, has been studied [12].

The below assumptions are accepted in computational modelling:

1. The problem is to be 2D considered;
2. Physical properties of areas are assumed to be constant;
3. Coils' sections are simulated with no discretion by individual turns;
4. Magnetic permeability of a magnetic core is taken as constant based on the properties of the material selected;
5. Motion of melt particles is disregarded when solving the EM problem.

With due account of the assumptions taken, the equation of the magnetic vector potential for the computational domain is written as follows [13, 14]

$$\Delta A=\mu\gamma\frac{\partial A}{\partial t}, \tag{1}$$

where A is a magnetic vector potential; μ is an absolute magnetic permeability; γ is a specific electric conductivity; t is time.

Dirichlet condition is applied to the normal component of the vector potential and Neumann condition is applied to the tangential component at the boundaries of the computational domain.

$$A_n=0; \qquad \frac{\partial A_\tau}{\partial n}=0. \tag{2,3}$$

2.2. Numerical setup

The total MMF in each coil section is set as the loading source. The waveform of MMF in coils is a trapezoidal alternating function which parameters are set by the rise/fall timet_{front} and by the time of the MMF being constant t_{const}. The MMF waveform depends on the non-dimensional parameter $\hat{\tau}=2t_{front}/T$ which is indicative of the current rise time to the peak value. Let's say that this value is a waveform parameter. Thus, parameters of the MMF curve are found as below

$$t_{front}=\frac{\hat{\tau}T}{2}; \qquad t_{const}=\frac{T}{2}-t_{front} \tag{3,4}$$

Therefore, when $\hat{\tau}\to 0$, MMF waveform represents bidirectional square pulses and, when $\hat{\tau}\to 1$, it represents triangular pulses. For meaningful comparison the peak MMF value varies with $\hat{\tau}$ so that the effective value remains constant

$$I_m n_{sec}=\frac{I_m n_{sec}}{\sqrt{2}}\sqrt{\frac{3}{3-2\hat{\tau}}}, \tag{5}$$

where I_m is a base current amplitude; n_{sec} is a number of turns at the coil section; I_m is an amplitude of the non-sine supply.

Picture 1 shows the diagram of the instantaneous current in the coil conductor I_m for various waveform values.

IOP Conf. Series: Materials Science and Engineering **950** (2020) 012015 doi:10.1088/1757-899X/950/1/012015

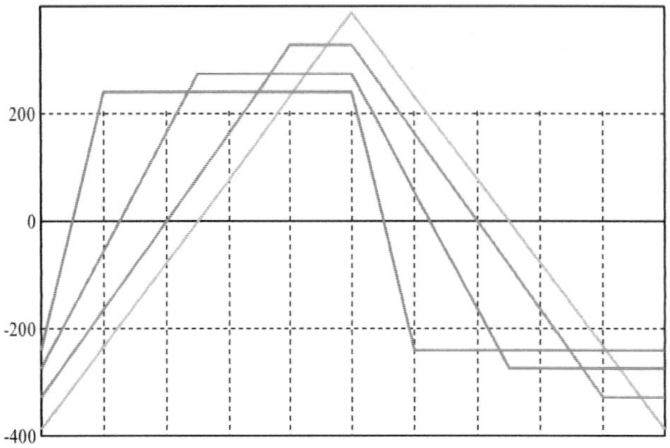

Picture 1. Diagram of the instantaneous current in the coil conductors for various waveform values.

Picture 2 shows the sketch of the computational domain with the key geometrical and physical parameters and the finite element grid. The computational domain includes melt 1, magnetic core 2, two-phase coil 3 and ambient air 4. The finite element mesh is automatically built with Maxwell facilities based on the adaptive algorithms [15]. Table 1 presents the main geometrical and power parameters of the computational model. Other geometrical dimensions are in parametric relations with the main ones. Thus, the depth of a tooth is automatically computed based on the number of coil sections and the gap in-between them.

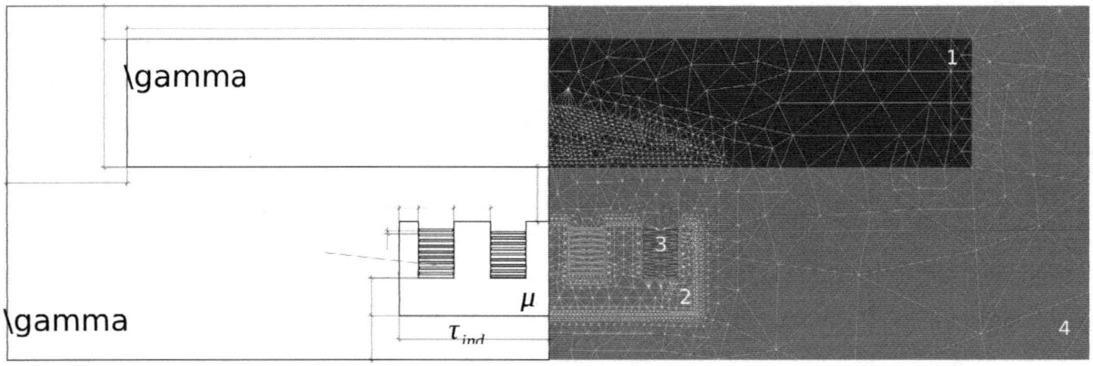

Picture 2. Geometry of the computational domain and the finite element grid.

Conductivity γ is given in [S/m]

IOP Conf. Series: Materials Science and Engineering **950** (2020) 012015 doi:10.1088/1757-899X/950/1/012015

Table 1. The main model parameters

Geometrical parameters [m]	
Pole pitch, τ_{ind}	1.1
Nonmagnetic gap, Δ_{nm}	0.4
Bath depth, H_{melt}	0.95
Bath length, L_{melt}	6.2
Height of the magnetic core yoke, h_{yoke}	0.275
Width of the end tooth, w_e	0.278
Width of the middle tooth, w_m	0.419
Gap between the coil sections, Δ_{sec}	0.022
Power parameters:	
Frequency, [Hz]	0.5, 1, 1.5
Base current in conductor, [A]	224
Phase shift between the coils' currents, [deg]	90

3. Results and discussion

3.1. Integral parameters

When studying MHD-stirrers' parameters, the tangential component of EM force is of the prime interest as this component is responsible for the heat-and-mass transfer within the furnace. When the inductor coil is supplied with sine current, it is reasonable to take the force magnitude as the base value in order to compare the influence of the supply current waveform on the electromagnetic force $F^{\sin}$.

The picture shows the dependence of the relative integral EM force on the waveform parameter. Thus, it is evident that when varying the supply current waveform from trapezoid to triangle with the effective value being the same, the integral value of the tangential force component will, at any time, be less if the inductor is sine supplied. Thereat the extreme value is observed in domain $\hat{\tau} \approx 0.6 \ldots 0.7$ for all the frequencies reviewed. This is attributable to the fact that the harmonic content is best matched to the sine function under this waveform parameter. In addition to the above, the more current waveform edges are, the lower tangential EM force component is.

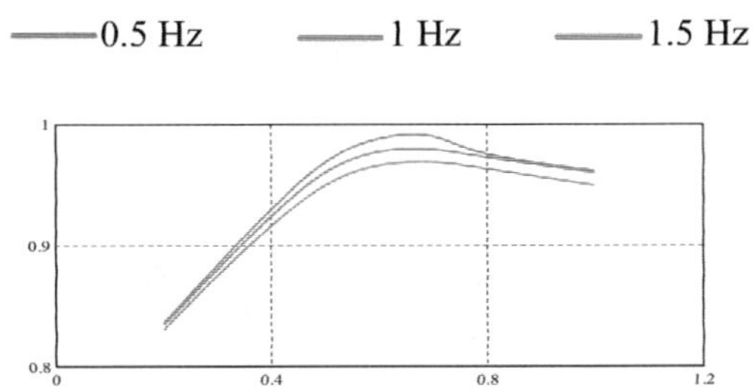

Picture 3. Dependence of the relative tangential EM force on the form factor.

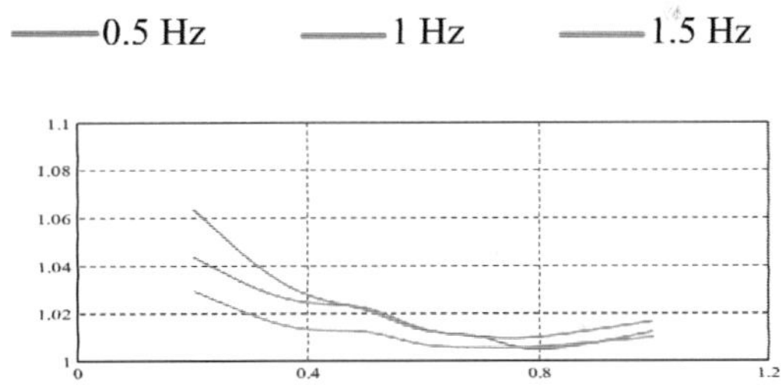

Picture 4. Dependence of the relative normal EM force on the form factor.

Therewith the integral value of the normal EM force component is slightly higher throughout the range $\hat{\tau}$ as compared with the inductor sine supply. Thus, with the frequency of 0.5 Hz, the integral normal EM force may increase by 5-7 percent when the inductor is supplied with trapezoidal current as compared with the sine supply. But the normal force component has no significant influence on the transient heat-and-mass transfer within the furnace. The work [16] states that the normal component provides only for the local vertical mass transfer with the general decrease of in-the-bath motion. On the other hand, such effect may be useful when installing an MHD-stirrer on the side wall of the furnace.

It also should be noted that the power consumed for coils' and metal preheat will change when the supply parameters are changed. The picture shows dependence of the tangential EM force relation to the active power consumed for coils' and metal preheat. Generally, the force-to-power increment is changed insignificantly throughout the reviewed range $\hat{\tau}$. It may just be noted that when current

IOP Conf. Series: Materials Science and Engineering **950** (2020) 012015 doi:10.1088/1757-899X/950/1/012015

approaches the rectangular waveform, the gain of the power consumed for preheat becomes more significant than the tangential EM force gain.

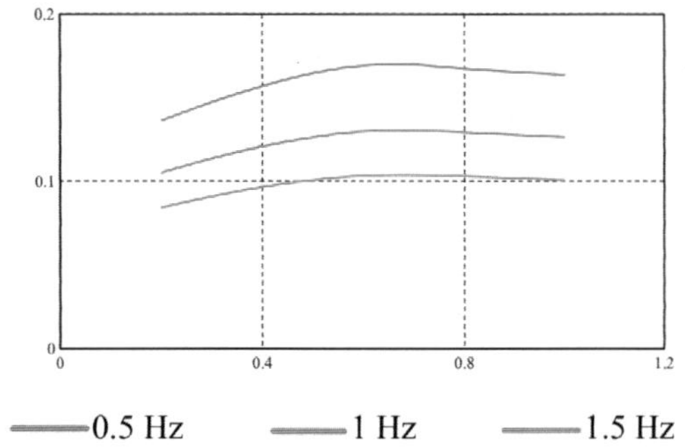

Picture 5. Dependence of the tangential EM force relation to the active power consumed for coils' and metal preheat on the waveform parameter.

3.2. Spatial and time distributions of electromagnetic forces

It is worth mentioning that in case of non-sine periodic inductor supply, the structure of the travelling magnetic field is disturbed whereby the spatial and time in-melt distribution of EM forces is much different from distribution of forces in case of the sine supply. Picture 6 shows the vector in-melt EMF distribution at different time with various inductor supply parameters (here T is the supply current oscillation period). When the inductor is sine supplied, the wave of electromagnetic force glides along the bath length in the magnetic field direction. When the inductor is non-sine supplied, the spatial and time EMF distribution is disordered and characterized by its reversal in direction at different melt points. Besides that, it bears mentioning the occurrence of the strong normal EM force component in the middle part of the inductor such component being of the negative direction (vector to inductor). Such EMF behavior may have a disastrous effect on the integrity of the refractory lining.

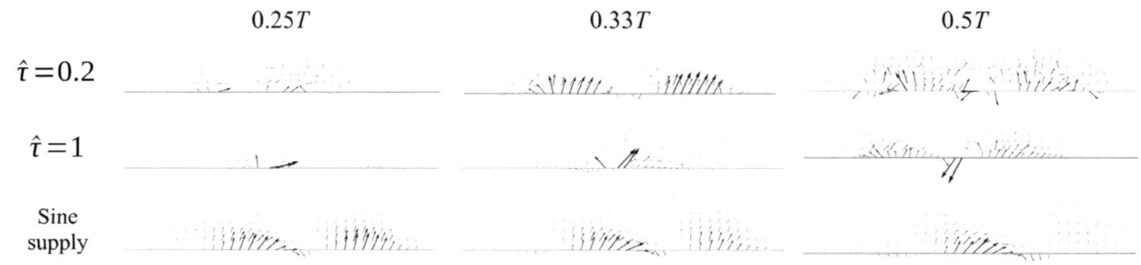

Picture 6. Vector in-melt distribution of EM forces over time under various inductor supply parameters.

Such spatial and time EMF in-melt distribution will generally have the adverse effect on the metal velocity, transient mass transfer and hence the mechanical efficiency. On the other hand, such pattern of changes opens up the potential for vibratory impacts in melt which, as stated above, also provides process advantages when producing multicomponent alloys.

EMF at any point of the melt is known to be sinusoidally changed in time if an MHD-stirrer is also sine supplied. The work [4] revealed that EMF pulses allow intensifying the heat-and-mass transfer in melt by increasing the kinetic pulse energy. Low-frequency EMF sine waves lead to the periodic fluctuations of metal velocity which intensify elimination of microinhomogeneity by additional local acceleration and increased turbulent viscosity due to the components' diffusion. Nevertheless low-frequency sine supply of the inductor does not provide for the sharp variation of the in-melt EM force whereat the amplitude of oscillations remains stable notwithstanding the supply parameters. Thus, such EMF behavior is advantageous which will ensure the mechanical impact on melt in order to remove microinhomogeneities.

Two bottom points (Picture 7) have been chosen to study the EMF behavior: point 1 is the middle point as relating to the inductor geometric symmetry (middle zone) and point 2 is in the area of the end inductor tooth (edge zone).

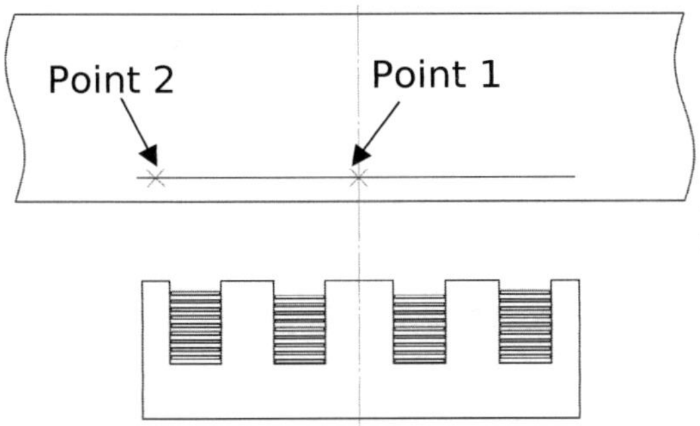

Picture 7. Points to study the EMF behavior in melt.

Picture 8 shows the instantaneous values of tangential EMF component at points 1 and 2 under various waveform parameter values. It can be seen that tangential EMF in the edge zone varies about the zero value which indicates that its period average in this zone is close to zero and, moreover, may be reverse in relation to the magnetic field direction. Such behavior of the tangential force component in the edge zone is attributed to the linear edge effect [17].
Tangential EMF component in the middle zone varies about some average value in the positive domain. Therewith the instantaneous EM force is close to the sine form when the waveform parameter is $\hat{\tau} \approx 0.4 \ldots 0.8$. With $\hat{\tau} = 0.2$ and $\hat{\tau} = 1$ the force is most suddenly changed, moreover, in case of triangular supply current the instantaneous tangential EM force also has the obvious triangular waveform.

$\hat{\tau} = 0.2$

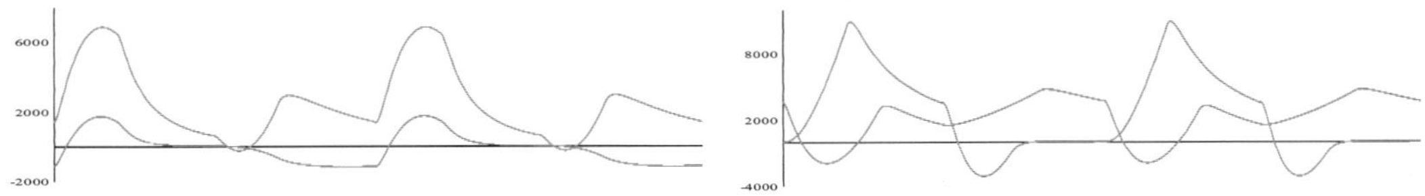

$\hat{\tau} = 0.4$

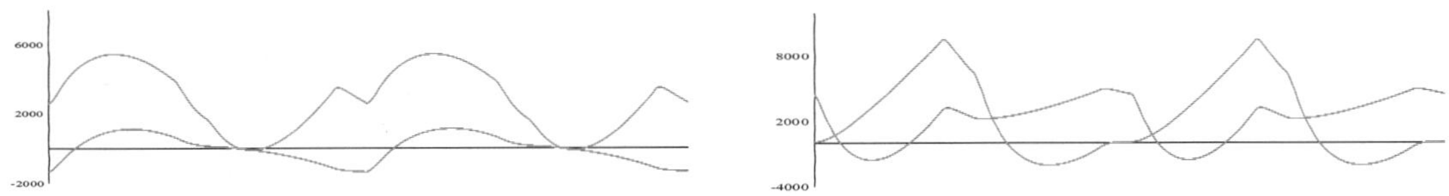

$\hat{\tau} = 0.6$

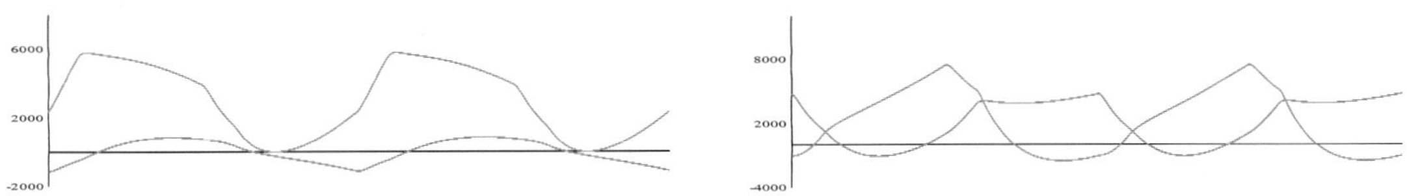

$\hat{\tau} = 0.8$

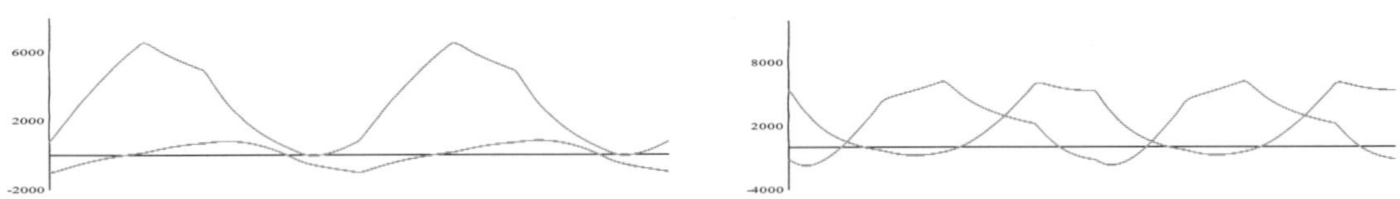

$\hat{\tau} = 1$

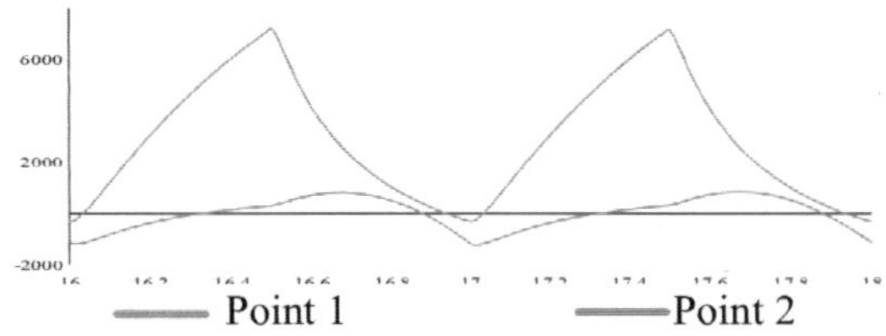

Picture 8. Diagram of the instantaneous tangential EM force in the middle zone

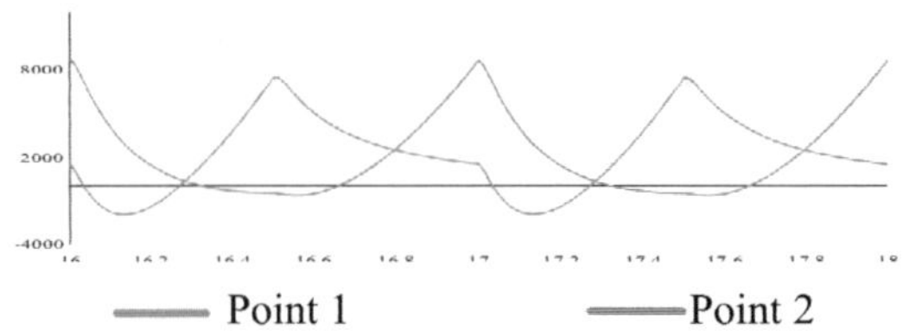

Picture 9. Diagram of the instantaneous normal EM force in the edge zone

Picture 9 shows the instantaneous normal EMF component at points 1 and 2 under various waveform parameter values $\hat{\tau}$. As regards the middle zone, normal force is most suddenly changed with $\hat{\tau}=0.2$. Beside the EMF behavior in time, the peak-to-peak force oscillations are also important for generation of the force impact required for the efficient homogenization and elimination of microinhomogeneities. The peak-to-peak EMF oscillations may be evaluated with the peak factor which corresponds to a peak-to-RMS relation

$$K_a = \frac{F_m}{F^{rms}}; \qquad F^{rms} = \sqrt{\sum_{n=1}^{N} \frac{F_n^2}{N}}, \qquad (6,7)$$

where F_m is the peak EMF value; F^{rms} is the RMS value of EM force; F_n is the instantaneous EMF for n time step; N is a number time steps over the period.

In the present case let us consider the EMF change function without the average value being taken into account. It is known that in this case for sine function $K_a^{\sin} \approx 1.41$. Picture 10 shows the diagrams of the peak factor dependence on the waveform parameter at the middle and edge points. It is apparent that dependence of the peak factor on the form factor is alike for all the reviewed frequencies of non-sine supply current and, moreover, the frequency has slight effect on the peak factor.

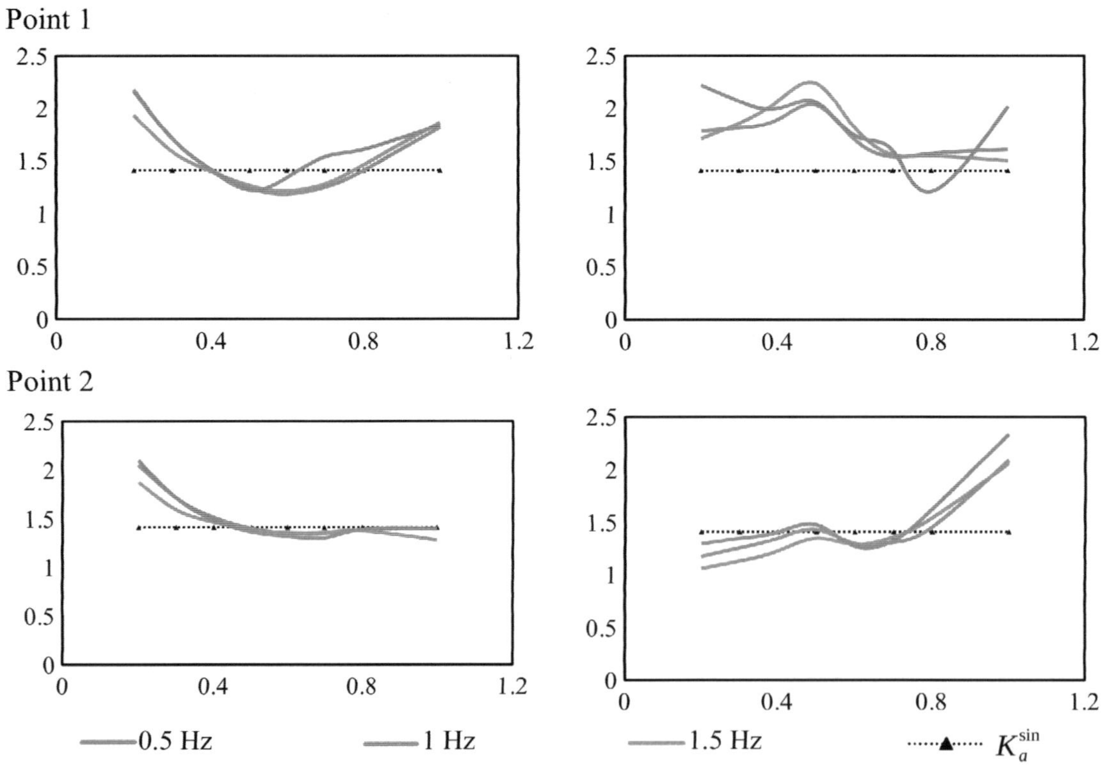

Picture 10. Dependence of the EMF peak factor on the form factor with various supply current frequencies.

The highest amplitude of tangential EMF component oscillation in the middle zone is ensured for the values $\hat{\tau}<0.4$ and $\hat{\tau}>0.8$, that is actually with almost rectangular supply current or triangular supply current. It is also noteworthy that peak oscillations of tangential force component within these values may be lower than when sine supplied. On the contrary, the highest amplitude of the normal component oscillations in the middle zone is provided at $\hat{\tau}\approx 0.4\ldots 0.5$.
Peak factor dependences for tangential and normal EMF component at Point 2 have no maximum. Peak oscillations of tangential component in this zone is increased as the current waveform approaches the squarewave. As regards the normal component, the peak oscillations are, on the contrary, increased as the supply current waveform approaches the triangle.

4. Conclusion

The influence of the waveform of the current supplying the MHD-stirrer inductor on the EMF behavior in melt has been assessed in this work. Several main conclusions can be made based on the findings:
1. When supply current waveform approaches the rectangle, the integral value of in-melt tangential force is decreased as compared with the value when the inductor is sine supplied.
2. The highest tangential EMF component is achieved when the inductor is supplied with trapezoidal current with the amplitude rise/fall time t_{front} being equal to approximately quarter-period.
3. As regards the non-sine current of the given waveform, it is expected that the transient mass transfer throughout the melt will decrease and hence the mechanical efficiency will also decrease. However the disordered spatial and time distribution of EM forces opens up the potential for the vibratory impact on the melt which may pose some process advantages for removal of microinhomogeneities.
3. The most sudden change and the highest amplitude of EMF fluctuations are ensured when triangle or trapezoid current is supplied with the minimum amplitude rise/fall time that is at $t_{front} \to 0$.

5. References

[1] Baake, E., et al. "MHD technologies in metallurgy." Intensive Course Specific IV. Study guide, Publishing house of ETU, St. Petersburg 232 (2013).
[2] Fdhila R, Sand U and Eriksson JE, et al. 2016 A stirring history. ABB Review.;3:45-48.
[3] Timofeev V and Khatsayuk M. 2016 Theoretical design fundamentals for mhd stirrers for molten metals //Magnetohydrodynamics (0024-998X). – Vol. 52. – №. 4.
[4] Musaeva D, Ilin V and Baake E, et al. 2015 Numerical Simulation of the Melt Flow in an Induction Crucible Furnace Driven by a Lorentz Force Pulsed at Low Frequency. Magnetohydrodynamics.
[5] Eckert S, Nikrityuk P and Rabiger D, et al. 2008 Efficient melt stirring using pulse sequences of a rotating magnetic field: Part I. Flow field in a liquid metal column. Metall Mater Trans B.
[6] Wang X. et al. 2009 A periodically reversed flow driven by a modulated traveling magnetic field: Part I. Experiments with GaInSn //Metallurgical and Materials Transactions B. – Vol. 40. – №. 1. – p. 82.
[7] Ma X, Yang Y and Wang B. 2009 Effect of pulsed magnetic field on superalloy melt //International Journal of Heat and Mass Transfer. – Vol. 52. – №. 23-24. – p. 5285-5292

[8] Guo-Jun C, Yong-Jie Z and Yuan-Sheng Y. 2013. Modelling the unsteady melt flow under a pulsed magnetic field //Chinese Physics B. – Vol. 22. – №. 12. – p. 124703
[9] Timofeev V, Lybzikov G and Khatsayuk M, et al. 2015. Magnetohydrodynamic stirrer liquid metal with a non-sinusoidal currents. Engineering & Technologies. 166-177
[10] Jia, Yonghui, et al. "Numerical study on action of HMF, PMF, DHMF, and DPMF on molten metal during electromagnetic casting." The International Journal of Advanced Manufacturing Technology 103.1-4 (2019): 201-217
[11] Patent RU 2680715
[12] web site: https://www.npcmgd.com/
[13] Landau, L. D. (Ed.). (2013). The classical theory of fields (Vol.2). Elsevier.
[14] Landau, L. D., Bell, J. S., Kearsley, M. J., Pitaevskii, L. P., Lifshitz, E. M., & Sykes, J. B. (2013). Electrodynamics of continuous media (Vol.8). elsevier.
[15] Ansys Maxwell 15.0 User's Guide 3D
[16] Maksimov, Aleksiy A., Maksim Y. Khatsayuk, and Viktor N. Timofeev. "Aspects of Energy Transformation and Energy Control Capabilities in Electric Machines with a Liquid-Metal Working Body." 2019 20th International Conference of Young Specialists on Micro/Nanotechnologies and Electron Devices (EDM). IEEE, 2019.
[17] Yamamura S., Ito H., Ishulawa Y. Theories of the linear, induction motor and compensated linear induction motor //IEEE Transactions on Power Apparatus and Systems. – 1972. – №. 4. – C. 1700-1710

Acknowledgments
This research was financially supported by the Russian Foundation for Basic Research, by the Government of the Krasnoyarsk Territory, by the Krasnoyarsk Territory Science Foundation as part of scientific projects No. 18-48-242013 entitled "Study of the effect of spatial and frequency-temporal distributions of electromagnetic forces in the melt on the nature of the MHD flow processes in liquid metal"

IOP Conf. Series: Materials Science and Engineering **950** (2020) 012016 doi:10.1088/1757-899X/950/1/012016

Intelligent complex security management system FEC for the Industry 5.0

N V Korneev[1,2,3]

[1] Gubkin Russian State University of Oil and Gas, Moscow, Leninsky Prospect, 65, Russian Federation

[2] Financial University under the Government of the Russian Federation, Moscow, Leningradsky Prospekt, 49, Russian Federation

[3] Email: niccyper@mail.ru

Abstract. Taking into account the digital transformation, the prospects for the development of intelligent complex security management system FEC are revealed. For the first time a definition of intelligent complex security management system FEC is given, also, its generalized structure is developed, considering the development of the Russian human-machine concept, as a complex information-energy system, on the basis of intellectualization of theoretical principles of human physiology and psychology. Its main elements are defined: a core, subsystems and components and mathematical dependencies implementing management functions are given. The core, subsystems and components are represented by the protected system at the level of security threat management in the form of a block-functional system scheme with elements of artificial intelligence that differ in the detail level. The mutual influence mechanism of separate subsystems and components is defined through simple, hybrid and complex threats. Besides, system threats of FEC were classified and their definitions were given. As an example, we consider the complex threat related to security of critical information FEC infrastructure. What is more, the basic intelligent management principles of complex FEC security are formulated, based on the block-functional system scheme with artificial intelligence elements and operator state security identification in real time. We obtained general equations of the output state variables vector, which is equivalent to the security model of the operator, vectors of output state variables in case of hybrid, complex, and system FEC threats.

1. Introduction

Rapid development of information technologies has led to the digital transformation [1]. Digital transformation can be considered as an integration process of innovative technologies and business processes. In the digital economy it required to change technologies, culture and principles for creating new products and services. Nowadays, The Industry 5.0 can be defined as the best groups, methods and tools for digital transformation. National artificial intelligence development strategy for the period up to 2030 (Decree of the President of the Russian Federation 10.10.2019 № 490) influenced on the Industry 5.0 development.

The artificial intelligence [2] – this is a dynamic system that is able to (without human intervention) create full-featured models displaying complex world phenomena and itself in this world; analyze an adequacy (correspondence) of different model variants in order to select the most accurate ones; generate variant predictions of expected consequences based on the selected models.

All of the above offer new opportunities for the information processing systems development. The structure of information processing systems changes fundamentally which is now based on distributed information-computing networks, connected to global data networks, convergent, hyper-converged, neuromorphic and quantum computing systems. At the same time, regulatory requirements are toughen, especially, in terms of complex object security: physical, economic, fire, informational, psychological, intellectual property security, technogenic, security against terrorism, ecological safety and power security. For the fuel and energy complex (FEC) – this is first and foremost Energy security doctrine of the Russian Federation (Decree of the President of the Russian Federation 13.05.2019 № 216), new version of the information security Doctrine (Decree of the President of the Russian Federation 05.12.2016 № 646), Federal law July 26, 2017 № 187-FL "On the Security of the Russian Federation Critical Data Infrastructure, Federal law July 21, 2011 № 256-FL "On the safety of fuel and energy complex facilities", Federal law July 27, 2006 № 149-FL "About information, information technology and data security ". In these documents, the priority is given to the fuel and energy complex facilities safety, including through continuous monitoring of object operation threats.

We consider the concept of the object in its broader notion. Modern management theory considers a person, the world around him and certain parts of this complex system as objects for studying the possible effective impact on them. Let us call the part of this world that interests us - an object of management, we will characterize the position of this part in the world by the term - an object state, purposeful change of object state and the way of achieving certain goals in it by the term-management. Thus, the object of management is any phenomenon of the world (or the subject itself) which is considered in terms of creating some purposeful movement. It is natural to call the final state of the object desired, i.e. which we are interested in and which defines the purpose of management as the conversion of an object to the desired state and the act of conversion as a process of management or just management.

In this case FEC as an object can include the oil, gas, coal and peat industries, electricity and heat supply that is fully corresponds to the energy security Doctrine of the Russian Federation. However, if considering complex object security, expand its use for solving national security problems.

Based on the above, it becomes obvious the essence of the complex security – ensuring all security components of the facility: physical, economic, fire, informational, psychological, intellectual property security, technogenic, security from terrorism, environmental security and energy security including new components.

In this case, the concept of complex security is more flexible и self-organizing for digital transformation of an object, and the theory of management does not change the fundamental principles, that give the opportunity for creating an appropriate object management system for the Industry 5.0 including artificial intelligence elements.

The new concept is not about replacing a person – EC (electronic computer), but in their mutual complement. This approach is described in detail in [3] as a new class of "systems with artificial intelligence elements".

Thus, intelligent complex security management system FEC is a class of managing human-machine systems providing all the fuel and energy complex safety components with elements of artificial intelligence or artificial intelligence.

Currently, the implementation of this approach is observed [2, 3] in all areas of life activity. Separate components of the complex security management system are being developed for FEC. For example: physical [4, 5], economic [6, 7], fire [4, 8, 9], informational [10, 11, 12], individual solutions for intellectual property security offer companies «SearchInform» and «InfoWatch», technogenic [13, 14, 15, 16, 17], ecological safety [18, 19, 20], power security [6, 21, 22], psychological, security against terrorism [4, 23] are unavailable.

It is easy to notice that there aren't any Russian solutions for the most subsystems and components, generally, there is no Russian intelligent complex security management system of FEC.

Thus, the generalized structure of the intelligent control system for complex security of FEC becomes obvious, based on the Russian concept of the human-machine system development.

2. Results and Discussion

The generalized structure of the intelligent control system for complex security of FEC, taking into consideration the Russian concept of the human-machine system development as a complex information-energy system, based on intellectualizing the theoretical principles of human physiology and psychology is shown in Figure 1.

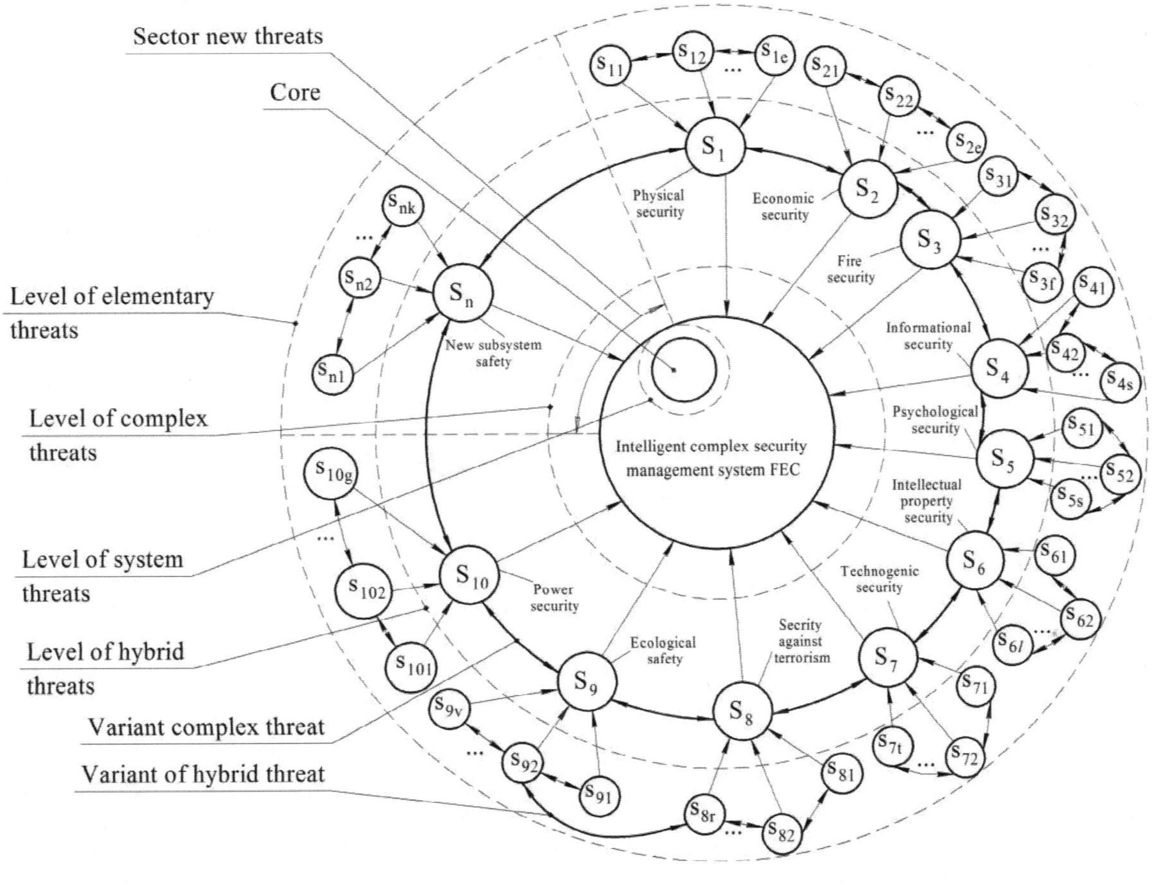

Figure 1. The general intelligent complex security management system of FEC

It is represented by a set of subsystems S_n ensuring all components of FEC security: physical (S_1), economic (S_2), fire (S_3), informational (S_4), psychological (S_5), intellectual property security (S_6), technogenic (S_7), security against terrorism (S_8), ecological safety (S_9), power security (S_{10}) including new elements (S_n) and their components: s_{1p}, s_{2e}, s_{3f}, s_{4s}, s_{5h}, s_{6l}, s_{7t}, s_{8r}, s_{9v}, s_{10g}, s_{nk}. Generally, the number of components in each subsystem is different. The core, subsystems and components – this are a representation of the protected system at the level of security threat management in the form of a block-functional scheme of the system with artificial intelligence elements (Figure 2) that differ in the detail level. For subsystems and components the features of the corresponding FEC security elements are taken into account. It should be noted that if considering the schemes of the system core, the operator of an intelligent integrated security management system of

FEC – is the president of Russia, and, the management system is an element of strategic planning in the sphere of ensuring the Russian Federation security.

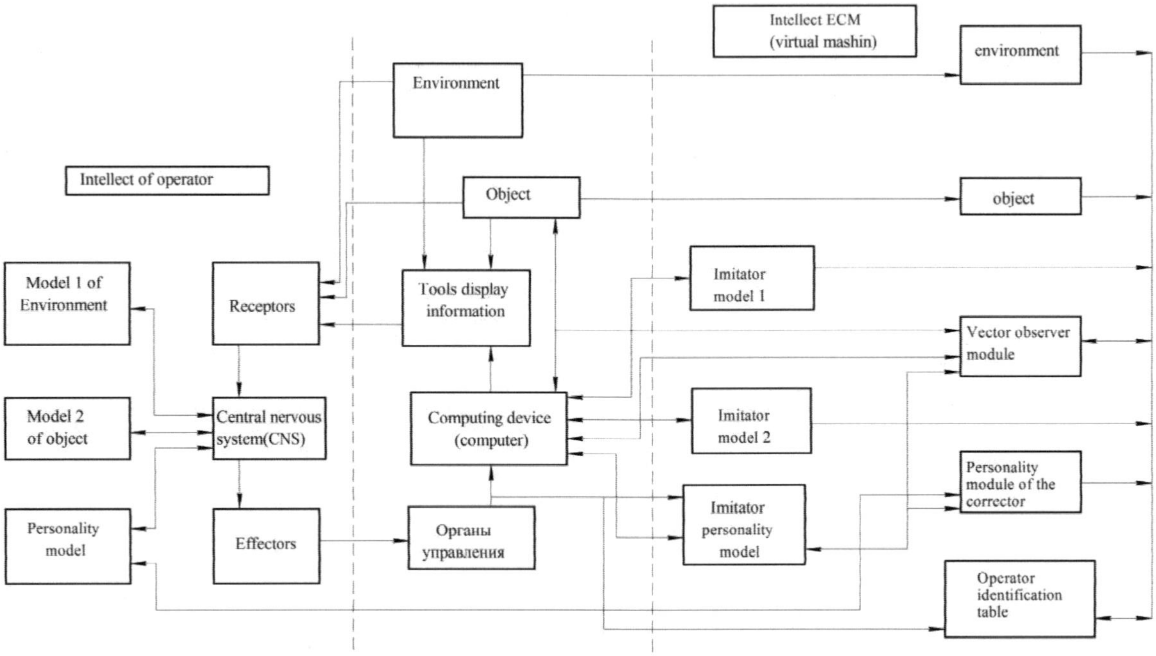

Figure 2. Block-functional system scheme with artificial intelligence elements

The scheme (Figure 2) shows the development of the block-functional scheme [3, 24] in the direction of supplementing the usual human-machine system with artificial intelligence elements. They are mainly related to the right side of Figure 2 under the pointer «Intellect ECM». The following computational models and modules are assigned to this system part: environment – «direct» model 1; the object – «direct» model 2; imitators: model 1 the operator of the environment; model 2 the object of the operator; personality model of the operator; personality module of the corrector; the vector observer module.

Model imitators 1 and 2, as well as the personality model of the operator of the corresponding security subsystem have a task to approximately repeat the security models that exist within the operator. Their tasks – explain and predict the behavior of the operator due to these factors. For example, the operator PCS thinks that a compressor is broken or the operator SIEM (Security Informational and Event Management) thinks that there is a cyber attack on the enterprise's SCADA system. The personality model displays his behavior when such a hypothesis appears. The imitators' task is to repeat the properties of internal models as closely as possible.

The difference between "direct" object model and the environment in the management system from the internal models of the operator and their imitators, i.e., the display of these models formed by the operator is as follows:

the direct model of the object is the result of its diagnosis and current monitoring in stationary conditions. In details, this model is significantly more accurate than the internal operator model (especially, if the operator is not highly qualified);

the direct model of the environment can be formed in the management system objectively (software and hardware), including maps of the current object operation area; communication and interrelation schemes of an object, atmospheric and operating conditions; the presence of dynamic and other types of interference when operating. If considering critical information infrastructure (CII) FEC – reports on current operation of potential CII objects, industrial safety data sheet of a hazardous industrial object, protection in the structure of potential CII objects;

the object and environment models of the operator are subjective and deformed by his personality model.

Generally, like the operator identification table, these modules have the character of handlers of accumulated information and devices for generating additional diagnostic actions and actions of an executive nature.

Direct display of the object state and environment based on diagnostic information about the object and the most complete environment available for the management system.

At the same time, we cannot assume that this information is identical to the equation given in the work [3, 24], for this reason:

$$\tilde{X}[k+1]=\tilde{\Phi}(\tilde{X},U,\tilde{F},t)\tilde{X}[k]+\tilde{\Gamma}(t)U[k]+\tilde{G}(t)\tilde{F}[k], \quad (1)$$

where $\tilde{X}[k+1], \tilde{X}[k]$ – the most accurate possible vectors evaluations of the object state and environment; $\tilde{\Phi}(\tilde{X},U,\tilde{F},t)$ – state transition function determined by the most accurately known parameters of the object state and environment; $\tilde{F}[k]$ – vector evaluation of direct environmental impacts; $\tilde{\Gamma}(t)U[k], \tilde{G}(t)\tilde{F}[k]$ – integral transformations of the most accurately represented controlling and disturbing influences.

Equations for the secondary object and environment display in the management system. The secondary display is the equation of the secondary models displaying the internal operator models in the management system. They can be recorded in a form similar to the given equation [3, 24] using the double dash sign at the top of the symbol to indicate the secondary model functions and variable estimates:

$$\bar{\bar{X}}[k+1]=\bar{\bar{\Phi}}(\bar{\bar{X}},U,\bar{\bar{F}},t)\bar{\bar{X}}[k]+\bar{\bar{\Gamma}}(t)U[k]+\bar{\bar{G}}(t)\bar{\bar{F}}[k],$$

(2)

where $\bar{\bar{X}}[k+1], \bar{\bar{X}}[k]$ – vector condition estimates of object state and environment in the secondary models; $\bar{\bar{\Phi}}(\bar{\bar{X}},U,\bar{\bar{F}},t)$ – the function of the secondary transition model, $\bar{\bar{F}}[k]$ – perturbation vector estimate in the secondary model; $\bar{\bar{\Gamma}}(t)U[k], \bar{\bar{G}}(t)\bar{\bar{F}}[k]$ – conversion of the secondary model impacts; $\bar{\bar{Y}}[k]$ – vector of the secondary models state variables: $\bar{\bar{Y}}[k]=\bar{\bar{C}}\bar{\bar{X}}[k]+\bar{\bar{\upsilon}}[k]$, где $\bar{\bar{\upsilon}}[k]$ – noise measurement vector; $\bar{\bar{C}}$ – estimator matrix of measurable variables.

Equations of the secondary personality model of the operator. This model is constructed similarly, taking into account the object functions and environment models that are already known:

$$\tilde{\tilde{Y}}[k+1]=\tilde{\tilde{\psi}}(\tilde{\tilde{Y}},t)\tilde{\tilde{Y}}[k]+\tilde{\tilde{\psi}}(\bar{\bar{X}},t)\bar{\bar{X}}[k+1]+\bar{\bar{\Gamma}}(t)(U[k])+\bar{\bar{G}}(t)\bar{\bar{F}}[k], \quad (3)$$

where $\tilde{\tilde{Y}}[k+1], \tilde{\tilde{Y}}[k]$ – operator's actions vectors based on its secondary model; $\tilde{\tilde{\psi}}(\tilde{\tilde{Y}},t)$ – transition function of the operator secondary personality model.

According to (3) the operator's actions are based on internal state and object state prediction and environment. Thus, the influence of the object state prediction and environment on the formation of the vector of operator's management considers separately. Predictive abilities of the operator for FEC are very important and require special testing and development.

Subsystems (Figure 2) may include separate components that are selected from the complex security subsystems based on the above concept of an object and solved security problem; for example, by the sequential decomposition of environment , object and models to the level of an object with elementary security threats.

IOP Conf. Series: Materials Science and Engineering **950** (2020) 012016 doi:10.1088/1757-899X/950/1/012016

It should be noted, that the issue of mutual interrelation of individual subsystems and components on each other is quite important and at the same time difficult to solve. In the Figure 1 links that characterize this effect are highlighted in bold. Such interaction is related to the mutual impact of individual subsystems and components on each other. One of the possible ways to account for this impact is related to the concepts of hybrid and complex threat.

Complex threat – a threat consisting of several different elementary threats, connected by means of certain synchronized mechanisms and not necessarily existing in one space.

Hybrid threat – a variation of a complex threat, which necessarily contains elementary threats that affect different areas of the protected system.

It is also worth noting that the existence of hybrid threats is closely related to the term "hybrid war" [25, 26, 27]. These are subtypes of complex threats and characterized by the property of forming and implementing the threat components not in a single space (for example, only in the physical) and in several spaces simultaneously (for example, in physical and information space).

The work [28] examined the practical problems of complex threats (CT) and the peculiarities of their formation and implementation. On that basis, proposed the approaches to ensuring security, which can most effectively be manifested in the presence of tools and methods for identifying the very fact of the presence of CT and determining the list of possible targets of CT. Implementation of these mechanisms can be achieved using big data analytics integrated from various sources, for example at the object level – FEC enterprise.

At the system core level, not only complex FEC threats are considered, but also FEC system threats that can be defined like this.

FEC system threat – this is a complex threat to the oil, gas, coal and peat industries, electricity and heat supply. In a system threat sets of complex threats are already subjects for consideration, selected on the basis of the above object concept and solved subsystem security problem and united, for example in a cluster. As an example of state systems, operating at the level of system threats – State system for detecting, preventing and eliminating the consequences of computer attacks, SIS FEC (State information system of the fuel and energy complex).

Further, as an example we will consider the complex threat connected with ensuring the security of the critical data infrastructure of the FEC. We will consider the FEC enterprise as an object. For the certainty in security tasks, assume that we know the basic threat model of the object (Figure 1) and includes information security threats (subsystems S_4), technogenic (S_7). The considered problem relates to complex threats, but we will take its variation – a hybrid threat that includes elementary threats: 1 information security threat and 1 technogenic safety threat. In short, assume that the environment in case of information security threat is formed by security events in SIEM (Security Informational and Event Management), technogenic safety – event report in PCS.

In this case, we will have two systems of block-functional structure (Figure 2). Using the example of these systems, we formulate the basic principles of intelligent management of complex FEC security based on the block-functional scheme of the system with artificial intelligence elements. This approach is based on identifying security states of the operator (Figure 2) in real time [3, 24].

1. Selection of the controlling element set used by the operator. For a considering task we assume that both operators have the ability to respond to a security event, using the computer mouse and entering data from the keyboard. It should be noted that in real conditions we can use advanced systems for actions control.

2. Defining the conditions when implementing event or set of events that are synchronized by a specific mechanism, for example time t can be considered as security threat. This problem has a lot of solutions. For example, the most well-known is the use of a rules set as a part of SIEM based on heuristic methods [29]. However, this option refers to an analytical approach of modeling that does not always adequately reflect reality, especially in the context of complex interaction processes in the real world.

The proposed intelligent complex security management system FEC (Figure 1) allows to realize not only analytical, but also informational approach based on modern data mining technologies and big data processing methods [28].

For a considering task we assume that both operators can respond to a security event directly using a computer mouse, but to respond to a security threat you must use the keyboard. This condition allows you to initially isolate a threat from security events in SIEM at the operator action level. This approach is justified because most of the events registered by SIEM are not threats and the operator scroll through them using the computer mouse. At the same time, for a threat such as domain control attack by selecting a software password RDP, the operator will be required to use the keyboard to create more complex password for RDP.

3. Identification of a functionally complete threat measurement system. According to modern concepts such measurement system contains:

1. Sensor of event 1 – Y_1, frequency of event 1 – Y_2.
2. Sensor of event 2 – Y_3, frequency of event 2 – Y_4.
3. Sensor of event 3 – Y_5, frequency of event 3 – Y_6
4. The sensor of the left key of a computer mouse – $U_1 = \xi_1$.
5. The sensor of the right key of computer mouse – $U_2 = \xi_2$.
6. Computer mouse scroll sensor – $U_3 = \xi_3$.
7. Keyboard activity sensor (buttons Enter) – $U_4 = \xi_4$.

In the list below we mentioned three event sensors. In real conditions their number is determined according to the analysis of the exploited vulnerability of the threat, i.e. a factor, based on the protected system properties or protection methods that are used when implementing a specific elementary threat. The dash sign means the sensor indicators as measurable coordinates: Y – measurable (output) state variables; U – control actions of the operator.

A significant number of identification variables follows from this set of sensors: for sensors number 4…7 the actions of the operator are also characterized by the direction (sign) of the impact which is essential for this class of systems. It should be noted that for other FEC objects measurable state variables and operator control actions can be similarly allocated.

It allows you to create the following table 1.

Table 1. Identification variables the sensors

Sensor number	Action direction	Impact direction	Impact
4	keypress	$\xi_{11} = \xi_{1+}$	$\xi_{121} = \mathrm{d}\xi_{11}/\mathrm{d}t$
4	button release	$\xi_{12} = \xi_{1-}$	$\xi_{122} = \mathrm{d}\xi_{12}/\mathrm{d}t$
5	keypress	$\xi_{21} = \xi_{2+}$	$\xi_{221} = \mathrm{d}\xi_{21}/\mathrm{d}t$
5	button release	$\xi_{22} = \xi_{2-}$	$\xi_{222} = \mathrm{d}\xi_{22}/\mathrm{d}t$
6	up	$\xi_{31} = \xi_{3+}$	$\xi_{321} = \mathrm{d}\xi_{31}/\mathrm{d}t$
6	down	$\xi_{32} = \xi_{3-}$	$\xi_{322} = \mathrm{d}\xi_{32}/\mathrm{d}t$
7	keypress	$\xi_{41} = \xi_{4+}$	$\xi_{421} = \mathrm{d}\xi_{41}/\mathrm{d}t$
7	button release	$\xi_{42} = \xi_{4-}$	$\xi_{422} = \mathrm{d}\xi_{42}/\mathrm{d}t$

Thus, $\xi[k]$ – vector of output (measurable in the control system) state variables, which is equivalent to the security model of the operator, can be represented as follows:

$$\xi[k] = \Lambda(V, U, t)V[k] + \nu[k], \tag{4}$$

where $\Lambda(V,U,t)$ – a vector-function, identifying the security state of the operator; $v[k]$ – a noise vector accompanying operator's action. In case of an analytical approach the vector-function is defined by an extended object space, based on differential equations in Cauchy form [24], information approach – data mining technologies and big data processing methods [29].

It is noteworthy that the output state variables vector for the second system of the block-functional structure will be similar (Figure 2), determining the threat to technotronic security. Then the record of the output state variables vector of the considered example in case of a hybrid threat will be:

$$\xi[k]_{hyb} = \xi^1[k] \cap \xi^2[k], \tag{5}$$

where the upper indexes belong to the considering block-functional structure systems (Figure 2).

Based on (Figure 1) the record of the output state variables vector in case of a complex threat will be:

$$\xi[k]_{com} = \bigcap_{S_1}^{S_n} \bigcup_{S_{11}}^{S_{nk}} \xi[k], \tag{6}$$

in case of a hybrid threat:

$$\xi[k]_{hyb} = \mathrm{var}\,\xi[k]_{com}. \tag{7}$$

The equations (6, 7) are valid if the object and environment are equivalent.

Based on (Figure 1) the record of the output state variables vector in case of a system threat will be:

$$\xi[k]_{sys} = \langle \triangleright\triangleleft\, \xi[k]_{com} \rangle, \tag{8}$$

where operator $\triangleright\triangleleft$ denotes a method for joining vectors of output state variables in the case of a complex threat, for example, clusterization - is a natural join.

3. Conclusion

Among the local models included in the structure (Figure 2) the personality model of the operator has the greatest non-stationarity. Studies of the current identification of this model parameters are a scientific complex security problem, where the tasks are: constructing procedure for identifying the object management style using the fuzzy set theory, identification procedures for an extended object threat model based on modern data mining technologies and big data mining methods, procedures of intelligent complex security management of FEC by replacing the operator with artificial intelligence, for example, in the form of a self-learning neural network or a neurograph [4]. We note that these tasks determine the need to consider dynamic evaluations which change according to conditions, and, above all - the state of the operator.

References

[1] Patel K and McCarthy M P 2000 *Digital transformation: the essentials of e-Business leadership* (New York: McGraw-Hill) p 144

[2] Korneev N V 2008 Printsipy postroeniya sovremennykh tekhnicheskikh sistem s elementami iskusstvennogo intellekta [Principles of construction actual technical systems with elements of artificial intellect] *Tekhnika mashinostroeniya* **2(66)** 2–7 (in Russian)

[3] Korneev N V 2008 Intellektual'naya sistema upravleniya transportnym sredstvom s uchetom lichnykh osobennostei voditelya [Intellect system control transport mashin based on driver's personal characteristics] *Tekhnika mashinostroeniya* **3(67)** 54–56 (in Russian)

[4] Korneev N V 2019 *IOP Conf. Ser.: Earth Environ Sci.* **224** 012021

[5] Mosolov A S and Grafova O V 2010 *Proc International Conference on Security of Information and Networks* (Rostov-on-Don: ACM Press and Digital Library) pp 134–136

[6] Cao J, Wang Y, Zhu C, Zhang Y, Guo C and Cao Y 2012 *Automation of Electric Power Systems* **2** 77–81

[7] Yu Y X and Luan W P 2009 *Zhongguo Dianji Gongcheng Xuebao / Proceedings of the Chinese Society of Electrical Engineering* **34** 1–8

[8] Danilov A I, Mel'kov S A, Solov'eva N V, et al 2019 *Russ. Meteorol. Hydrol.* **44** 300–304

[9] Thompson P and Marchant E 1995 *Fire Safety Journal* **24(2)** 131–148
[10] Markov A S, Barabanov A and Tsirlov V 2018 Periodic Monitoring and Recovery of Resources in Information Systems *Probabilistic Modeling in System Engineering* ed A Kostogryzov (London: IntechOpen) chapter 10 p 278
[11] Korneev N V 2013 *Algoritmicheskie i programmnye metody i sredstva otsenki al'ternativnykh proektov zashchity sistemy obrabotki informatsii predpriyatiya na osnove mnogokriterial'nogo analiza* [Algorithmic both program methods and tools estimation of alternative projects of the guard data reduction system of firm on the basis of the multicriteria analysis] (Togliatti: Volga Region State University of Service Press) p 116 (in Russian)
[12] Korneev N V and Korneeva Yu V 2014 Audit sistemy menedzhmenta informatsionnoi bezopasnosti [Audit of system management information security] *Standarty i Kachestvo* **7(925)** 60–64 (in Russian)
[13] Kershenbaum V, Grigoriev L, Kanygin P and Nistratov A 2018 Probabilistic Modeling Processes for Oil and Gas *Probabilistic Modeling in System Engineering* ed A Kostogryzov (London: IntechOpen) chapter 3 p 278
[14] Oliveira L E S and Alvares A J 2016 *Procedia CIRP* **53** 198–205
[15] Wei L, Chuipin K, Qiang N, Jingguo J, Xionghui Z 2020 *Robotics and Computer-Integrated Manufacturing* **61** 101842
[16] Peng L, Tianliang H, Chengrui Z 2011 *Procedia Engineering* **15** 840–847
[17] Korneev N V and Yanitskiy A I 2019 *Russ. Electr. Engin.* **90** 696–701
[18] Gerardo A, Vibeke S May-Britt E, Jukka T and Adrian T *Sustainability* **11(17)** 4611
[19] Smolenskaya N M Smolensky V V, Korneev N V 2018 *IOP Conf. Ser.: Earth Environ. Sci.* **121** 052009
[20] Smolenskaya N M, Korneev N V 2017 *IOP Conf. Ser.: Earth Environ. Sci.* **66** 012016
[21] Kovalev A 2017 *Energy manager* **1** 3–5
[22] Karasev Y, Konev V, Evlanov M and Krykov I 2014 *Automation and IT in energy* **11(64)** 23–33
[23] Young C 2014 *The Science and Technology of Counterterrorism* (Oxford: Butterworth-Heinemann) p 512
[24] Korneev N V, Kustarev Yu S and Morgovsky Yu Y 2008 *Teoriya avtomaticheskogo upravleniya s praktikumom* [Theory automatic control with workshop] (Moscow: Publishing house Academia) p. 224 (in Russian)
[25] Davis Jr J R 2015 *The Three Sword Magazine* **28** 19–25
[26] Hunter E and Pernik P 2015 *The challenges of hybrid warfare* (Tallinn: International Centre for Defence and Security) p 7
[27] Mälksoo M 2018 *European security* **27(3)** 374–392
[28] Korneev N V and Merkulov V D 2019 Intelligent analytics for complex security *Secure Information Technologies* (Moscow: Bauman Moscow State Technical University Press) pp 215–219
[29] Markov A S, Matveev V A, Fadin A A and Tsirlov V L 2016 *Vestn. Mosk. Gos. Tekh. Univ. im. N.E. Baumana Priborostr.* **1** 98–111

Issues of induction sorting of scrap and waste of non-ferrous metals

M E Zyazev[1], E S Lyampasova[1], Z O Abdullaev[2] and A Yu Konyaev[3]

[1] Master Student, Chair of Electrical Engineering and Electrotechnological Systems, Ural Federal University, Yekaterinburg, Russia
[2] Graduate Student, Assistant, Chair of Electrical Engineering and Electrotechnological Systems, Ural Federal University, Yekaterinburg, Russia
[3] Doctor of Technical Scienses, Professor, Chair of Electrical Engineering and Electrotechnological Systems, Ural Federal University, Yekaterinburg, Russia

E-mail: zyacho72@gmail.com

Abstract. Nowadays, one of the main economic advantages is using secondary raw materials for the production of useful materials for further use. The greatest environmental and economic effect provides the use of secondary non-ferrous metals. Eddy-current separation based on interaction force of a magnetic field with eddy currents induced by this field in conducting particles, is the most effective for the recycling of non-ferrous metals. The paper is devoted to the problem of non-ferrous metals sorting. The research results of the number of traveling magnetic field separators designs are described. The ways of improving separation efficiency are shown. Eddy-current separators can be used as stand-alone solutions, and in combination with shredder and classifying systems, in a variety of recycling plants and at metallurgical enterprises for non-ferrous metals refinement.

Keywords: separation, induction sorting of metals, research results.

1. Introduction

One of the trends in the development of the economy, which allows solving environmental and economic problems, is an increase in the production of secondary non-ferrous metals. When using secondary metals the need for mineral raw materials, energy consumption and emissions of pollutants into the biosphere are reduced. At the same time, thorough the sorting of collected scrap, which should ensure the production of high-quality secondary alloys with an admixtures level not higher than that of primary metal alloys, is becoming increasingly important. The uncontrolled mixing of a scrap of non-ferrous metals and their alloys in combination with the melt purification constraints during remelting creates big problems in obtaining high-quality alloys from secondary raw materials [1-4]. Therefore, the development and creation of technologies for the collection and processing of scrap and non-ferrous metal waste, as well as related technological equipment, are relevant.

Electrodynamic (eddy-current) separation is the most often method used for non-ferrous metals recovery from solid waste and also separating non-ferrous metals from each other [5-9]. Eddy-current separation is based on the force interaction of a magnetic field with eddy currents induced in conducting particles using a time-varying field. An electromagnetic force ejects metal particles out of the stream of materials. A traveling or rotating magnetic field can be generated by three-phase linear inductors or rotating inductors based on permanent magnets [5, 9]. The most widespread, are separators based on permanent magnets [5-7]. At the same time, the possibilities of separators, based on three-phase linear inductors, are not well understood [5, 8-9].

The article presents the results of the research of eddy-current separators based on linear induction machines.

2. Decided technological problems

One of the fast-growing large-tonnage types of solid waste is automobile scrap. Therefore the problems of disposal of end-of-life vehicles require increased attention. The main commercial product of auto recycling is steel scrap. At the same time, in the automotive industry, there is a tendency to increase the share of light alloys (primarily aluminum) in the design of cars. For example, in the cars of the world's leading automobile companies, released in 2010, the share of aluminum reaches 15–20% (by weight) and its subsequent growth is forecast to be up to 30% [10–12]. The increase in the share of aluminum is accompanied by the expansion of the list of alloys used in the automotive industry. Therefore, in auto recycling, it is advisable to solve three consecutive technological tasks:

- the selection of non-ferrous metals from the flow of non-metals after shredding processing and magnetic separation;
- separation of non-ferrous scrap by types of alloys (aluminum, copper, zinc);
- division of aluminum scrap into groups and grades of alloys.

Similar technological problems also arise at metallurgical plants in the preparation of mixed non-ferrous scrap for melting [13].

Another fast-growing group of metal-containing waste is electronic scrap. Such waste can be a significant source for the extraction of not only non-ferrous but also precious metals [9, 14-15]. One of the important problems in the processing of crushed electronic scrap is the separation of aluminum and copper alloys. The collection of copper alloys in a separate group facilitates the subsequent separation of precious metals.

A significant proportion of waste containing non-ferrous metals (primarily copper and aluminum) is scrap and cable waste. In the manufacture of cables, technically pure highly conductive metals are used, as well as high-quality insulating materials (polymers, rubber). Powerful cables, in most cases, have a protective sheath (made of lead, aluminum or steel tape). One of the common technologies for processing cable scrap is its crushing, followed by separation of metal fractions and insulating materials [16-17]. The most important separation task is to improve the quality of copper or aluminum concentrates. Removal of impurities from such concentrates allows the secondary use of highly conductive metals for the production of cable products.

The literature analysis shows that electrodynamic separation is most often used to solve these technological problems.

3. Objects of research

The presented technological tasks can be divided into two types:

- the extraction of non-ferrous metals from the flow of non-metals;
- sorting of non-ferrous metals by types and brands of alloys.

The tasks of the first type can be solved by using electrodynamic separators, schematically shown in figure 1. Both versions of separators are widely used in practice. The result of separation is most often a collective metal concentrate.

The separator shown in figure 1-a is characterized by low power consumption and can operate at increased frequencies of the magnetic field. At the same time, such a separator has a more complex structure and requires the supply of material in a thin layer, which causes a number of limitations. Separators based on three-phase linear induction machines (LIM) are easily integrated into complete

IOP Conf. Series: Materials Science and Engineering **950** (2020) 012017 doi:10.1088/1757-899X/950/1/012017

production lines and have great productivity. However, they operate at a frequency of 50 Hz, which makes it difficult to separate metal particles with a particle size of less than 40 mm.

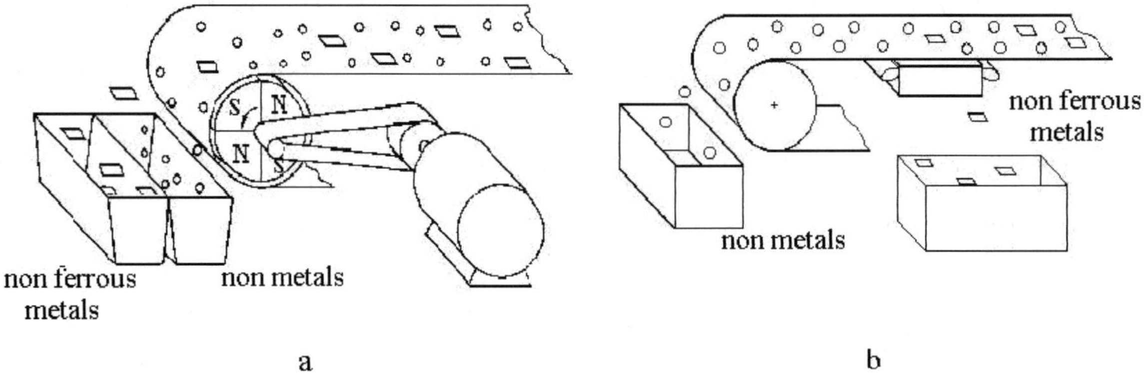

Figure 1. Electrodynamic separators with magnetic rotor (a) and with linear induction machine (b)

For induction sorting of non-ferrous metals, the most important indicator is the selectivity of the separation of metals, which is difficult to ensure when feeding materials along the conveyor belt. Great opportunities to control the trajectories of particles of different metals and to increase the selectivity of separation appear when materials are fed into the working area along an inclined plane. An example of such a separator is shown in figure 2. To increase the electromagnetic forces, it is possible to use a LIM with a double-sided inductor. In this case, the size of the working gap is determined by the size of the separated particles.

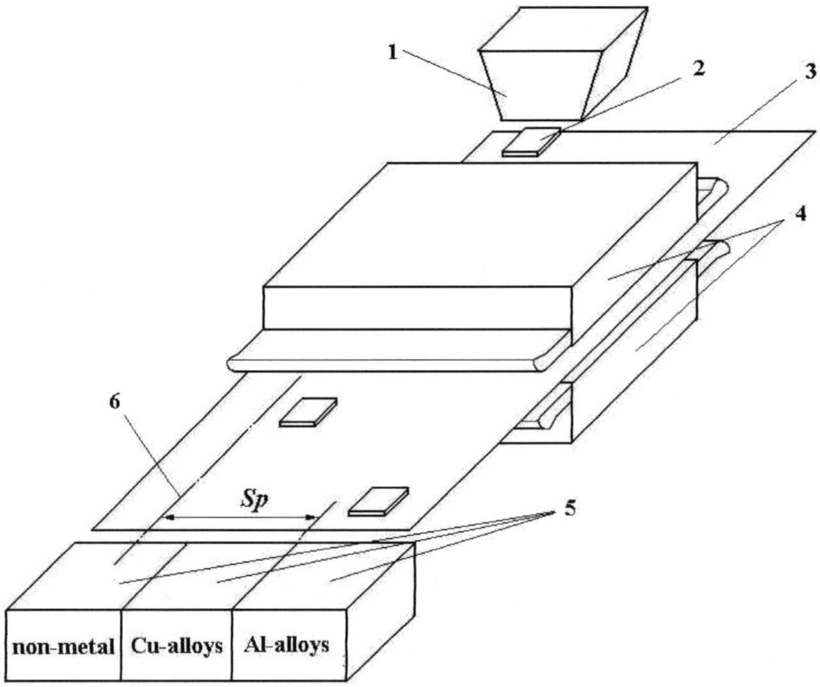

Figure 2. Electrodynamic separator with double-sided linear induction machine:
1 – feed hopper; 2 – particle; 3 – feed plane; 4 – primary LIM;
5 – collection bins; 6 – feed line; *Sp* - splitter position

To increase the selectivity of metal separation, the separated materials are fed along a feed line in a narrow stream, the width of which is determined by the particle size.

In the research laboratory of the Ural Federal University, all types of facilities, shown in figures 1 and 2, were created. This allows a variety of studies of separation processes to solve various technological problems.

4. Research results

4.1. Automobile scrap

A study of the processes of induction sorting of non-ferrous metals, represented in automobile scrap, was carried out on samples of aluminum alloys (cast or wrought), of magnesium and zinc alloys. At the first stage, the specific electromagnetic forces were evaluated (the ratio of the electromagnetic force to the particle mass, N/kg or m/s^2). Figure 3 shows the dependences of such forces acting on plates with dimensions of 40×40×3 mm in a double-sided linear machine (pole pitch 0.18 m, air-gap 0.06 m) on the current in the separator. Experimental forces were obtained for currents from 0 to 40 A (at a rated current of laboratory installation of 30 A). The calculations were performed for currents up to 60 A, which can be achieved in an industrial installation with forced air cooling.

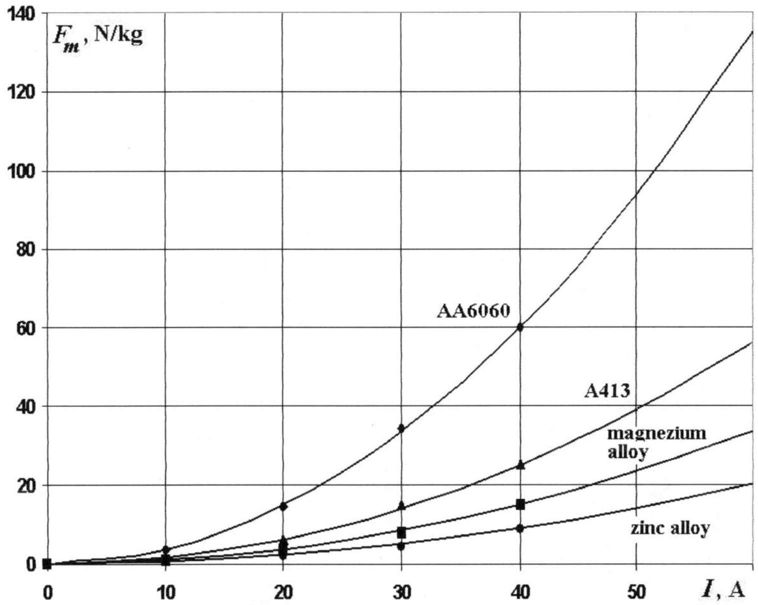

Figure 3. The dependence of the specific electromagnetic force on the LIM current: calculated (lines) and experimental (points)

A significant difference in forces for different alloys indicates the possibility of sorting. This is confirmed by testing the electrodynamic separator when feeding metals on an inclined plane. The total deflections of the metal particles from the feed line B at the outlet of the installation (at the end of the inclined plane) were measured depending on the distance from the input edge of the inductor to the starting point L_0. The measurements were performed at the rated currents of the installation (power consumption 1 kW). Figure 4 shows the dependences for two alloys in the range of particle sizes from 20 to 40 mm. It can be seen that there is a choice of the location of the flow splitter, which causes a collection of different alloys in different receivers. For example, at L_0 = 0.1 m, the distance from the feed line to the splitter can be Sp = 0.1 m.

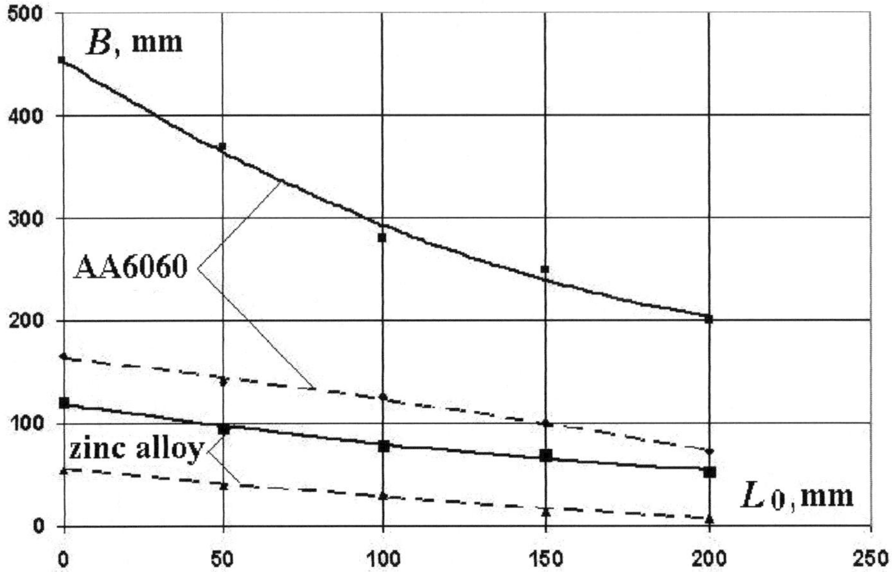

Figure 4. Path deflections of particles of different alloys from the feed line, depending on the distance from the input edge of the inductor to the starting point for different particle size: 40 × 40 mm - solid lines, 20 × 20 mm - dotted lines

4.2. Electronic scrap

For the separation of electronic scrap, an installation was used with feeding the material along an inclined plane based on a two-sided linear induction machine (power consumed by the inductor P = 750 W, pole pitch 0.066 m, air-gap 0.020 m). During the tests, a large fraction of crushed electronic scrap with a particle size of 8 to 20 mm was used. The characteristics of the scrap sample are shown in table 1. The main purpose of separation is the separation of particles of aluminum alloys from copper alloys.

Table 1. Mass and volume indicators of the processed sample electronic scrap

Fractional composition	Mass part		Volume part	
	g	%	cm^3	%
Aluminum alloys	273.00	60.67	101.11	79.83
Copper alloys	155.00	34.45	17.41	13.74
Other	22.00	4.88	8.14	6.43
Total	450	100	126.66	100

Figure 5 shows the dependences of the technological parameters of the separator on the position of the material flow splitter Sp achieved in one of the series of tests. Further studies show that separation indicators can be further improved.

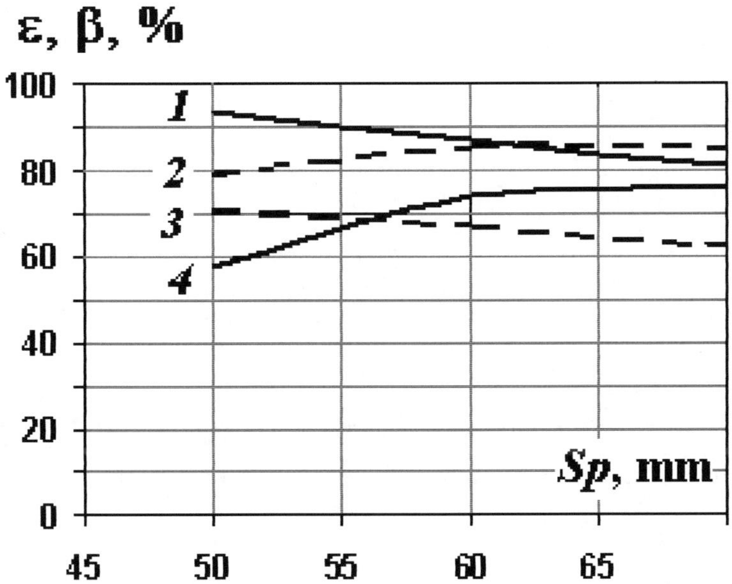

Figure 5. Separation indicators (by weight): recovery (solid lines) and grade (dotted line): 1, 2 - for aluminum alloys (in concentrate); 3, 4 - for copper alloys (in tails)

4.3. Cable scrap

When processing crushed cable scrap, electrodynamic separation is used to improve the quality of the base metal concentrate (copper or aluminum). For testing, we used a setup with feeding material along an inclined plane based on a two-sided linear induction machine (power consumed by the inductor P = 900 W, pole pitch 0.066 m, air-gap 0.01 m). Samples of crushed cable scrap provided by enterprises consisted of particles with a particle size of 3 to 8 mm. The compositions of the samples of the starting aluminum concentrate are shown in table 2, and the source of copper concentrate in table 3.

Table 2. Morphological composition of the sample of the starting aluminum concentrate

Material	Mass, g	Mass part, %
Aluminum	478.8	95.76
Copper	19.2	3.84
Insulation	2.0	0.40
Total	500.0	100.00

Table 3. Morphological composition of the sample of the starting copper concentrate

Material	Mass, g	Mass part, %
Copper	131.1	87.4
Lead	17.9	11.9
Insulation	1.0	0.7
Total	150.0	100.00

Some test results are shown in figures 6 and 7.

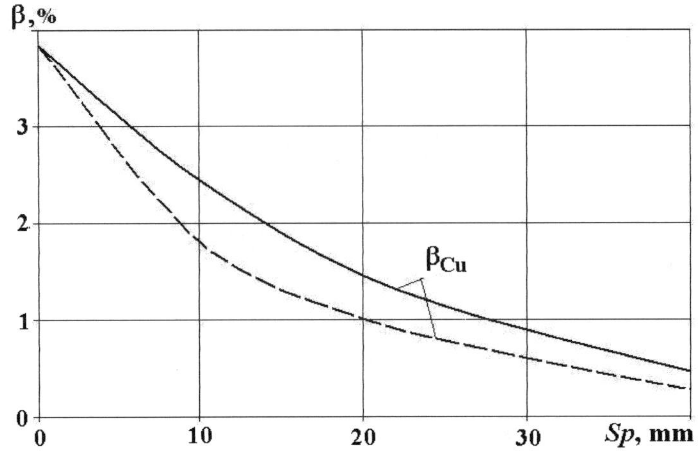

Figure 6. Reducing the proportion of copper in aluminum concentrate

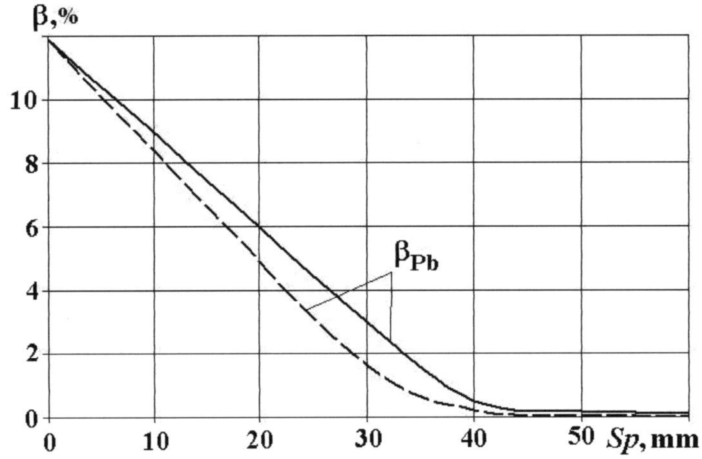

Figure 7. Reducing the proportion of lead in copper concentrate

5. Conclusion
Summarizing all the facts, the performed research has shown the possibility of efficient separation of non-ferrous metals in solving various technical problems in installations based on double-sided linear induction machines when feeding material on an inclined plane.

References
[1] Environmental modelling of aluminium recycling: A Life Cycle Assessment tool for sustainable metal management / D. Paraskevas , K. Kellens, W. Dewulf, J.R. Duflou // Journal of Cleaner Production, 2015, volume 105, 15 October 2015, pp. 357-370. DOI:10.1016/j.jclepro.2014.09.102

[2] Gaustad G., Olivetti E., Kirchain R. Improving aluminum recycling: A survey of sorting and impurity removal technologies. Resources, Conservation and Recycling, 2012, issue 58, pp.79– 87. DOI:10.1016/j.resconrec.2011.10.010

[3] Tatarkin A.I., Romanova O.A., Dyubanov V.G., Dyushin A.V., Bryantseva O.S. Trends and prospects for the development of metal recycling. [In Russian]. Ecology and Industry of Russia, 2013, No. 5, pp. 4-10.

[4] Ovsyannikov B.V. Manufacturing of deformed products, using scrap and wastes of aluminium alloys. [In Russian]. Tsvetnye metally, 2014, No 5 (857), pp. 66-70.

[5] Smith Y.R., Nagel J.R., Rajamani R.K. Eddy current separation for recovery of non-ferrous metallic particles: A comprehensive review. Minerals Engineering, 2019, issue 133, pp. 149–159. DOI:10.1016/j.mineng.2018.12.025

[6] Lungu M., Rem P.C. Separation of small nonferrous particles using and inclined drum eddy-current separator with permanent magnets. IEEE Transaction on Magnetics, 2002, volume 38, No. 3, pp. 1534–1538. DOI:10.1109/20.999128

[7] Settimo F., Belivacqua P., Rem P. Eddy current separation of fine non-ferrous particles from bulk streams / Physical Separation in Science and Engineering, 2004, volume 13, No. 1, pp. 15–23. DOI: 10.1080/00207390410001710726

[8] Konyaev A.Yu., Bagin D.N. Modeling an Electrodynamic Separator Based on a Linear Inductor. Russian Electrical Engineering, 2018, volume 89, No. 3, pp. 168–173. DOI:10.3103/s1068371218030100

[9] Obvintseva E.Yu., Konyaev A.Yu. (2017). Linear induction machines for electrodynamic separation of non-ferrous metals. 2017 IEEE Conference of Russian Young Researchers in Electrical and Electronic Engineering (EIConRus). DOI:10.1109/elconrus.2017.7910871

[10] Component- and alloy-specific modeling for evaluating aluminium recycling strategies for vehicles / R. Modaresi, A.N. Lovik, D.B. Muller. Journal of Metals, volume 66, issue 11, 2014, pp. 2262-2271. DOI: 10.1007/s11837-014-0900-8

[11] Aluminium flows in vehicles: enhancing the recovery at end-of-life / F. Passarini, L. Ciacci, A. Santini, I. Vassura, L. Morselli // Journal of Material Cycles and Waste Management, volume 16, issue 1, February 2014, pp. 39-45. DOI:10.1007/s10163-013-0175-0

[12] Cui J., Roven H. Recycling of automotive aluminium // Transaction of non-ferrous metals society of China, 2010, issue 20, pp. 2057-2063. DOI:10.1016/S1003-6326(09)60417-9

[13] Kercher S. A., Webb M. Scrap processing by eddy current separation techniques. Resources and Conservation, 1982, volume 8(1), pp. 61–74. DOI:10.1016/0166-3097(82)90053-0

[14] Cui J., Forssberg E., Mechanical recycling of waste electric and electronic equipment: a review. Journal of Hazardous Materials, 2003, volume 99 (3), pp. 243–263. DOI:10.1016/S0304-3894(03)00061-X

[15] Kumari R., Karthaka P.P., Rajan A.P. Electronic waste- a journey from global menace to wealth generation by its effective management strategy (Review). Research Journal of Pharmacy and Technology, 2019, volume 12, issue 2, pp. 848-858. DOI:10.5958/0974-360X.2019.00146.X

[16] Fletcher D., Gerber R., Moore T. The electromagnetic separation of metals from insulators. IEEE Transactions on Magnetics, 1994, volume 30 (6), pp. 4659-4661. DOI:10.1109/20.334181

[17] Overview of the recycling technology for copper-containing cables / Liquan Li, Gongqi Liu, Dean Pan, Wei Wang, Yufeng Wu, Tieyong Zuo. Resources, Conservation & Recycling, 2017, issue 126, pp. 132–140. DOI:10.1016/j.resconrec.2017.07.024

IOP Conf. Series: Materials Science and Engineering **950** (2020) 012018 doi:10.1088/1757-899X/950/1/012018

Liquid metal free surface dynamics in rotating permanent magnet stirrer

V Dzelme, A Jegorovs and A Jakovics

Institute of Numerical Modelling, University of Latvia, Jelgavas street 3, Riga, Latvia

E-mail: valters.dzelme@lu.lv

Abstract. We study liquid metal stirring by rotating permanent magnets in a laboratory-scale rectangular glass container. The main goal is numerical model validation using experimental free surface shape data. We find reasonable agreement between the experiments and coupled liquid metal magnetohydrodynamics simulations. Since the surface tension forces are not dominant here, free surface is deformed mainly by the dynamic pressure of the bulk flow. Therefore, we can conclude that not only the free surface profile is similar to experiments, but the bulk flow must be also very similar.

1. Introduction

Liquid metal free surface dynamics is an important aspect in many applications, such as metal casting, stirring etc. Numerical modelling can provide very useful information regarding surface stability under different operating conditions of the system. However, numerical results usually require experimental validation.

Rotating permanent magnet-induced liquid metal flows have been studied in the past, however, acceptable agreement between experiments and simulations regarding flow velocity has not been obtained.

A popular method of liquid metal flow velocity measurements is the ultrasound Doppler velocimetry (UDV). Rotating permanent magnet-induced flow has been studied in [1], where reasonable quantitative agreement between UDV and simulations has been achieved, citing vortical instabilities as a possible reason for some qualitative differences. UDV measurements are known to be sensitive to presence of oxides on the container walls and probe [2], as well as temperature variations [3], so the results may not be consistent.

Another arguably very powerful tool of investigating liquid metal flows is the neutron radiography reported in [4], where the authors have used small tracer particles for neutron beam contrast. Assuming that the particles do not affect the flow directly, flow velocity distribution can be obtained from image sequences using particle image velocimetry, for example. There have been efforts of processing the obtained images and comparing the velocities to simulations, such as in [5], but the results, while qualitatively good, are quantitatively inconclusive, with the high concentration of particles in some areas as the main problem.

The goal of this study is to validate numerical results of rotating permanent magnet-induced liquid metal flow. To achieve this, we take a step back from multi-magnet systems with complicated flow patterns and create a very simple configuration - a single rotating magnet located next to a rectangular liquid metal container. In such a case, the flow pattern is basically a single vortex (apart from possible small secondary vortices which are usually unavoidable in turbulent flows). Rotation direction of the magnet is such that an upward liquid metal jet is induced at the wall closest to it which consequently deforms the free surface.

2. Experiments

Experimental setup is shown in Fig.1. All experiments are conducted at room temperature of approximately 20°C. The magnet is a nickel-plated NdFeB cylinder magnetized perpendicular to its axis (N52, remanence $B_r = 1.42 - 1.47T$) enclosed in a non-magnetic shell fixed to a shaft connected to a motor.

Magnet rotation frequency is varied using a frequency converter of the motor up to 50Hz in the present configuration. The frequency is set via a computer program.

Liquid metal is Galinstan with density 6440 kg/m^3, viscosity 2.4 mPa·s, electrical conductivity 3.46 MS/m, surface tension coefficient 0.535 N/m and melting point 11°C. Initial experiments were performed with pure gallium (melting point 29°C), but it was difficult to keep it from solidifying during experiments. To prevent free surface oxidation and avoid complicated wetting phenomena, we pour HCl solution on the surface.

The main experimental measurement tool is video recording. Free surface shape is extracted from the images using a self-written *python* script. To achieve better contrast, we use a bright background behind the container, which then makes the liquid metal dark. Examples of raw as well as postprocessed images will be shown in the results section.

Figure 1. Experimental setup

3. Numerical model

The present problem involves coupled electromagnetism and two-phase flow, in other words, we are dealing with liquid metal magnetohydrodynamics (MHD). Numerical solution of such problems is quite time-consuming task. Recently developed open-source solution EOF-Library [6][7] efficiently couples *Elmer* (finite element method) [8] and *OpenFOAM* (finite volume method) [9]. In this work, we use these programs and solve electromagnetics in *Elmer* and two-phase flow in *OpenFOAM*. Liquid metal shape data is transferred from *OpenFOAM* to *Elmer* and electromagnetic forces back from *Elmer* to *OpenFOAM*. All simulations are 3D.

Scheme of the system (magnet and liquid metal volume) is shown in Fig.2. Distance from the magnet edge to liquid metal is fixed at 7 mm and the bottom of the magnet coincides with the bottom of liquid metal.

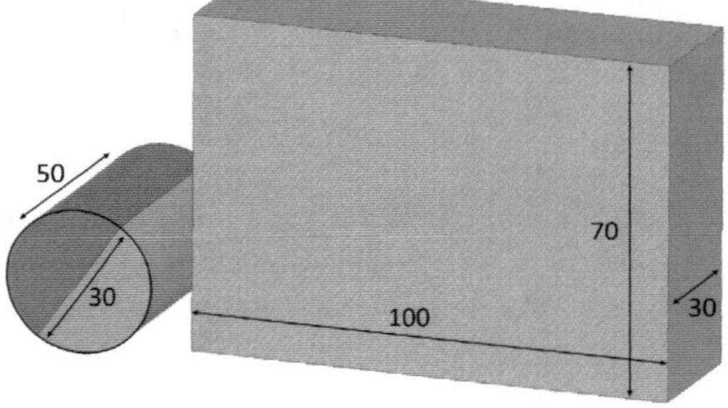

Figure 2. Numerical model scheme with dimensions in millimeters

3.1. Dimensionless parameters

Relevant dimensionless parameters in this system are Reynolds number $Re = \frac{\rho U L}{\mu}$, magnetic Reynolds number $Re_m = \mu_0 \sigma U L$ and Weber number $We = \frac{\rho U^2}{\gamma}$, where ρ - fluid density, U - characteristic flow velocity, L - characteristic length, μ - viscosity, μ_0 - vaccuum permeability, σ - electrical conductivity, γ - surface tension coefficient. Typical values of these in this case are $Re \approx 10^5$, $Re_m \approx 10^{-2} - 10^{-1}$ and $We \approx 1 - 100$.

Large Re means that the flow is fully turbulent and turbulence models are needed.

Small Re_m means that the fluid flow velocity $\vec{v}$ doesn't affect the external magnetic field much, i.e. the magnetic field of the $\sigma\vec{v} \times \vec{B}$ current is negligible. This would mean that fluid flow and magnetic field can be decoupled and $\vec{v}$ can be neglected in the electromagnetics simulation. This does not mean that the $\vec{v} \times \vec{B}$ term can be dropped from the Lorentz force, though.

Large We means that surface tension forces are much weaker than fluid inertia. This is very important here. It allows us to propose that the free surface deformation is caused and affected almost exclusively by the fluid flow or, in other words, the dynamic pressure $p_v = 0.5\rho v^2$. This is the main point of simulation validation in this paper - if the simulated free surface profile is the same as in experiments, we can conclude that the flow velocity must also be the same.

3.2. Electromagnetics

Using potential formulation, i.e. writing magnetic flux density as $\vec{B} = \nabla \times \vec{A}$, where $\vec{A}$ - magnetic vector potential with $\nabla \cdot \vec{A} = 0$, relevant electromagnetic field equations in this problem are the following:

$$\nabla \times \left(\nabla \times \vec{A}\right) = \mu_0 \vec{j} + \mu_0 \nabla \times \vec{M} \tag{1}$$

$$\vec{j} = \sigma \left(-\frac{\partial \vec{A}}{\partial t} + \vec{v} \times \left(\nabla \times \vec{A}\right)\right) \tag{2}$$

$$\nabla \cdot \vec{j} = 0 \tag{3}$$

where μ_0 - vacuum permeability, $\vec{j}$ - current density, σ - electrical conductivity, $\vec{v}$ - fluid velocity, $\vec{M}$ - magnetization. Equation (1) is Ampere's law and is the main equation solved to obtain magnetic field distribution. Equation (2) is the Ohm's law and Eq. (3) is the current continuity.

Electromagnetism is solved in frequency domain by expressing all oscillating fields using complex exponents, e.g. $\vec{A} = \vec{A}_0 cos(\omega t) = Re\{\vec{A}_{0C} e^{i\omega t}\}$, where subscripts 0 and C denote amplitude and a complex field, respectively, ω is angular frequency, t is time. In the complex form, the time-dependent exponents drop out of equations which are then solved for amplitudes.

The magnet rotation is modelled as a rotating magnetization vector and in complex representation that means we must set amplitudes of real and imaginary parts of magnetization. For counter-clockwise rotation time-dependent form is $M_x = \frac{B_r}{\mu_0} cos\,(\omega t)\,, M_y = \frac{B_r}{\mu_0} sin\,(\omega t)$ and in complex form this is $M_{x,re} = \frac{B_r}{\mu_0}$, $M_{y,im} = \frac{B_r}{\mu_0}$.

Result of the electromagnetic simulation in *Elmer* is the complex amplitude of $\vec{A}$ from which current density $\vec{j}$ and magnetic field $\vec{B}$ can be calculated, which are then sent to *OpenFOAM*. The fluid dynamics simulation is transient and we convert the complex amplitudes back to time-dependent sinusoidal form.

Two important simplifications in electromagnetics model must be mentioned here. Firstly, the fluid velocity is neglected in the electromagnetic simulation, but the $\vec{v} \times \vec{B}$ term is considered in fluid flow simulation. This approximation is justified by the small magnetic Reynolds number. Secondly, electromagnetic simulation in the whole coupled simulation process is done only once at the initial stage with undeformed free surface. This is justified by noticing that the rotating magnet is relatively far from the free surface and so the induced forces for the deformed free surface shape would be almost the same. This reduces simulation time considerably.

Electromagnetics model consists of the whole system - liquid metal, magnet and air around them. Container is non-conductive and non-magnetic, so it is modelled as air. There is only one boundary condition - $\vec{A} = 0$ on the external boundary of the domain, which means that the magnetic field is parallel to the boundary.

3.3. Fluid flow

The incompressible two-phase flow is governed by the Navier-Stokes equations with body forces:

$$\frac{\partial \rho \vec{v}}{\partial t} + (\vec{v} \cdot \nabla)\,\rho \vec{v} = -\nabla p + (\mu + \mu_t)\Delta \vec{v} + \vec{f} \tag{4}$$

$$\nabla \cdot \vec{v} = 0 \tag{5}$$

where ρ - fluid density, $\vec{v}$ - velocity, μ - viscosity, μ_t - turbulent viscosity, p - pressure, $\vec{f}$ - body forces. To account for turbulence effects and calculate μ_t, the $k - \omega\ SST$ [10] model is used.

To model the two immiscible phases (liquid metal and HCl solution (assumed to be pure water)), we use the Volume of Fluid method [11], implemented in the *interFoam* solver, which allows solving only one set of Navier-Stokes equations for both phases. This method introduces additional equation:

$$\frac{\partial \alpha}{\partial t} + (\vec{v} \cdot \nabla)\,\alpha = 0, 0 \le \alpha \le 1 \tag{6}$$

with $\alpha = 0$ for one fluid and $\alpha = 1$ for the other. The interface between the fluids is usually assumed to be in numerical cells where $\alpha = 0.5$. Density and viscosity are then

$$\rho = \alpha \rho_1 + (1 - \alpha)\rho_2 \tag{7}$$

$$\mu = \alpha \mu_1 + (1 - \alpha)\mu_2 \tag{8}$$

where indices 1 and 2 designate fluid 1 and 2, respectively. Using VOF model, surface tension forces are implemented as a continuum surface force

$$\vec{f_s} = \gamma \kappa \nabla \alpha \tag{9}$$

where $\kappa = -\nabla \vec{n}$ - surface curvature, $\vec{n}$ - surface normal, γ - surface tension coefficient. Note that even though surface tension forces are not dominant in this particular case, we cannot neglect surface tension completely because we need to set contact angle different from 90 degrees and the surface must be rounded near the container walls as in experiments.

As mentioned in the previous subsection, the $\sigma\vec{v} \times \vec{B}$ term is included in the fluid flow simulation. Since this is a part of electric current density, it must obey the continuity equation $\nabla \cdot \vec{j} = \nabla \cdot (\sigma\vec{v} \times \vec{B}) = 0$. Since such an equation may not hold as written, additional correction ϕ is introduced and equation

$$\nabla \cdot (\sigma\vec{v} \times \vec{B} - \nabla\phi) = 0 \tag{10}$$

is implemented in *OpenFOAM* to find ϕ. In the end, the total current density used for Lorentz force calculation is $\vec{j} = \vec{j}_{elmer} + \sigma\vec{v} \times \vec{B} - \nabla\phi$. Body force in equation (4) is then

$$\vec{f} = \gamma\kappa\nabla\alpha + \rho\vec{g} + (\vec{j}_{elmer} + \sigma\vec{v} \times \vec{B} - \nabla\phi) \times \vec{B} \tag{11}$$

where $\vec{j}_{elmer}$ represents the current density received from *Elmer* that does not include the $\vec{v} \times \vec{B}$ or $\nabla\phi$ terms.

Fluid flow model consists of only the liquid metal and some water above it, enough for free surface deformation. Boundary conditions are $\vec{v} = 0$, $\frac{\partial p}{\partial n} = 0$ on container walls and $\frac{\partial \vec{v}}{\partial n} = 0$ (for outflow) or $\vec{v} = 0$ (for inflow) and static pressure $p_s = 0$ on the top boundary. Boundary condition for α is a fixed gradient corresponding to 160 degree contact angle (the value is estimated experimentally) on walls and $\alpha = 0$ (for inflow) or $\frac{\partial \alpha}{\partial n}$ (for outflow) on the top boundary.

4. Results

Let us first look at typical Lorentz force and velocity distributions in this system, shown in Fig.3. As expected from the magnet rotation, the force is directed mainly upwards and so is the flow in that zone, deforming the free surface.

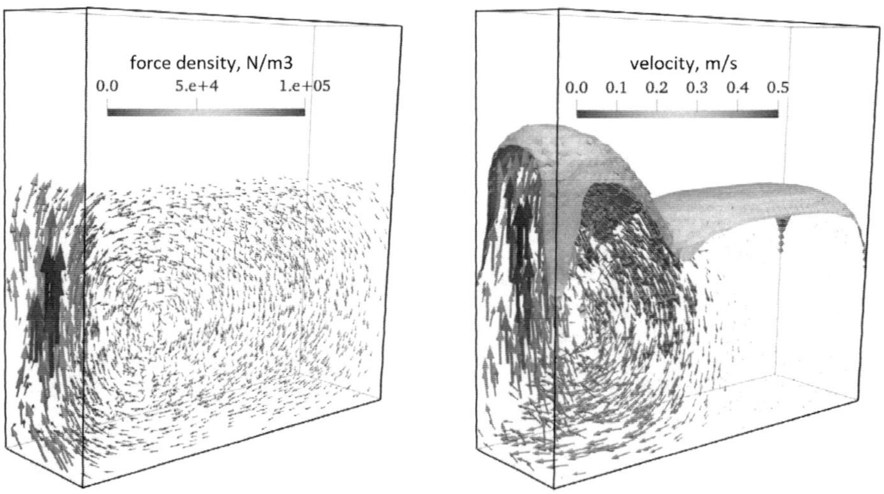

Figure 3. Characteristic Lorentz force density (left) and velocity (right) vectors.

Fig.4 shows raw images from experiments without magnet rotation. The right image is with a bright background to make postprocessing easier.

The main result of interest here is the free surface shape and deformation height. Since the free surface oscillates quite considerably, especially in experiments, it is reasonable to look at the time-averaged results. Time-averaged free surface profiles are shown in Fig.5 for different magnet rotation frequencies. Standard deviation of experimental data is 0.3 mm at 5 Hz and

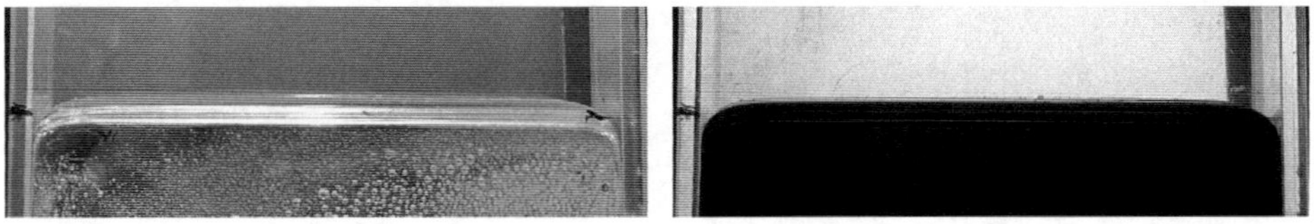

Figure 4. Comparison of different lighting conditions: normal light (left), dark room with bright background (right).

1.7 mm at 20 Hz. Note that the vertical axis scales are different for each plot. The rectangular bump on the right side for experimental data is due to a black tape present in the background (see Fig.4).

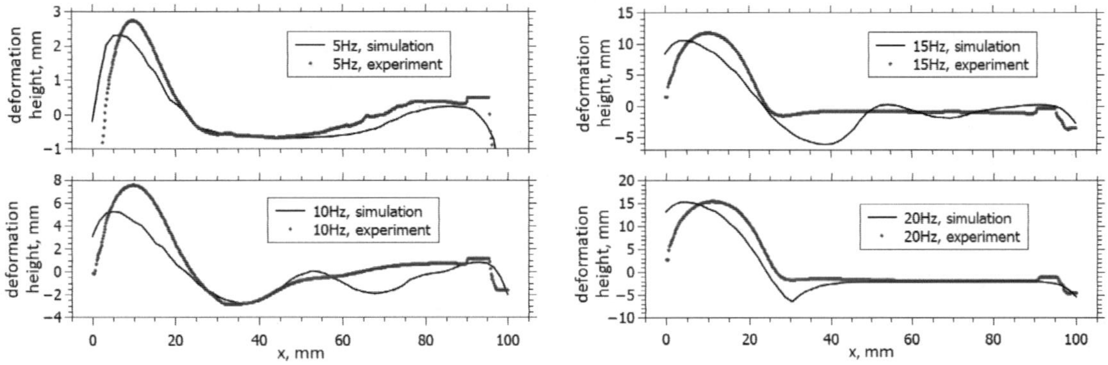

Figure 5. Time-averaged free surface profiles for different magnet rotation frequencies.

Similarities and differences between simulations and experiments are immediately clear. There is one pronounced maximum near the magnet, followed by a minimum and an increased height on the other end. A quantitative difference is that the maximum surface height is smaller in simulations than in experiments for all considered frequencies. This can simply be due to turbulence model - such two-parameter models tend to overestimate turbulent viscosity [12] leading to increased flow damping. Another reason could be that the distance between the magnet and melt is not matched absolutely exactly. Since the magnetic field decreases rapidly by increasing distance to the magnet and the force is proportional to the magnetic field squared, even a small distance error can lead to considerable induced force difference.

The main qualitative difference is the presence of a local maximum in the middle of container in simulations. This is due to unphysical water flow above the free surface. Test simulations with water flow damping (we multiply the velocity $\vec{v}$ with the volume fraction α which is 0 in water and 1 in liquid metal) confirms this - the surface is then flatter, more resembling the experimental one. However, such artificial second phase damping can lead to slight primary phase damping, so another solution to this problem must be sought.

As already mentioned above, the free surface oscillates. To demonstrate that, Fig.6 shows simulated free surface height approximately at the maximum position in time.

Two distinct oscillations can be distinguished here. The smallest one has frequency equal to double the magnet rotation frequency or equal to the Lorentz force frequency. The large oscillations are of hydrodynamic nature - surface waves and oscillations determined by material properties and container geometry.

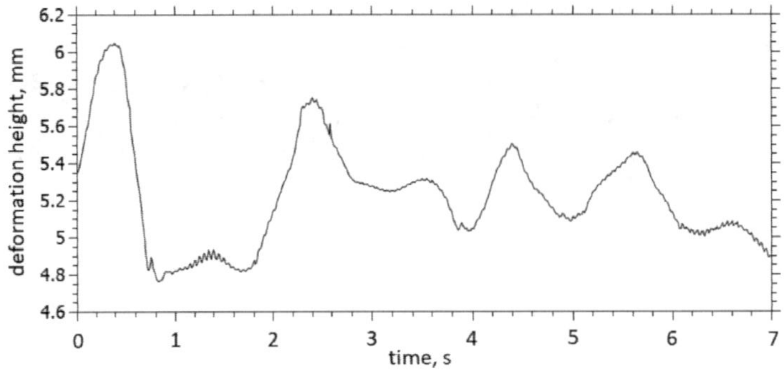

Figure 6. Simulated free surface height oscillations at 10Hz magnet rotation frequency.

5. Conclusions

The simulated free surface shape is similar to the experimental one for different magnet rotation frequencies. As mentioned above, differences can arise because of turbulence model, some system parameter differences (distance from magnet to melt), as well as unphysical flow in the fluid above the liquid metal.

Further numerical work will focus on the solution to the unphysical secondary phase flow above the melt. Improved turbulence models, such as Large Eddy Simulation, which have been shown in literature to describe liquid metal free surface dynamics more realistically, will also be employed. Experimental improvements should include more precise control of crucial system parameters. Video camera with higher frame-rate will be used to capture the faster free surface oscillations.

Acknowledgements

This work was supported by the ERDF project No. 1.1.1.1/18/A/108 "Development of numerical modelling approaches to study complex multiphysical interactions in electromagnetic liquid metal technologies".

References

[1] Dzelme V, Jakovičs A, Venčels J, Köppen D, and Baake E. *IOP Conf. Series: Materials Science and Engineering*, 424:012047, 2018.
[2] Eckert S, Cramer A, and Gerbeth G. *Velocity Measurement Techniques for Liquid Metal Flows*, pages 275–294. Springer Netherlands, Dordrecht, 2007.
[3] Losev G, Khalilov R, and Kolesnichenko I. *IOP Conference Series: Materials Science and Engineering*, 208:012022, 2017.
[4] Ščepanskis M, Sarma M, Vontobel P, Trtik P, Thomsen K, Jakovičs A, and Beinerts T. *Metallurgical and Materials Transactions B*, 48:1045, 2017.
[5] Dzelme V, Ščepanskis M, Geža V, Jakovičs A, and Sarma M. *Magnetohydrodynamics*, 52:461, 2016.
[6] EOF-Library, https://github.com/jvencels/EOF-Library.
[7] Venčels J, Raback P, and Geža V. *SoftwareX*, 9:68, 2019.
[8] Elmer FEM, https://www.csc.fi/web/elmer.
[9] OpenFOAM, https://openfoam.org.
[10] Menter F R, Kuntz M, and Langtry R. *Proceedings of the 4th international symposium on turbulence, heat and mass transfer*, 2003.
[11] Heyns J and Oxtoby O. *6th European Conference on Computational Fluid Dynamics*, 2014.
[12] Spitans S, Baake E, Nacke B, and Jakovičs A. *Metallurgical and Materials Transactions B*, 47:522–536, 2016.

IOP Conf. Series: Materials Science and Engineering **950** (2020) 012019 doi:10.1088/1757-899X/950/1/012019

Machine learning application for magnetohydrodynamic pump research

N V Tarchutkin, I A Smolyanov

Department of Electrical Engineering and Electrotechnology Systems, Ural Federal University, Yekaterinburg, Russia

Nik_tar@mail.ru

Abstract. The article is devoted to finding out best machine learning model used for the analysis of magnetohydrodynamic pump. The machine learning model was created on the basis of data obtained as a result of numerical simulation of the unit using COMSOL Multiphysics. The paper compares errors of output data collected by using various machine learning methods for out-of-sample data.

1. Introduction

Machine Learning (ML) is a form of predictive analytics that uses a set of training data to develop a mathematical model to extrapolate conclusions from input data. At the moment, machine learning is used as a research tool in thermoelectricity [1], construction [2], chemistry [3], engineering [4] and other scientific fields. This popularity is due to the ability to analyze large, complex and streaming data and find in them valuable information that allows to make predictions based on input data. However, they also have a disadvantage associated with a lack of understanding of the physical basis of the solution. In addition, for the machine learning approach to be successful in presenting the physics of the problem, a large and statistically reliable dataset must be used.

2. Formulation of the problem

The paper compares various ML algorithms used to analyze the data obtained during the simulation of the induction pump in the COMSOL Multiphysics software package (figure 1). The model is based on the geometric parameters of a real installation (table 1).

The data used for ML are obtained at various current amplitudes and current frequencies of the pump windings. Used libraries: scikit-learn (ML algorithms), pandas (data extraction), plotly (data visualization).

The objective of this work is to compare the error of the ML algorithms using out-of-sample data.

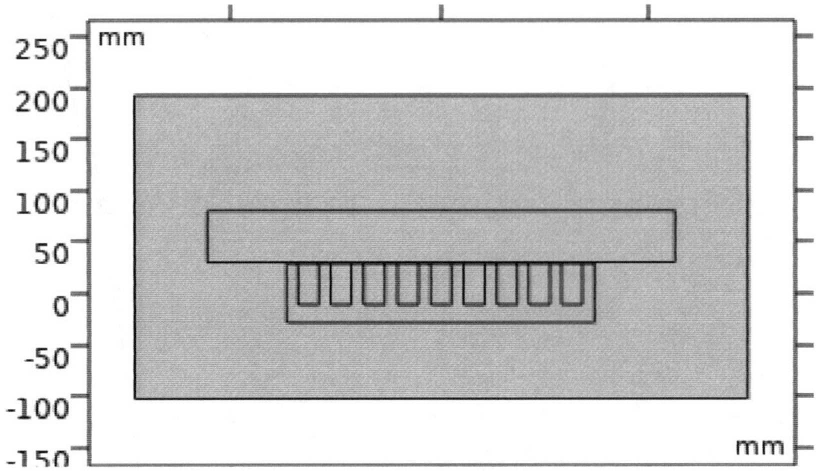

Figure 1. Scheme of the simulated induction pump.

TABLE 1. Initial data.

Description	Unit	Value
Inductor length	mm	295
SE length	mm	450
The number of turns in the slot	-	135
Number of slots	-	9
Inductor thickness	mm	58
Inductor width	mm	77
SE width	mm	78
Rectangular slot height	mm	40
Rectangular slot width	mm	20
Tooth width	mm	11.5
SE height	mm	50
Gap between the inductor and the SE	mm	1

3. Overview of algorithms

At the moment, the scikit-learn library contains several algorithms. Each of them receives a set of input and output data, based on which the dependencies between the variables are detected. The following is a description of the functions used in this paper.

4. Linear Regression

The model is based on the ordinary least squares method. The dependence of the output variable on the input is given by the equation:

$$y = \omega_0 + \omega_1 x_1 + \omega_2 x_2 + \ldots + \omega_n x_n, \tag{1}$$

where *n* is a number of input variables, $\omega_1, \omega_2, \ldots, \omega_n$ are the slope parameters, $x_1, x_2, \ldots, x_n$ are an input variables, and ω_0 is an intercept.

The model selects the coefficients ω_i in such a way as to minimize the residual sum of squares between the predicted values of y and the actual values at the points from dataset [5]:

$$\min_{\omega} \|X\omega - y\|_2^2, \tag{2}$$

where $\|\cdot\|_2$ is the Euclidean norm.

5. SGD Regressor

Stochastic Gradient Descent (SGD) Regressor builds a plane using the method specified by the loss variable which can take the following values:

- loss="squared_loss": ordinary least squares;
- loss="huber": Huber loss;
- loss="epsilon_insensitive": linear Support Vector Regression;
- loss="squared_epsilon_insensitive": same as previous, but becomes squared loss past a tolerance of epsilon.

In this work model uses a method based on Huber loss function [5]. The same equation as in the case of Linear Regression is used to create the plane, but the choice of coefficients is aimed at minimizing the sum of Huber losses for all points of dataset. Losses for each point are calculated as follows:

$$L_\delta(a) = \begin{cases} \frac{1}{2}a^2 & |a| \le \delta \\ \delta\left(|a| - \frac{1}{2}\delta\right) & otherwise \end{cases}, \tag{3}$$

where a is the difference between actual and predicted values, δ is the slope of the linear part of the function.

6. Ridge

The model uses the least squares method with L2 regularization:

$$\min_{\omega} \|X\omega - y\|_2^2 + \alpha\|\omega\|_2^2, \tag{4}$$

where α is the regularization strength ($\alpha \ge 0$) [5].

7. Lasso

Least Absolute Shrinkage and Selection Operator (Lasso) uses the ordinary least squares method with L1 regularization [5]. Optimization is performed according to the following formula:

$$\min_{\omega} \frac{1}{2n}\|X\omega - y\|_2^2 + \alpha\|\omega\|_1, \tag{5}$$

where $\|\cdot\|_1$ is the taxicab norm (or Manhattan norm).

8. Lasso Lars

The model uses the ordinary least squares method with the least-angle regression (LARS) algorithm. Optimization is performed according to the same formula as for the Lasso method.

9. ElasticNet

The model uses the least squares method with both L1 and L2 regularizations [5]. Optimization is performed according to the following formula:

$$\min_{\omega} \frac{1}{2n}\|X\omega - y\|_2^2 + \alpha\rho\|\omega\|_1 + \frac{\alpha(1-\rho)}{2}\|\omega\|_2^2, \tag{6}$$

where ρ is the component of L1 regularization ($0 \leq \rho \leq 1$).

10. Passive Aggressive Regressor

The model analyzes the data set points in turn and adjusts the values of slope parameters and intercept by using ε–insensitive hinge loss function:

$$l_\varepsilon(\boldsymbol{\omega};(\boldsymbol{x},y)) = \begin{cases} 0 & |\boldsymbol{\omega}\cdot\boldsymbol{x} - y| \leq \varepsilon \\ \left||\boldsymbol{\omega}\cdot\boldsymbol{x} - y| - \varepsilon\right| & otherwise \end{cases}, \tag{7}$$

where ε is the permissible error value.

The initial values of the coefficients are taken equal to 0. Then, the loss function is calculated alternately for each data point. If the value of the loss function is 0, the coefficients remain unchanged, otherwise the coefficients are adjusted according to the formula [6]:

$$\boldsymbol{\omega}_{i+1} = \boldsymbol{\omega}_i + sign(y_i - \hat{y}_i)\frac{l_i}{\|\boldsymbol{x}_i\|_2^2}\boldsymbol{x}_i. \tag{8}$$

11. Theil-Sen Regressor

The model calculates the slopes of the planes, each of which passes through three points that are not on the same line and do not have the same X coordinates. The result is a plane with a slope determined by the spatial median of the calculated slopes.

The intercept is determined from the condition of minimizing the sum of the distances from dataset points to the plane [7].

12. Support Vector Machines

With linear approximation, the model solves the problem of minimizing the objective function [8]:

$$\min_{\omega,\omega_0,\xi,\xi^*} \frac{1}{2}\|\boldsymbol{\omega}\|_2^2 + C\sum_{i=1}^{n}(\xi_i + \xi_i^*). \tag{9}$$

The following conditions are met:

$$y_i - \boldsymbol{X}_i\omega - \omega_0 \leq \varepsilon + \xi_i, \tag{10}$$

$$\boldsymbol{X}_i\omega + \omega_0 - y_i \leq \varepsilon + \xi_i^*, \tag{11}$$

$$\xi_i, \xi_i^* \geq 0, i = 1, \ldots, n. \tag{12}$$

13. K-Nearest Neighbors

To determine the value of a variable at a specific point, the model calculates the average value of the variable for the nearest points from the training data. In this work k=4. An example of predicted point with its nearest neighbors is shown in figure 2.

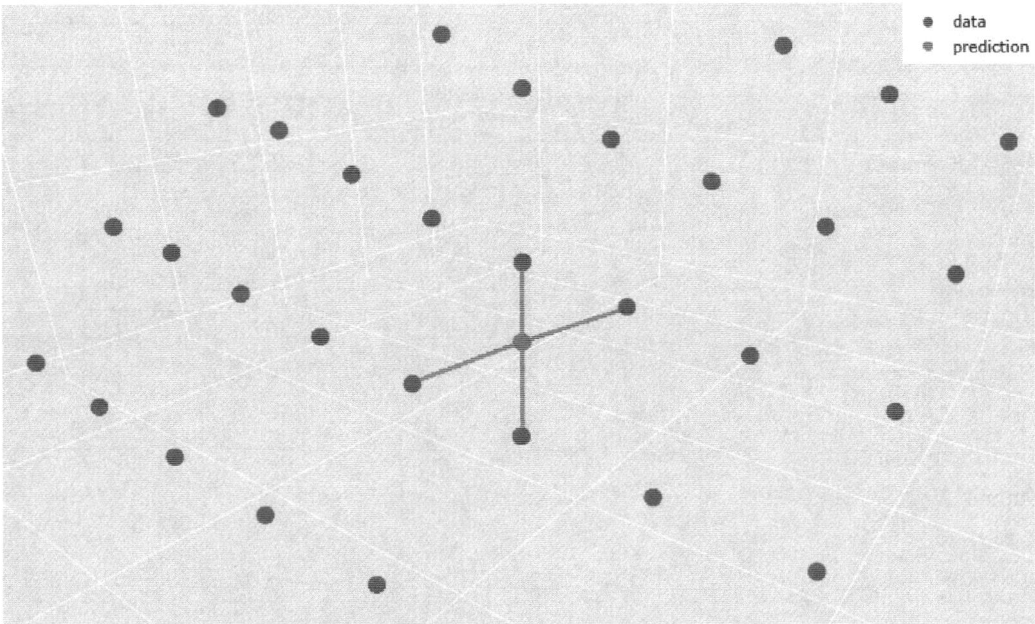

Figure 2. Prediction of output value based on KNN.

14. Decision Trees

The model creates an "if" condition tree, depending on the fulfillment of which the function will be assigned one of the values of the output variable obtained from the training data (figure 3).

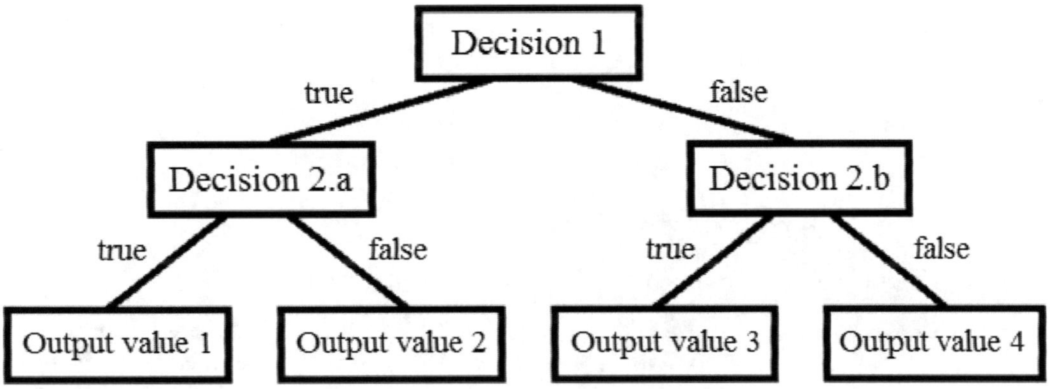

Figure 3. Example of Decision Tree.

15. Analysis of results

Comparison of algorithms is carried out according to the relative error of the results. To train the ML models, 80 results of solving the problem using COMSOL Multiphysics were used, while the verification was carried out using the other 63 solutions. Root-mean-square error (RMSE) and relative error were calculated. For some models dataset was scaled in order to increase accuracy. The calculation results are presented in table 2.

Table 2. Errors of ML algorithms.

Model	Variable			
	B, mT		v, mm/s	
	RMSE	Relative error	RMSE	Relative error
Linear Regression	5.23	4.81	2156.38	241.7
SGD Regressor	9.36	12.65	3677.42	96.0
Ridge	5.24	4.81	2156.17	241.4
Lasso	5.24	4.81	2156.38	241.7
Lasso Lars	5.51	4.4	2156.26	241.5
ElasticNet	5.77	5.13	2150.19	233.5
Passive Aggressive Regressor	13.33	15.72	2156.89	273.3
Theil-Sen Regressor	7.25	6.41	2635.02	138.9
Support Vector Machines	4.30	4.96	2121.62	194.8
K-Nearest Neighbors	3.75	2.76	141.39	5.89
Decision Trees	16.77	19.75	1065.26	28.3

The noticeable difference in the errors of the same methods for different results is explained by the fact that the dependence of the magnetic flux density B on the amplitude and frequency of the current is close to linear, while for velocity v the dependence is nonlinear. Figures 4-6 show the set of input data used to create the model (green) and the results of machine learning (blue). It can be seen that the result of approximating strongly nonlinear dependences by most models cannot be called reliable.

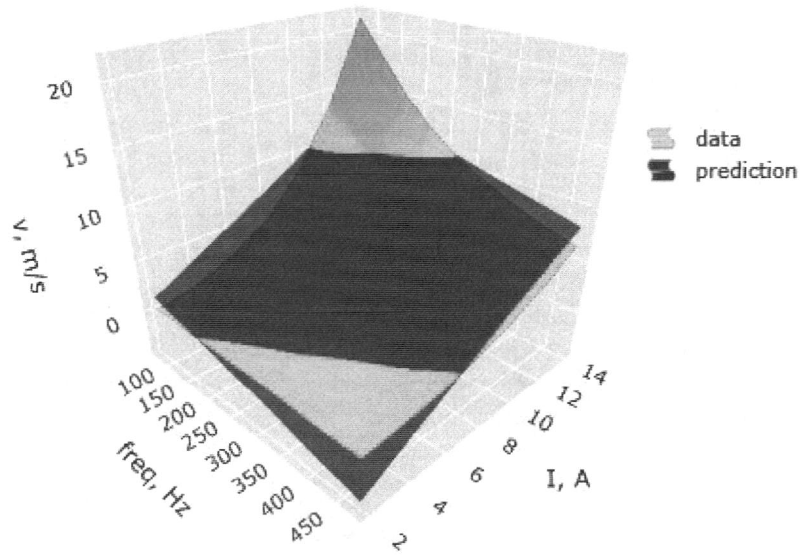

Figure 4. Dependence of the average velocity of the SE metal (m/s) on the frequency and amplitude of the inductor windings current (Linear Regression).

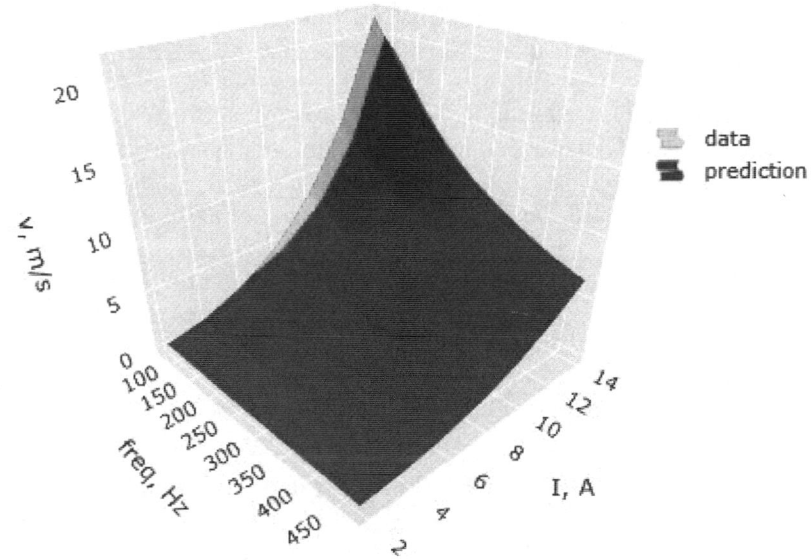

Figure 5. Dependence of the average velocity of the SE metal (m/s) on the frequency and amplitude of the inductor windings current (K-Nearest Neighbors).

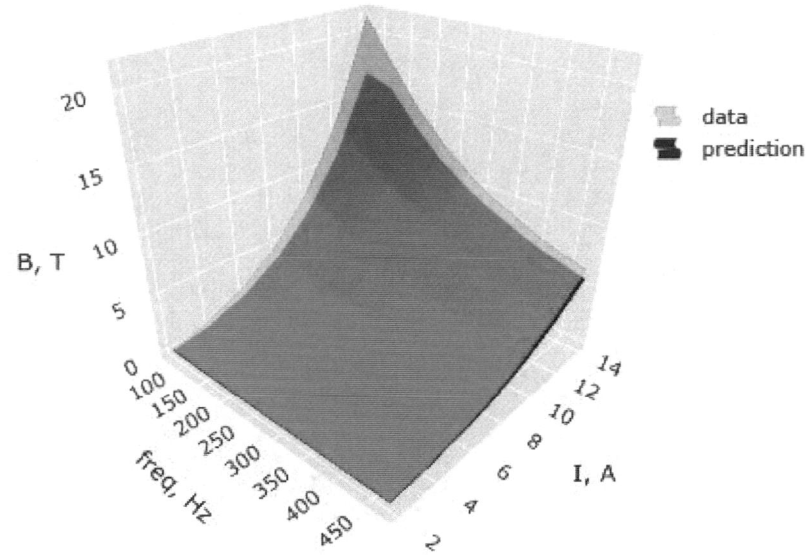

Figure 6. Dependence of the maximum magnetic flux density of the SE metal (T) on the frequency and amplitude of the inductor windings current (Decision Trees).

It should be noted that in the training data set, neighboring points are located in the nodes of the rectangular grid, while all test data points are located at the same distance from nearest training data, which significantly improves the quality of the KNN model. When examining data with an uneven distribution, it is necessary to use a KNN models with different values of the nearest neighbors. The quality of models is also affected by the number of training data points.

16. Conclusion

Based on the data presented, most of the methods considered in the work have high accuracy in predicting linear dependencies, while the K-Nearest Neighbors and Decision Trees methods show the best results with strongly non-linear dependencies.

References

[1] Chen L, Tran H, Batra R, Kim C and Ramprasad R 2019 Machine learning models for the lattice thermal conductivity prediction of inorganic materials. *Computational Materials Science* **170** 109155

[2] DeRousseau M A, Laftchiev E, Kasprzyk J R, Rajagopalan B and Srubar III W V 2019 A comparison of machine learning methods for predicting the compressive strength of field-placed concrete. *Construction and Building Materials* **228** 116661

[3] Rizkin B A and Hartman R 2019 Supervised machine learning for prediction of zirconocene-catalyzed a-olefin polymerization. *Chemical Engineering Science* **210** 115224

[4] Swischuk R and Allaire D 2019 A Machine Learning Approach to Aircraft Sensor Error Detection and Correction. *Journal of Computing and Information Science in Engineering* **4** no. 1

[5] 1. Supervised learning – scikit-learn 0.21.3 documentation (2019). Available at: https://scikit-learn.org/stable/supervised_learning.html (accessed 25 November 2019).

[6] Crammer K., Dekel O., Keshet J., Shalev-Shwartz S. and Singer Y. 2006 Online Passive-Aggressive Algorithms. *Journal of Machine Learning Research* **7** pp. 551–585

[7] X. Dang, H. Peng, X. Wang and H. Zhang 2008 Theil-Sen estimators in a multiple linear regression model. *Tech. Rep.*

[8] Smola, A. J. and Schölkopf, B. 2004 A tutorial on support vector regression. *Statistics and Computing* **14**(3) pp 199–222.

IOP Conf. Series: Materials Science and Engineering **950** (2020) 012020 doi:10.1088/1757-899X/950/1/012020

Mathematical modeling of electromechanical characteristics of linear electromagnetic and induction-dynamic motors

Sidorov O. Yu. [1], Sarapulov F.N. [2], Tomashevsky D.N. [3]

[1] Ural Federal University, Nizhny Tagil, Russia, sidorov-ou-62@yandex.ru
[2] Ural Federal University, Russia, sarapulovfn@yandex.ru
[3] Ural Federal University, Russia, 2036969@mail.ru

Abstract. The calculations of the electromechanical characteristics of the pulse action linear motor (PALM) with cylindrical symmetry are performed. The case of the motor power from the capacitor bank of a given capacitance is considered. The model of "field-circuit" was used for electromechanical characteristics modeling. The electromagnetic field was calculated by finite difference method taking into account the displacement of the secondary element (SE). The problem of circuit theory was solved together with the field problem (via flow coupling) in order to determine temporal dynamics of the current and voltage. The current and voltage in the primary circuit of the PALM, the force in the SE of the PALM, the value of the SE displacement was calculated for each time point. It is shown that the electromechanical characteristics depend on the design and material of the SE.

Keywords: electromagnetic field, vector potential, induction-dynamic motor, the electromechanical characteristics, pulse action linear motor (PALM), cylindrical symmetry model.

1. Introduction

Recently electric devices of pulse action with forward movement of the working element are intensively developed. Such devices must be direct drive, since they are used to create a large forces.

Most suitable in this case are two types of PALM – electromagnetic motor and induction-dynamic motor (IDM). Depending on the SE stroke length, the SE end speed, the number of cycles per unit of time the effectiveness of aforementioned types of engines is different. To select the type and design of PALM it is necessary to have the methods for calculating their characteristics, as well as quantitative estimates of their parameters. The aforementioned quantitative estimates are impossible to obtain without learning the peculiarities of distribution of magnetic flux in the PALM. The electromagnetic and induction-dynamic linear motors impulse actions with cylindrical symmetry are considered.

2. Relevance

Issues of experimental and theoretical studies on the application of electromagnetic PALM of are the subject of numerous works [1-5, 10-14]. Technological processes, e.g. in the construction, mining and metallurgical industries require the implementation of short-term shock effects, which can be implemented by using of the electromagnetic PALMs.

3. Statement of the problem

In this study, the task of mathematical modeling of electromechanical processes occurring during operation of the electromagnetic PALMs with cylindrical symmetry when they are powered by a capacitor bank.

4. A mathematical model

For the solution of the field task cylindrical coordinate system was used, and the cylindrical symmetry of considered model exists.

In this case, the system of equations of the electromagnetic field can be reduced to one equation [6, 7]

$$-\frac{\partial}{\partial r}\left(\frac{A}{\mu r}+\frac{1}{\mu}\frac{\partial A}{\partial r}\right)-\frac{\partial}{\partial z}\left(\frac{1}{\mu}\frac{\partial A}{\partial z}\right)+\beta\frac{\partial A}{\partial z}+\alpha\cdot\left(\frac{A}{r}+\frac{\partial A}{\partial r}\right)+j\cdot\omega\cdot\gamma\cdot A=J. \quad (1)$$

Here $\beta=-\gamma\cdot w$ and $\alpha=-\gamma\cdot v$; γ - is the specific conductivity of the material; v, w - are components of the velocity of the SE; A, J - are φ–components of the vector potential of the electromagnetic field and density of the external current, respectively.

Components of magnetic induction are described by equations

$$B_r=-\frac{\partial A}{\partial z};\quad B_y=\frac{A}{r}+\frac{\partial A}{\partial r}.$$

The value of the induced current is described by equation

$$J_{in}=-\gamma\cdot\frac{\partial A}{\partial t}.$$

Ponderomotive force, acting on the SE is described by equation

$$F=F_y=B_r\cdot J_{in}.$$

To solve the equations of the electromagnetic field (1) the finite difference method, which took into account displacement of the SE and the induced currents in the SE of is applied.

Further, the PALM power from the capacitor bank is considered. His capacity is 1600 mkF and the initial voltage is 1000 V. The resistance of the circuit was equal to 0.125 Ohms and inductance 0,021 H. The equation of an electric condition of the primary circuit would be

$$\begin{cases} U_c=R\cdot i+\dfrac{d\Psi}{dt} \\ i=C\cdot\dfrac{dU_c}{dt} \end{cases}. \quad (2)$$

Here *R, C, i,* U_c, Ψ – the resistance, the capacitance, the current in the circuit, the voltage on the capacitor and flux linkage, respectively.

The integration of the system of differential equations (2) was performed by the Runge-Kutt method (see e.g. [8]).

The joint solution of the field problem (1) to determine the vector potential and then Ψ, as well as the circuit problem (2) to find *i* and U_c allows to trace the dynamics of changes of the electromechanical characteristics of the motor and the SE displacement.

According to Newton's second law the displacement *y* of a secondary element (armature, his weight is *m*) under the force of *F* over time *t* was determined by integrating the equations corresponding to

$$m\frac{d^2y}{dt^2}=F.$$

IOP Conf. Series: Materials Science and Engineering **950** (2020) 012020 doi:10.1088/1757-899X/950/1/012020

5. The simulation results

Further, the model of armor electromagnetic PALM [9] is considered. Winding is located in the cavity between the core and the armature, it has 780 turns. The PALM has two working gap value δ.

The experimental dependences of the tractive force $F(\delta)$ at a constant value current of 5 A is shown in Fig.1. It also shows the calculated curves obtained with and without the saturation of the stator and armature. You can see that taking into account the saturation of the ferromagnetic parts leads to a significant decrease of tractive force in the region of small gaps (about 2 times) and improved convergence of experimental results and the results of the calculations. It should be noted that at small gaps the stator, upper and lower parts of the armature are the most saturated as a consequence of their smaller thickness compared to the central part of the armature.

Further, the calculation of electromechanical characteristics of an IDM is considered. It is assumed that the IDM power from the capacitor bank. The capacitor bank initial voltage and the active electric resistance of the primary circuit are known. The design scheme of the IDM is shown in Fig.2. The mass of the armature is 50 kg.

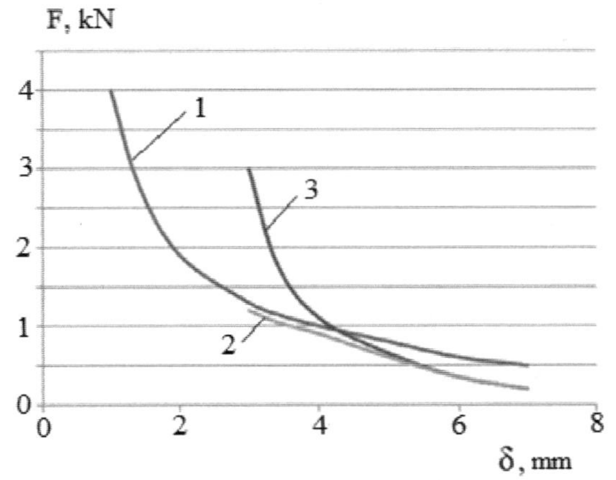

Figure 1. The dependence of the tractive force from the value of the air gap δ. 1 - experimental data [9]; 2, 3 – the results of the calculations with and without the saturation respectively

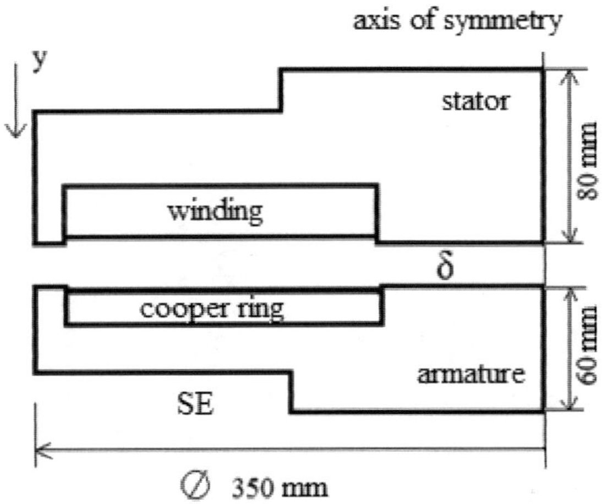

Figure 2. The design scheme of the IDM

A study of the influence of the copper ring thickness in the secondary element on the electromechanical characteristics i, U_c, y, F IDM was performed. The calculation results are shown in Fig.3-6. The basic version in these figures is identified by the number «0». It corresponds to the thickness of the copper rings of 6 mm. The rest material of the secondary element is a steel with a relative magnetic permeability of 10 and a specific conductivity of about 5 times lower than that of copper. The number «1» corresponds to a secondary element of non-magnetic steel with a conductivity of approximately 5 times lower than that of copper. The number «2» corresponds to the basic version, but with increased to 8 mm thickness of the copper ring. The number «3» corresponds to the basic version, but with reduced up to 4 mm thickness of the copper ring.

According to the results of the study using the SE without cooper ring leads to a slower decrease in voltage (Fig.3), to a smaller the displacement of the SE (Fig.4), to a less force on the SE (Fig.5), to a smaller current in the primary winding (Fig.6). Increasing the thickness of the copper ring results in larger force on the SE (Fig.5) and correspondingly to larger displacement of the SE (Fig.4). Reducing the thickness of the copper ring leads to the opposite effect.

IOP Conf. Series: Materials Science and Engineering **950** (2020) 012020 doi:10.1088/1757-899X/950/1/012020

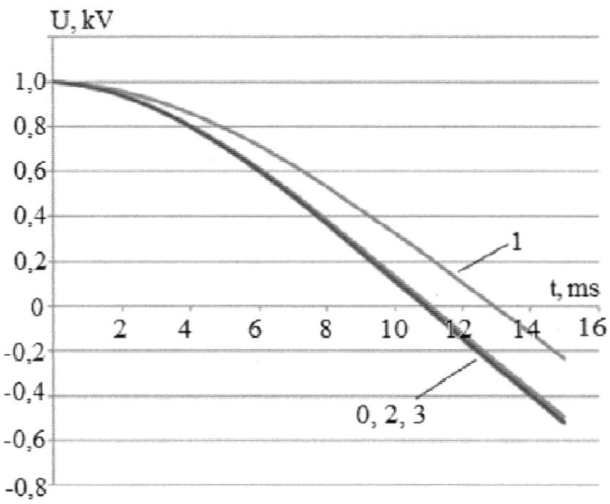

Figure 3. The time dependence of the voltage in the primary winding

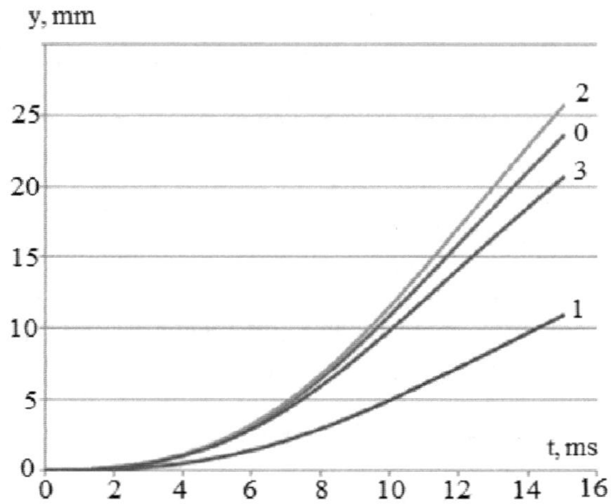

Figure 4. The time dependence of the SE displacement for different designs

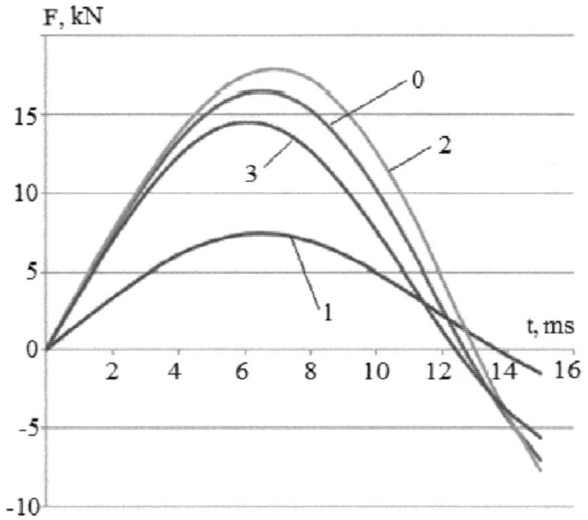

Figure 5. The time dependence of forces acting on the SE

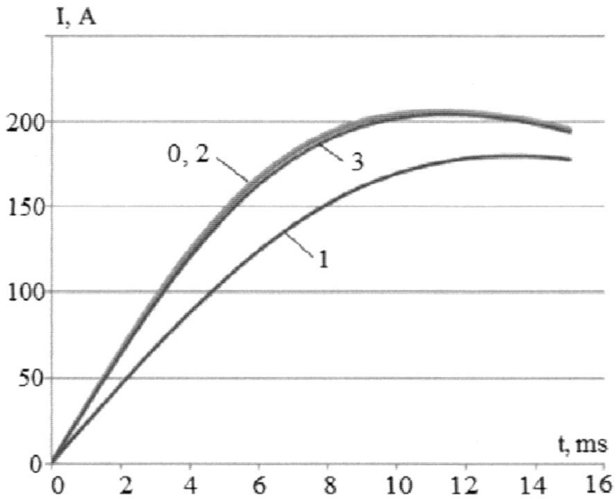

Figure 6. Time dependence of current in the primary winding

6. Conclusions

1. The mathematical model "field-circuit" to describe the electromechanical characteristics of the PALM is constructed.
2. It is shown that the increasing of the SE copper ring thickness leads to increase of force acting on the SE from the magnetic field created by the primary winding.
3. The design of the SE has a significant influence on the electromechanical characteristics of the PALM.

Reference

[1] Tomashevsky D.N. Modeling of the linear motor pulse action / D.N. Tomashevsky, A.N. Coshkin // Electrical engineering. – 2006. – №1. – pp. 24-27.

[2] Ugarov G.G., Neiman V.Yu. Trends in the development and use of handheld percussion machines with electromechanical energy conversion / G.G.Ugarov, V.Y. Neiman // Proceedings of the universities. Electromechanics. – 2002. – №2. – pp. 37-43.

[3] Moshkin V. I. Comparison of magnetic cycles of the pulsed linear electromagnetic motor with the power losses accounting in the coil // Bulletin of the Tomsk Polytechnic University. – 2012. – Vol. 321. – №4. – pp. 93-96.
[4] Neumann, A. L., Skotnikov A. A., Neumann, Y. V. Study of electromagnetic heating of the engine in transient conditions // Proceedings of the universities. Electromechanics. – 2012. – №6. – pp. 50-54.
[5] Neumann L., Neumann V. Yu. Shabanov, A. S., A simplified calculation of the electromagnetic impact actuator in intermittent operation mode // Electrical engineering. – 2014. – №12. – pp. 50-53.
[6] Sidorov O. Yu. finite-difference modeling of the characteristics of the axisymmetric induction devices / Sidorov O. Yu.V. A. Semenov, S. F. Sarapulov, Proceedings of the universities. Electromechanics. – 2001. – №1. – pp. 32-35.
[7] Sidorov O. Yu., Methods of finite elements and finite differences in electromechanics and Electrotechnology / O. Yu. Sidorov, F. N. Sarapulov, S. F. Sarapulov. – M.: Energoatomizdat, 2010. – 331 p.
[8] Kireev V. I., Panteleev A. V. Numerical methods in examples and problems. M.: Higher school, 2008. – 480 p.
[9] Ryashentsev N. P. Ugarov G. G., Levitin A. V. Electromagnetic presses. Novosibirsk: Nauka, 1989. – 210 p.
[10] Electrotechnological virtual laboratory: A textbook /F.N. Sarapulov, S.F. Sarapulov, D.N. Tomashevsky and others. Yekaterinburg: Ural State Technical University, 2003. – 233 p.
[11] Sarapulov F.N., Tomashevsky D.N., Padukov M.P., Research of induction-dynamic motor in electric drives of translational motion / Proceedings of the VI international Symposium ELMASH-2006. – Moscow: MA "Interelectromash", 2006. – Vol. 2. – pp. 116.
[12] Sarapulov F.N., Tomashevsky D.N., Petrov I.S., Simulation of linear impulse motors using the Simulink application / III international STC "Electromechanical and electromagnetic energy converters and controlled electromechanical systems". Yekaterinburg: Ural State Technical University, 2007. – pp. 235.
[13] Tomashevsky D.N., Koshkin A.N., Ponikarovsky V.A., Simulation of the mechanical part of pulsed shock systems in the SIMULINK application / Proceedings of the XII International conference "Electromechanics, electrical technologies, electrical materials and components-FEM-2008". Crimea, Alushta, 2008. – pp. 170.
[14] Tomashevsky D.N., Koshkin A.N., Magnetic-pulse device with linear motors / Industrial power engineering. – 2010. – №5. – pp. 45-47.

IOP Conf. Series: Materials Science and Engineering **950** (2020) 012021 doi:10.1088/1757-899X/950/1/012021

Mathematical modeling of mutual operation of superconductive devices in autonomous power engineering

L I Chubraeva[1], S S Timofeyev[1] and V A Lazerko[1]

[1]Institute of Electrophysics and Electric Power Engineering of the Russian Academy of Sciences 191186, Russia, St. Petersburg, 18 Palace Embankment

Abstract. The paper presents results of mathematical modeling of mutual operation of electromechanical energy converters: high temperature superconductive (HTSC) synchronous generators and motors. It includes modes of operation of synchronous alternator and synchronous motor, mutual operation of alternators of comparable rating in an autonomous energy installation, mutual operation of synchronous HTSC motor and semi-conductive converter, influence of semi-conductive converter on the loss value in HTSC armature windings. In all cases specific features of HTSC electrical machines and their parameters and peculiarities of HTSC windings are accounted for with imposed restrictions on the current and magnetic flux variation velocity.

1. Introduction

Investigations of electro-technical devices, comprising elements with high-temperature superconductors (HTSC) display maximum economical effect in HTSC complexes and not in individual units. A complex may operate with air-liquefier, supplying all the installation with LN_2. Our team took part in the development of a number of autonomous electric power engineering complexes, intended for different purposes and including several HTSC devices. Their investigations revealed problems, associated with mutual operation of HTSC electrical machines or HTSC devices and semi-conductive converters. Presented below is the general view of autonomous energy installation (Figure 1,a) with synchronous alternator, comprising HTSC saddle-shape armature winding (Figure 1,b) and synchronous motor of a disc type with circular HTSC armature coils (Figure 1,c). These electrical machines and their mutual operation served the base for mathematical modeling, presented below.

IOP Conf. Series: Materials Science and Engineering **950** (2020) 012021 doi:10.1088/1757-899X/950/1/012021

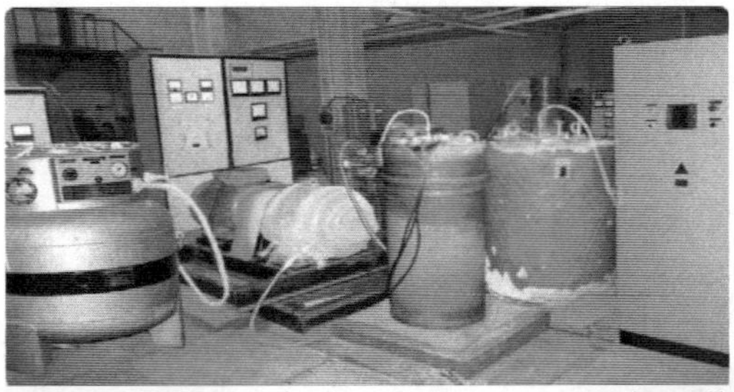

Figure 1. General view of autonomous energy installation (a),

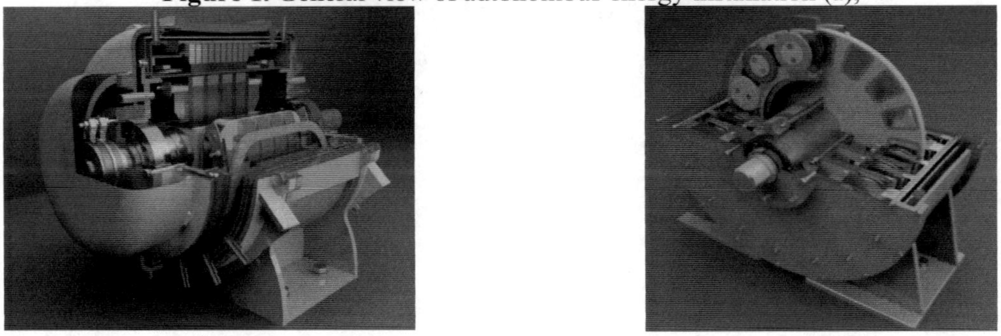

b c

Figure 1. Synchronous alternator, comprising HTSC saddle-shape armature winding (b), synchronous motor of a disc type with circular HTSC armature coils (c).

2. Mutual operation of HTSC synchronous alternator and synchronous motor

Let us consider peculiarities of parallel operation of SC synchronous generator and SC synchronous motor of similar rating without a voltage regulator. Both machines are non-saturated and non-salient pole.

It should be noted that in contrast with conventional synchronous machines the value of synchronous reactance may be varied in a wide range from 0.1 to over 2.5 p.u. [1] without affecting the size and weight parameters. Therefore presented below is the way to obtain maximum rating and overload capacity in HTSC machines, operating in parallel.

Figure 2 presents the simplified diagrams of both alternator and motor (x_G и x_M – generator and motor synchronous reactances; E_{0G} и E_{0M} – vectors of EMF, induced by the rotor flux in armature windings, δ – load angle, φ – the angle between the voltage and current vectors) [2].

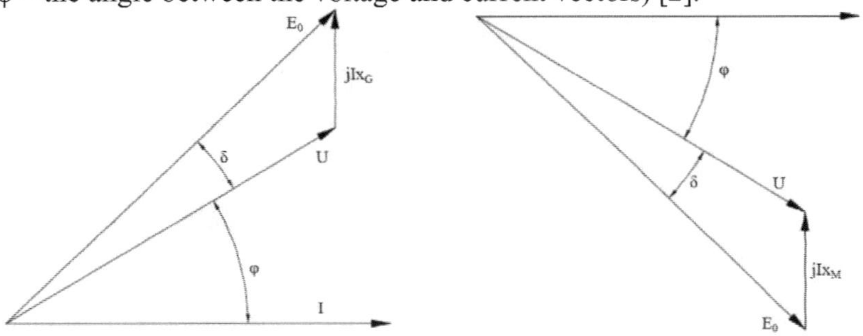

a b

Figure 2. Simplified vector diagrams of SC synchronous generator (a) and motor (b).

Let us accept that superconductive (SC) synchronous generator and SC synchronous motor are connected by SC cable without intermediate resistance. Then according to vector diagrams (Figure 3)

$$E_{0G} = U + jIx_G, \tag{1}$$

$$E_{0M} = U + jI\,x_M. \tag{2}$$

The following parameters may be obtained from the diagram (Figure 3) [3];

$$\frac{1}{2}E_{0G}E_{0M}\sin\delta = \frac{1}{2}UI(x_G + x_M)\cos\varphi, \tag{3}$$

$$P_G = P_M = mUI\cos\varphi, \tag{4}$$

$$P_G = P_M = \frac{mE_{0G}E_{0M}}{x_G + x_M}\sin\delta. \tag{5}$$

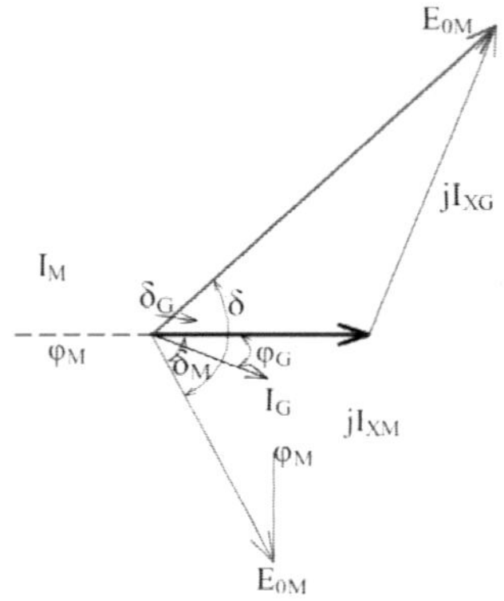

Figure 3. Vector diagram for parallel operation of synchronous alternator and synchronous motor

Maximum rating is achieved at sinδ=1, that is, at δ=π/2:

$$P_{G\max} = P_{M\max} = \frac{mE_{0G}E_{0M}}{x_G + x_M}, \tag{6}$$

The overload capacity is determined by the safety coefficient of the current in HTSC armature windings, which usually equals 1,5. Therefore the overload capacity is not to exceed this value.

Accounting for the possibility to vary the synchronous resistance on design stage, it is possible to choose the optimal value for both the alternator and motor with HTSC windings to obtain the maximum rating and overload capacity value with δg+δm=π/2 (6).

It should be noted that unlike parallel operation with the network of infinite power the rating of individual synchronous alternator or motor is determined not only by its excitation current value, but by the exaltation of other machines operating in parallel with it. The presented vector diagram permits to determine the values of the voltage, the armature current and cosφ. In case δ=π/2 the expressions are very simple [3].

If the voltage changes and the frequency of rotation remains constant, the following vector diagram may be used for calculations (Figure 4).

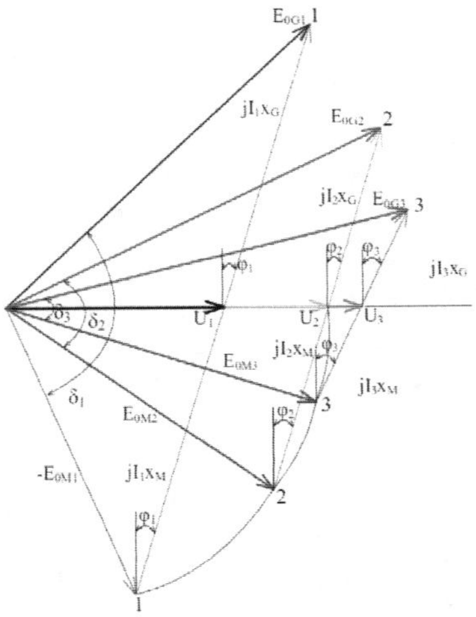

Figure 4. Vector diagram for parallel operation of synchronous alternator and motor with constant frequency of rotation and varying voltage value.

The presented in Figure 4 modes of operation show, that the voltage increase results in variation of maximum rating and overload capacity with certain increase of the load angle from $\delta=\delta_1$ to $\delta=\pi/2$ and then decrease to δ_2 and δ_3.

3. Parallel operation of synchronous alternators of similar rating

Let us consider 2 similar synchronous alternators. It is well-known: to transfer the load from one generator to the other at mutual rating P=const, U=const and f=const it is necessary to accelerate the rotor of the alternator I and to decelerate the rotor of the alternator II (with f=const) [4, 5]. As the rating of the generator I increases due to load angle increase, the synchronizing rating will tend to return the rotor to the initial position. The velocity of the process depends on the steepness of the synchronizing power density curve. The synchronizing rating equals:

$$P_S = \frac{mE_0U}{x_d}\cos\delta. \tag{7}$$

Vector diagram in Figure 8,a shows the initial mode of operation of both alternators. During the load transfer vectors E_{0I} and E_{0II} move around the arc of a circus (Figure 8,b). The load angle and the armature current of the alternator being loaded increase and respectively the same parameters of the unloaded alternator decrease. The velocity of armature current variation in both alternators must correspond to the current change velocity di/dt, permitted for the superconductor applied. The important moment of the process lies in the fact that the current variation in HTSC armatures may be relatively slow due to these restrictions. The rating of the first generator increases, the rating of the second generator decreases according to the preliminary specified program.

Both alternators are to have equal values of cosφ, therefore the rate of EMF increase of the first generator and its decrease for the second generator are to keep the voltage value constant. It is necessary to vary the excitation currents as well (Figures 8,b,c). The rate of excitation current variation may be limited in case the rotor contains HTSC windings.

As a result, to transfer the rating from one alternator to the other it is necessary to accelerate the rotor of the alternator being loaded along with the increase of the excitation current and to slow the rotor of the generator being unloaded with subsequent decrease of its excitation current. The rate of these changes must keep the frequency and the voltage constant.

IOP Conf. Series: Materials Science and Engineering **950** (2020) 012021 doi:10.1088/1757-899X/950/1/012021

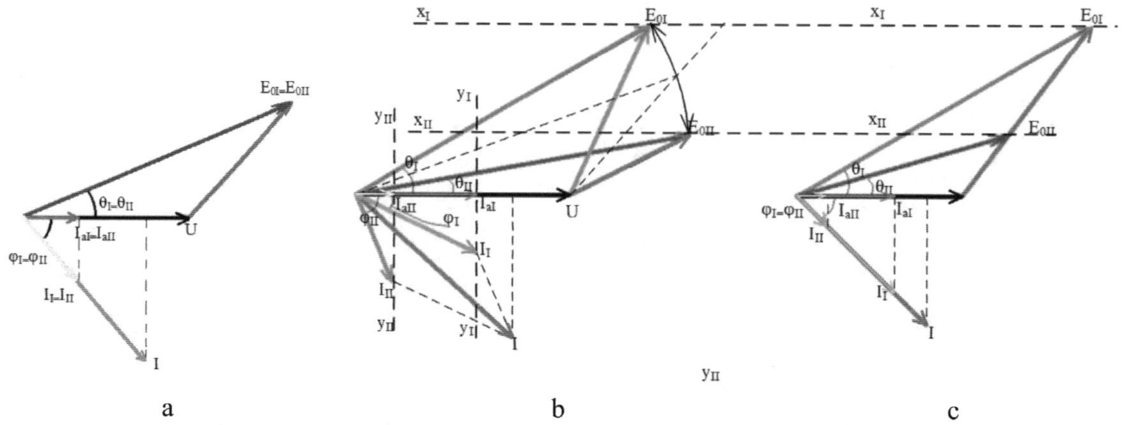

Figure 5. Vector diagrams presenting the load transfer between the alternators.

The synchronizing rating curve steepness depends on the parameters, obtained during the process of HTSC alternator designing: synchronous reactance x_d and number of phases m. The curves are presented in Figure 9. The lower is synchronous reactance and the higher is the phase number, the faster is the stabilizing process.

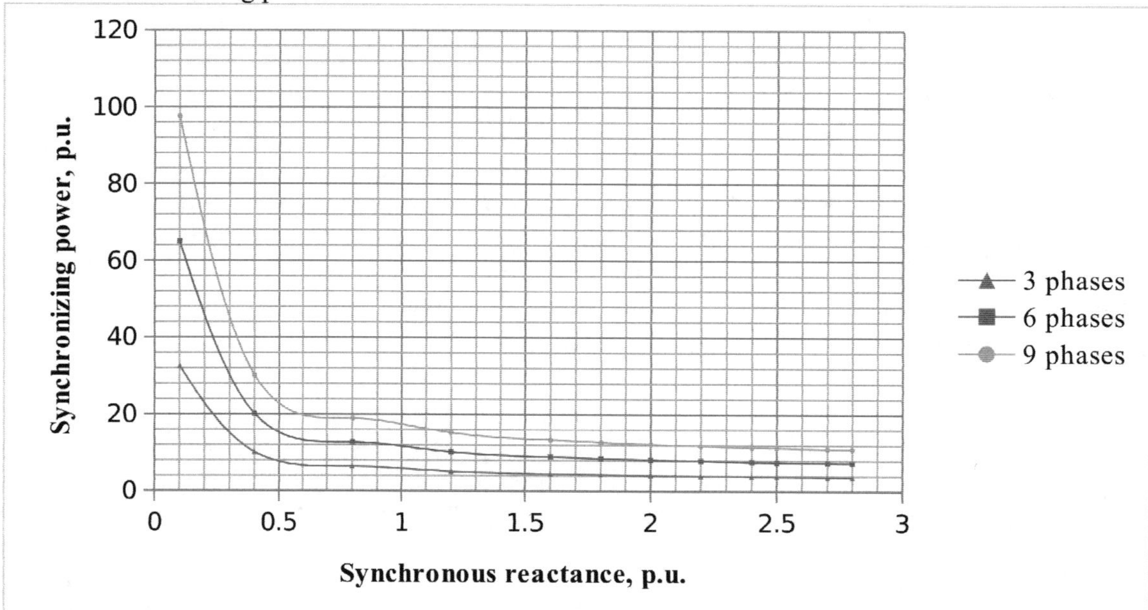

Figure 6. Dependence of synchronizing power density on synchronous reactance and phase number.

It may happen that the synchronizing process will be relatively fast due to low synchronous reactance or increased phase number, but the process of the rating transfer will take increased time because of restrictions, imposed by the HTSC wire applied for windings manufacturing.

4. Mathematical model of HTSC motor operating with frequency converter and voltage invertor

The model was developed for the motor with permanent magnets on the rotor and HTSC armature windings fed from the frequency converter with voltage inverter. The developed functional model of the motor control is presented in Figure 10.The task for the shaft frequency of rotation is preliminary being received by the unit permitting to establish the desirable intensity of its growth and then by proportional integral frequency of rotation controller (PI-controller). The signal of rotor frequency is obtained by the PI-regulator as well. Depending on the disagreement the signal from the exit of PI-

regulator is being sent to the unit of current calculation. The same unit receives the signal from the rotor position sensor. The unit forms the task for the phase A,B, and C currents with reference to the current rotor position.
The maximum motor torque value is obtained when the angle between the flux vector and overall current vector equals p/2. Zero angle value corresponds to the agreement of the above mentioned vectors. The data forms the base for the vectors of motor current calculations.
In the next unit the tasks for phase currents are being sent to the channel, forming the voltages of armature winding. Then the tasks for currents are being transformed into the pulses of the power switch of the invertor control in the units of pulse width modulation. Modeling of the motor modes of operation was based on application of simulation package Power System Simulink (MATHLAB). The model is based on standard blocks of the package.

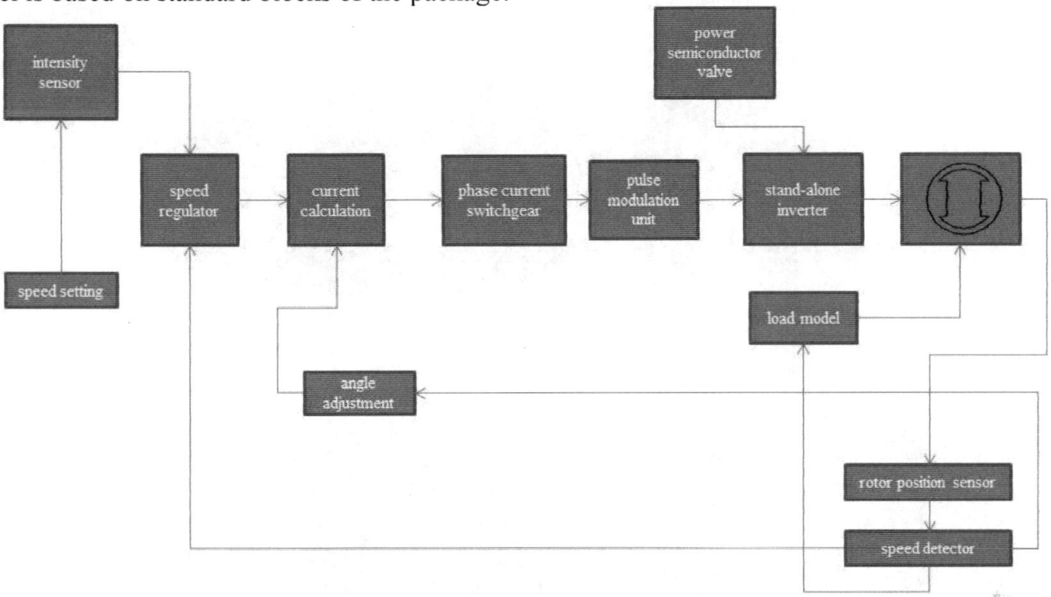

Figure 7. Structural scheme of synchronous motor.

The model used for calculation is not presented as it is relatively huge. Figures 11-12 present the calculation results for the motor start by screw characteristics $n = n_n$ (graphs for Ws_m - velocity, M_m - torque, I_{abc} - phase current, U_{ab} – motor linear voltage).
There is witnessed a certain armature current value jump. In our case it does not exceed the safety coefficient with reference to the superconductor critical current. But it should be accounted for to exclude transition of the superconductor to the normal state.

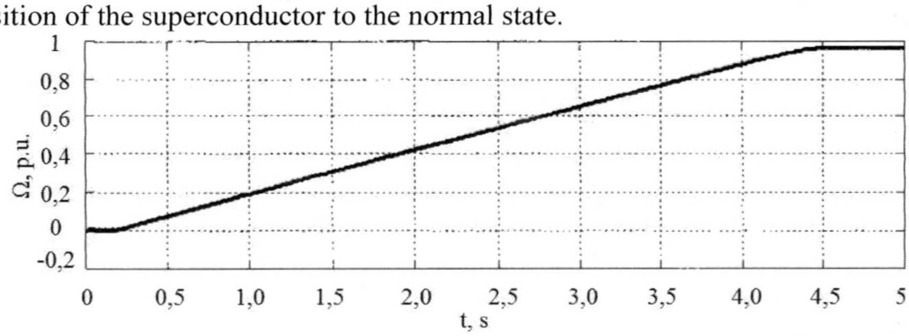

a

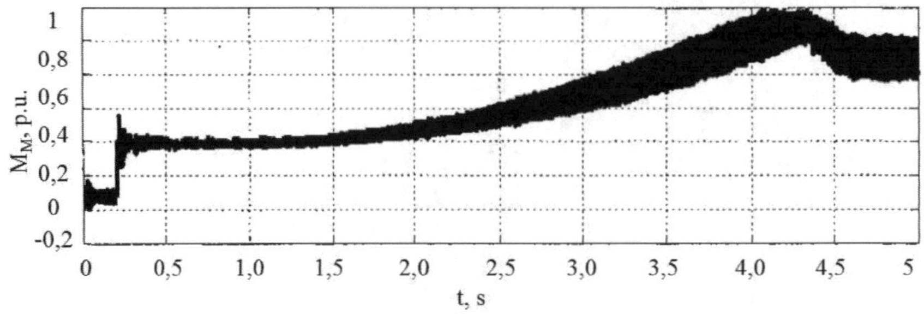

b

Figure 8. The calculation results for the motor start by screw characteristics.

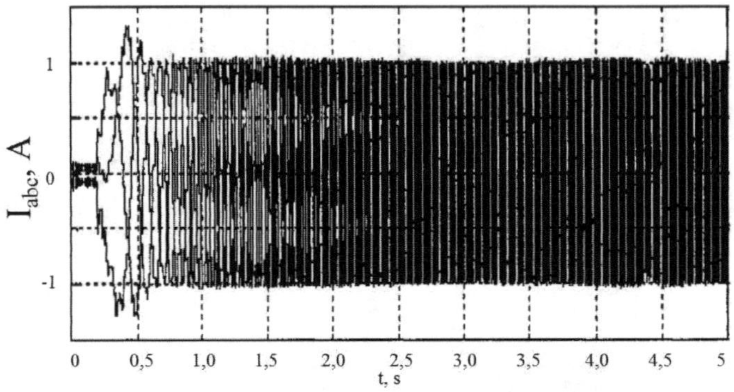

a

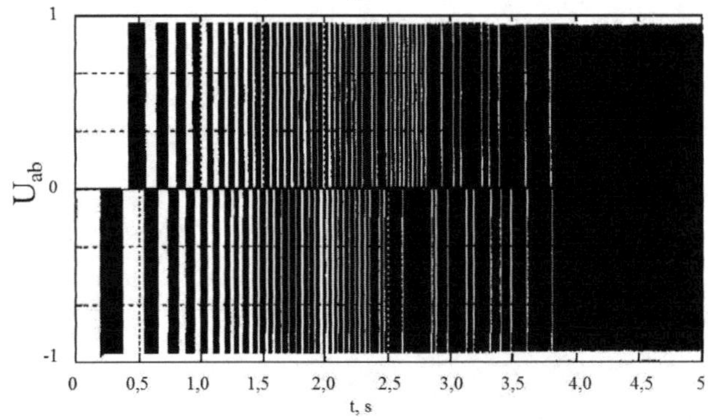

b

Figure 9. The calculation results for the motor start by screw characteristics.

5. Modelling of influence of semi-conductive converter on the level of loss in HTSC armature winding

It is well-known that discreteness of operation of semi-conductive converter elements results in appearance of high harmonics in current and voltage curves. In the model the converter is accomplished as a three-phase bridge circuit with f smoothing filter and current invertor. Results of modeling of operation of HTSC windings with a simi-conductive converter for the no-load of the converter and rated load are presented in Figures 13-14 (a – phase current curves, b –their spectral content). Spectral content, corresponding to the no-load mode, contains except $k = 6n \pm 1$ harmonics of the order $k = 2$, $k = 3$, $k = 4$, $k = 6$. As the load increases, the even harmonics start to disappear and the 3-d harmonic decreases with reference to the 1-st one.

Numerical and relative values of current harmonics from mains supply for the no-load mode are presented in Table 1.

IOP Conf. Series: Materials Science and Engineering **950** (2020) 012021 doi:10.1088/1757-899X/950/1/012021

Table 1. Numerical and relative values of current harmonics for the no-load mode

K	1	2	3	4	5	6	7	11
I_k,A	12,1	1,2	0,6	0,25	7,09	0,25	2,4	1,6
I_k, p.u.	1,0	0,1	0,05	0,02	0,58	0,02	0,2	0,135

Numerical and relative values of current harmonics from mains supply for the rated mode are presented in Table 2.

Table 2. Numerical and relative values of current harmonics for the rated mode

K	1	3	5	7	11
I_k,A	82,0	1,48	14,8	6,71	1,48
I_k, p.u.	1,0	0,018	0,18	0,08	0,018

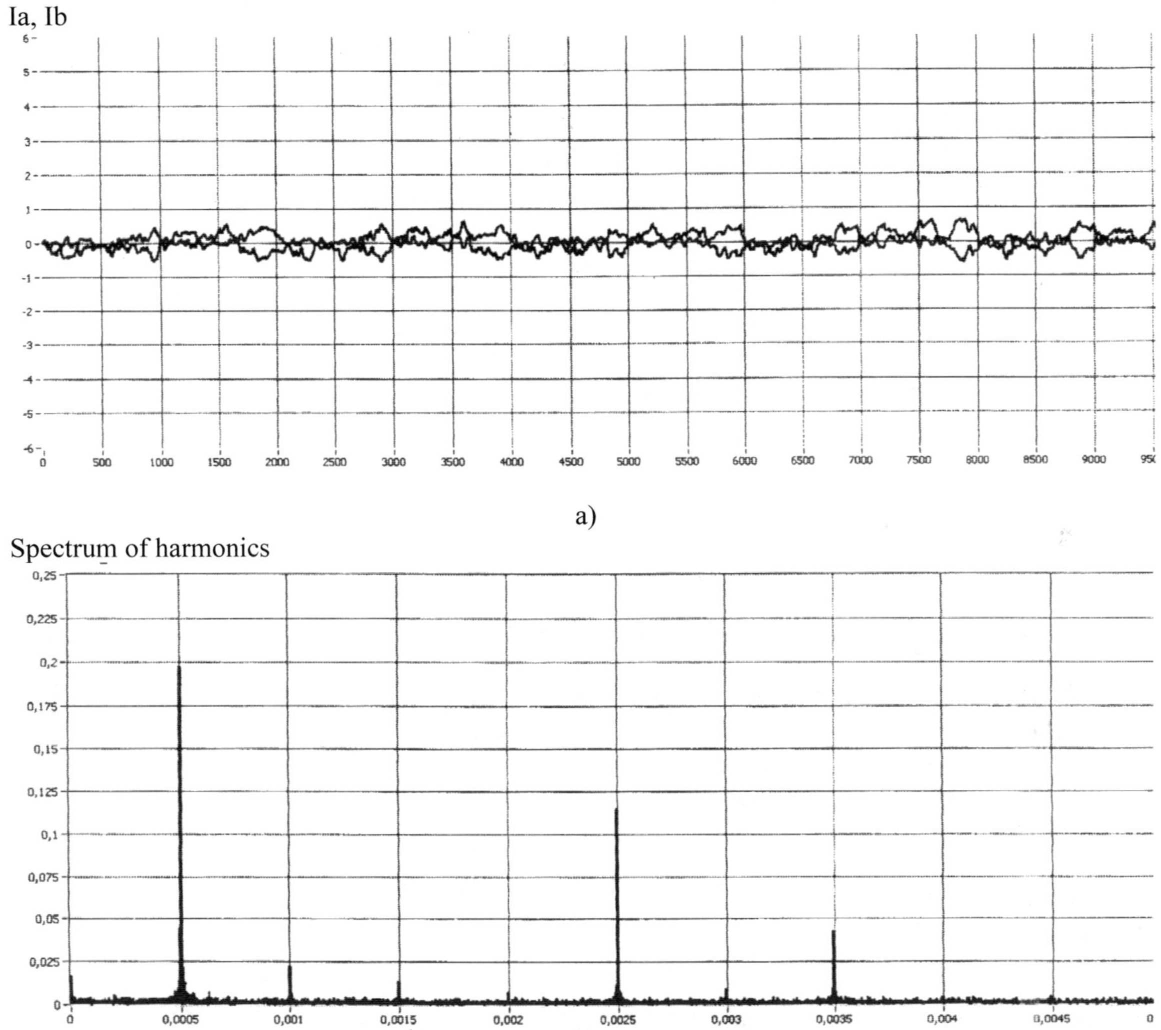

Figure 10. Operation mode simulation results of the converter at no-load duty.

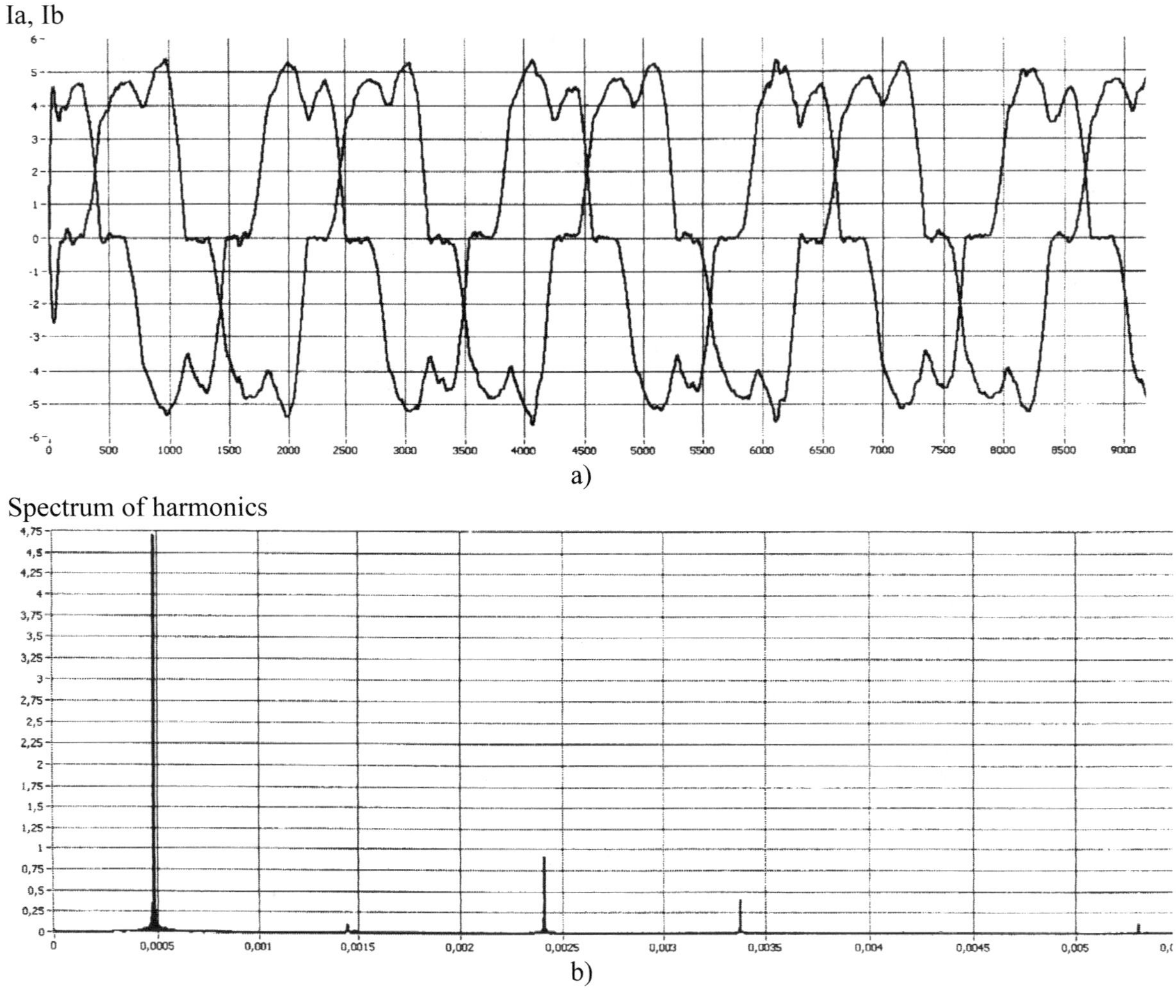

Figure 11. Results of modeling the rated operating mode of the converter.

Therefore the results of modeling show that the high harmonic content will not result in substantial loss in HTSC armature windings under load but it can cause increased losses in the no-load mode. The comparison of the loss values is to be compared for each particular case of mutual operation of HTSC synchronous motor and a semi-conductive converter. Therefore the results of modeling show that the high harmonic content will not result in substantial loss in HTSC armature windings under load but it can cause increased losses in the no-load mode. The comparison of this loss values is to be compared for each particular case of mutual operation of HTSC synchronous motor and a semi-conductive converter.

6. Conclusions

The results of modeling of mutual operation of HTSC synchronous alternators and synchronous motors in autonomous power installations show that they can fulfill the main requirements demanded the electrical equipment modes of operation, but there exist a lot of limitations, imposed by the type of superconductor applied, permissible current variation and impossibility to exceed the critical current or magnetic flux density to exclude transition of the superconductor to normal state.

References

[1] Chubraeva L I 1991 *Generators of nonconventional design (research calculation methods)* (Leningrad: Science. Leningrad Branch) p 246
[2] Vajnov A I 1968 *Electrical machines* (Leningrad: Energy) p 768
[3] Zimin V I 1960 *Ship electrical machines of alternating current* (Moscow: Military publishing house of the USSR Ministry of Defense) p 463
[4] Mezin E K *1985 Ship electrical machines. Textbook* (Leningrad: Shipbuilding) p 320
[5] Kitaev E and Greftsev N 1960 *Course of genera-l electrical engineering*. (Leningrad: Publication of the fifth revised and supplemented state union shipbuilding publishing house)

Acknowledgements.
The researches are being supported by the Program № 7 of Presidium of RAS «New developments in promising areas of energy, mechanics and robotics».

IOP Conf. Series: Materials Science and Engineering **950** (2020) 012022 doi:10.1088/1757-899X/950/1/012022

Method for modeling the modes of induction heating of turbine blades

A R Lepeshkin[1,4], A B Kuvaldin[2], C A Lepeshkin[2], Z Guanghua[4], V V Nazarov[3] and O I Ilinskaya[1]

[1]Moscow Aviation Institute - National Research University, 4 Volokolamskoe Shosse, Moscow, 125993, Russia
[2]National Research University «MPEI», Russia, Moscow, 111250, Russia
[3]Institute of Mechanics of Lomonosov Moscow State University, 1 Michurinsky Ave., Moscow, 119192, Russia
[4]Northwestern Polytechnic University, 127 West Youyi Road, Beilin District, Xian, 710072, China

E-mail: lepeshkin.ar@gmail.com

Abstract. In this paper there are offered calculated and experimental methods of simulating the high-speed modes of induction heating and designing the new inductor to provide the preset nonuniform thermal and thermostress state for cooled and uncooled blades of gas-turbine engine by changing the inductor width (specific power of induction heating). The peculiarity of this technique is the possibility of conducting the thermal cyclic tests of gas-turbine-engine parts with remaining the guaranteed minimal constant clearance between inductor and blade surface that reduces the probability of temperature-field distortion in changing the blades and promotes the efficiency enhancement of induction heating. Obtained results were used to conduct the tests of blades of gas-turbine-engine turbine. Carried out tests at this set with using the high-speed modes of induction heating helped to reveal quickly required changes in the blade structure.

1. Introduction

Rotor blades of turbines are one of the most important components which determine the service life of aircraft gas-turbine engines [1-5]. Therefore, the problems of providing their service life are of very importance. The good proportion of blade damages in the aircraft gas-turbine engines is bound up with the thermal variable loads arising in engine operating. The rise of such loads is explained by the large nonuniformity of gas-flow temperature along the blade height and the nonuniformity of heat transfer on the blade surface. Thermocyclic tests of the rotor blades with using the localized high-frequency induction heating the dangerous section of blade in which the strength margin is minimal provide the simulation of operating temperatures and loads. Under high enough frequency the induction heating allows to simulate quite satisfactorily the gas heating and because of contouring the inductor the nonuniform temperature field is obtained in accordance with preset one [3, 6]. In this paper the method for modeling the modes of induction heating of turbine blades of aviation engines on the test rig is developed.

2. Method for modeling the modes of induction heating of turbine blades, results and discussion

The calculated and experimental methods of simulating the high-speed modes of induction heating and designing the inductor to provide the preset nonuniform thermal and thermostress state for cooled and uncooled blades of gas-turbine engine by changing the inductor width (specific power of induction heating) are offered. The objective of test conducting was as follows: design-technological strength development of blade which provided while simulating in the dangerous section the operational thermomechanical loading the increase in service life 2 to 3 times. The studying program included: tests of series-produced blades and ones with thickened edges; assessment of acting the different variants of protective coatings; study of effect of stress redistribution in the blade during disk buckling (umbrella effect) on the cyclic life (in this connection, for tests the blades with biased center of gravity were prepared). There were tested seven variants of blades: new series-produced blades; series-produced blades completed ones' service life; new series-produced blades strengthened by microballs; blades with coating of titanium nitride; blades with "composite" enameled coating; blades with thickened edges. For each variants three blades were studied. With taking account of operating-thermal-state calculating data and experimental data under operating conditions there were made high-speed modes of heating at the thermocyclic tests. The temperature range in thermocycling was equal to 200 – 800 °C at the temperature rate: 100 – 400 K/s. Thermocycle parameters are shown in figure 1 and the preset temperature field of blade is given in figure 2,a.

On the basis of calculation the maximal value of mechanical load for simulation of axial tension in blade under the action of centrifugal forces was determined. Allowing for that the mechanical-load range was assumed in the following limits: 2000 – 25000 N. The test method of blade at the test rig was the determination of high-speed heating modes and designing the new profile inductor to provide the preset temperature field, permissible maximal temperature differences between the blade surface and core and thermostress state of blade in its dangerous section.

This task of induction heating the blade up to preset temperature field was solved in the following formulation:

$$\Phi = \sum_{i=1}^{i=m} (t_{preset,i} - t_i(p_i(h_{\text{ind},i})))^2 \Rightarrow \min \tag{1}$$

on the target functional (1) there were imposed the inequalities of linear type:

$$((\sum_{i=1}^{i=m} p_i)/m) \cdot S < \eta_{\text{g}} \cdot p_{\text{g}}^{\max}\text{;}\ \ h_{\text{ind,min}} < h_{\text{ind},i} < h_{\text{ind,max}}\text{;}\ \ \delta_{\min} < \delta_i < \delta_{\max}\text{;}\ \ \Delta t_i < \Delta t^{\text{permis}} \tag{2}$$

where i is the number of heating-zone portion, m is the number of heating portions, t_i is the calculated temperature of part surface, $t_{preset,i}$ is the preset temperature of part surface, p_{i} is the specific power of inductor, $h_{\text{ind},i}$ is the inductor width, $h_{\text{ind},min}$ is the minimal width of inductor due to maximal value of current, $h_{\text{ind},max}$ is the maximal width of inductor determined by conditions of local heating, Δt_i is the temperature gradient, $\Delta t^{permis.}$ is the permissible temperature difference (thermostress limitations), $p_{\text{g}}^{\max}$ is the power of high-frequency generator (HFG), η_{g} is the efficiency of high-frequency generator, S is the area of heated surface of blade, $\delta_{\min}$ is the minimal value of clearance, $\delta_{\max}$ is the maximal value of clearance.

Relationships of temperature to specific power of heating given in (1) were determined as a result of solving the differential equation of heat conductivity including the test conditions at the rig. This task was solved with using computer; as a result of its solving there were determined the powers of HFG4-10/0,44: while heating the blade from $\tau_{\min}$ to $\tau_{\max}$ the power was 6.5 kW, at the mode of keeping it was 1.5 kW, the new inductor profile (figure 2,b) and the clearance values between inductor and blade surface. The rig tests of blades were carried out by the modes of thermocycling (figure 1,c), thermostress state (figure 1,d) and hydraulic force loading (figure 1,a). Loading device ran in synchronism with the thermal cycle: $\tau_{\min} - \tau_{\max}$. Operating modes of HFG4-10/0,44 during tests are shown in figure 1,b. In the process of tests the thermal state over the contour of dangerous section of blades was monitored with chromel-alumel thermocouples. Control of cooling-air flow allowed to

IOP Conf. Series: Materials Science and Engineering **950** (2020) 012022 doi:10.1088/1757-899X/950/1/012022

select the predesigned parameters of thermocylce and to shorten the test periods. The distribution of surface thermostresses (figure 3) over the dangerous section of blade at the maximal mode and the change of thermostresses (figure 1,d) in the middle portion of blade surface during heating from τ_{min} to τ_{max} were obtained with using the computing program of finite-element method.

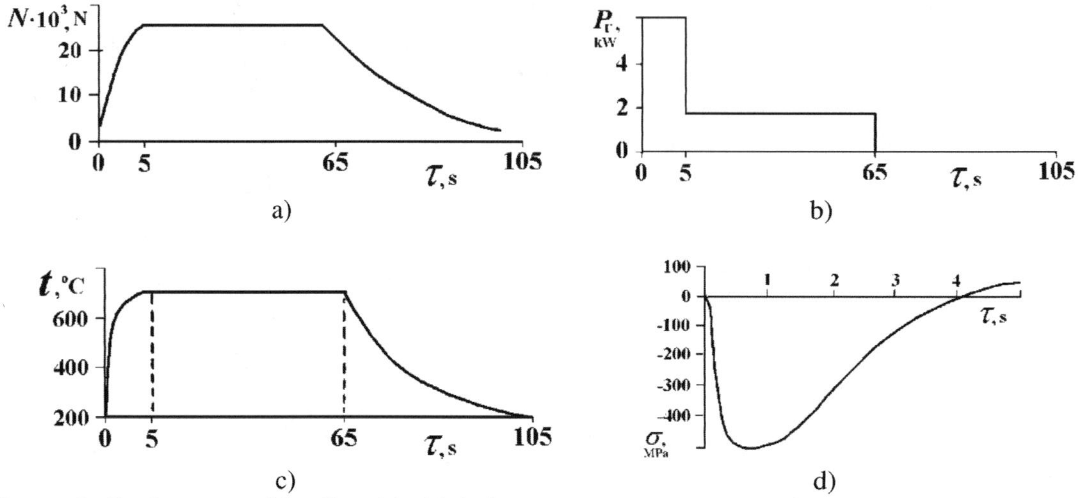

Figure 1. Cyclograms of loading (a), high-frequency generator operation (b), temperature changing (c) and thermostresses (d) on portion of blade surface

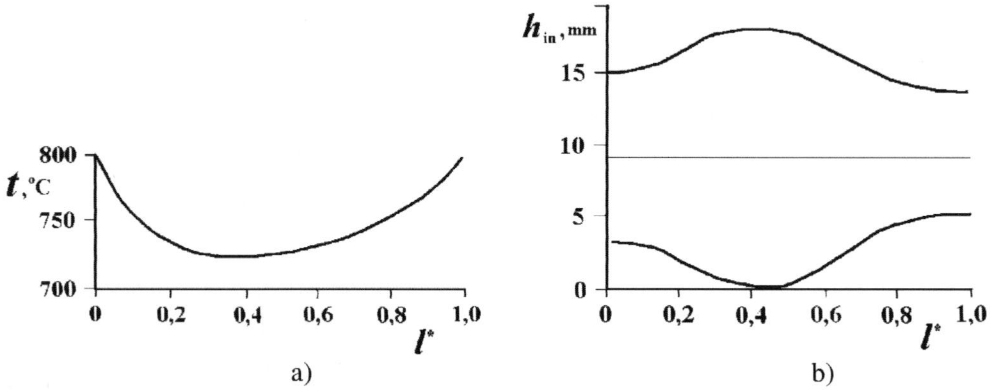

Figure 2. Temperature distribution (a), inductor profile (b) and thermostress distribution (c) under thermocyclic tests

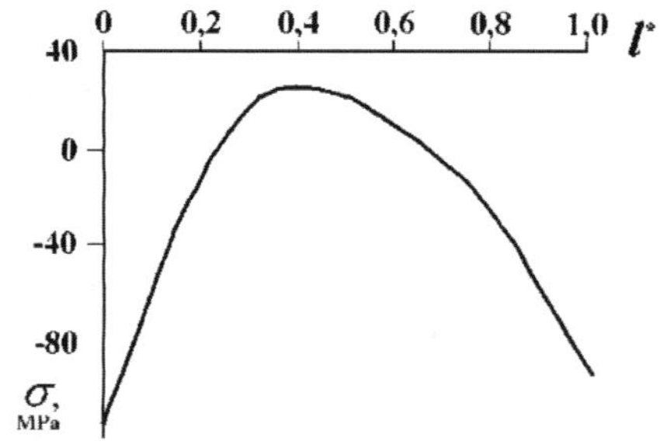

Figure 3. Thermostress distribution under thermocyclic tests

Numerical simulation of induction heating and thermostress state of blade according to experimental data made it possible also to reveal the maximal difference in simulation of thermostresses in the predetermined section of blade from the surface heating by value of 10 MPa at the time instant of 0.5 s. This value was 2.5% of the maximal thermostresses observed in the heating cycle (figure 1,d), therefore, this value can be neglected. Hence it follows that the high-speed modes of induction heating with the current frequency of 440 kHz simulate accurately enough the thermostress state corresponding to operating conditions. Analysis of test results showed that the life of blades completed ones' service life is one-fifth as many as new ones. Use of protective coatings on the series-produced blades did not lead to increasing in service-life. It was revealed that the structural changes as increase in edge thickness and bias of center for tensile-force applying provided the increase in service life. The average value of life under cyclic thermoloading the blades with reinforced edges gave sixfold rise of service life. Therefore, to develop the engine with predetermined service life the blades of improved structure (with reinforced edges) were used effectively. That made it possible to shorten essentially the engine developing periods. Carried out tests at this set with using the high-speed modes of induction heating helped to reveal quickly required changes in the blade structure.

3. Conclusion

In this paper there are offered calculated and experimental methods of simulating the high-speed modes of induction heating and designing the inductor to provide the preset nonuniform thermostress state for cooled and uncooled blades of gas-turbine engine by changing the new inductor width. The peculiarity of this technique is the possibility of conducting the thermal cyclic tests of gas-turbine-engine parts with remaining the guaranteed minimal constant clearance between inductor and blade surface that reduces the probability of temperature-field distortion in changing the blades and promotes the efficiency enhancement of induction heating. Obtained results were used to conduct the tests of blades of gas-turbine-engine turbine. Carried out tests at this rig with using the high-speed modes of induction heating helped to reveal quickly required changes in the blade structure. Developed program package of calculation and inductor structures can be used during tests and technological processes of heat treatment for different parts.

References

[1] Lepeshkin A R, Ilinskaya O I and Nazarov V V 2017 Investigations of thermal barrier coatings of turbine parts using gas flame heating *J. Phys.: Conf. Ser.* **899** 072002

[2] Lepeshkin A R, Nazarov V V, Ilinskaya O I and Kartavtsev A S 2019 Modelling the stressed state of thermal protective coating of turbine blades taking into account action forces from the gas flow de based on finite element analysis // *J. Phys.: Conf. Ser.* **1359** 012101

[3] Kuvaldin A B and Lepeshkin A R 2019 High-speed modes of induction heating and thermal stress in products. *Monograph.* M.: Infra-M. 282

[4] Li B, Fan X, Li D and Jiang P 2017 Design of thermal barrier coatings thickness for gas turbine blade based on finite element analysis *Mathematical Problems in Engineering.* **ID 2147830**

[5] Lepeshkin A R 2017 Technique of study of thermal difffusivity of metallics in different directions in a field of centrifugal accelerations and forces *J. of Engineering Thermophysics.* **26**(1) 1-7

[6] Fedin M A, Kuvaldin A B, Kuleshov A O, Zhmurko I Y and Kondrashov S S 2019 Theoretical investigation of induction heating of a two-layer conducting body by the example of an induction crucible furnace with a conductive crucible // *IOP Conf. Ser. Earth an Environmental Science.* November

IOP Conf. Series: Materials Science and Engineering **950** (2020) 012023 doi:10.1088/1757-899X/950/1/012023

Mixed Inverter Regulation Method for Induction Heating

Ilias Abdulkhakov [1], Yu Perevalov [1]

[1] Department of Electrotechnology and Converter Engineering, Saint-Petersburg Electrotechnical University (LETI), 197376 St.Petersburg, prof. Popov 5, Russia

E-mail: yyperevalov@yandex.ru

Abstract. One of the commonly used methods of voltage source inverter regulation for induction heating is a frequency method of regulating the output power of the inverter, based on the frequency selectivity of the resonance load circuit. With frequency regulation at low values of the Q-factor of the resonance circuit, deep regulation of the output power is impossible. As a result, other methods of regulation are desirable. The article discusses the issue of using a mixed pulse and frequency phase or latitude regulation method under the obligatory conditions of the soft switching mode.

1. Introduction

Transistor resonance voltage source inverters are widely used as power supply sources for induction heating installations. The most common method for regulating power in these devices is frequency regulation method with inductive misalignment of the inverter, as a result of which the soft commutation mode of semiconductor power devices is achieved. Using this method is not suitable for deep adjustment of the output power - the operating frequency must vary over a wide range, which is not always possible at low values of the Q-factor of the load circuit, as the operating frequency range of the inverter is limited and the heating technology requirements are specific [1].

In order to expand the regulating range of inverter output power, it is possible to use other regulation methods: amplitude, phase, pulse and width, pulse and code and their combinations, including frequency [2]...[5]. As a general principle, the use of these methods significantly reduces the inverter efficiency due to the deterioration of the switching modes of the power keys resulting in the decrease in reliability and added complexity of the converter circuit, which in turn affects the reliability and cost. Thus, this article provides a study on the efficiency in terms of the depth of adjustment, use of frequency and phase and frequency and latitude methods for regulating the inverter output power under the obligatory and sufficient conditions for ensuring the soft switching mode [6] of power keys and without complicating power part of the circuit.

2. Problem

To study the operation of the voltage source inverter in the LtSpice package, a computer model shown in Figure 1 was developed. The model consists of: constant voltage source V5, the output characteristics of which simulate the output characteristics of a 3-phase rectifier with a low-pass filter; bridge inverter on field-effect transistors VT1-VT4, controlled from pulsed voltage sources with adjustable time and amplitude parameters Vg1-Vg4 with gate resistances R1-R4; resonance load circuit consisting of series-connected effective resistance of inductor R5 and the inductance of inductor L1, as well as the compensation capacitance of inductor C1.

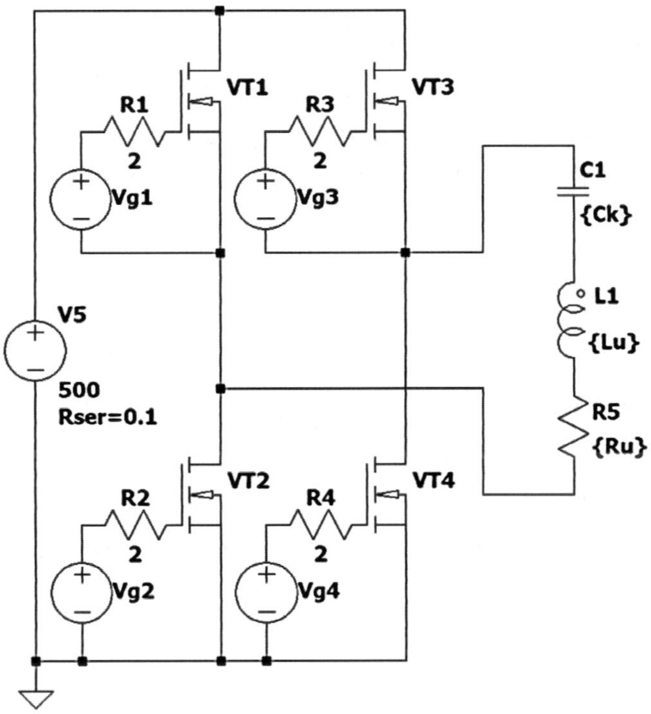

Figure 1. The computer model

Based on the calculations, a dependence (Figure 2) for minimum permissible relative pulse duration D with pulse and latitude regulation and the maximum permissible phase shift ϕ with phase and pulse regulation depending on phase ϕ between output current and output voltage on the first harmonic of the inverter, the output phase ϕiu1 is in turn regulated by the operating frequency of the inverter, with the invariant parameters of the resonance load circuit, was developed. This dependence is easily confirmed analytically based on the calculation of the resulting time interval between the output current and voltage on the first harmonic when the frequency changes, within which the switching of the semiconductor power keys should occur (in the limit to the left of the current zero point), and, based on this, it becomes clear that this dependence remains unchanged at different Q-factors of the load circuit. It should be noted that the presented dependence is achievable provided that the reactive power of the circuit is localized in the inverter. This condition is achieved automatically when using phase and frequency dependence, and when using the latitude and frequency regulation, oscillations in the circuit are damped and therefore the soft switching condition ceases to be met, due to a change in the relative pulse duration of the control pulses of transistors of the same level (upper or lower) only or one half-bridge (arm) of the inverter, if this condition is not met in the dead time interval with narrowing time (relative pulse duration) of the control pulses of the transistors because of the counter emf effect on the circuit through the bypass transistor diodes. Also, the dependence shown in Figure 3 does not take into account the switching time of transistors and the dead time between control pulses of opposite diagonals. In actual implementation, for example, when adjusting a product or simulating using models of the corresponding transistors, the dependence shown in Figure 2 can be reconstructed several percent higher from the indicated one (for both ordinates) taking into account the dead time of control pulses, transistors switching delay, as well as the fulfillment of the conditions of the soft switching mode.

To illustrate, at point ϕiu1 = 44.289 °, a calculation with the following parameters of the load circuit: Lu = 200 μH; C1 = 50 nF; Ru = 6.325 Ohm - was carried out. To establish the selected value of the phase deviation, the operating frequency of the inverter based on the calculation of the first harmonic should be approximately (excluding current harmonic distortion) equal to 52.846 kHz. Figure 3 shows timing diagrams explaining the operation of the inverter (Figure 1) at the specified operating point with the set parameters of the load circuit: Figure 3a - transistor control pulses VT2 (Vg2) and VT4 (Vg4), corresponding to curves 1 and 2; Figure 3b – transistor current VT2 (ID(VT2)); Figure 3c – transistor current T4 (ID(VT4)); Figure 3d - output current (I(Ru)) and voltage (Vout) of the inverter corresponding to curves 2 and 1.

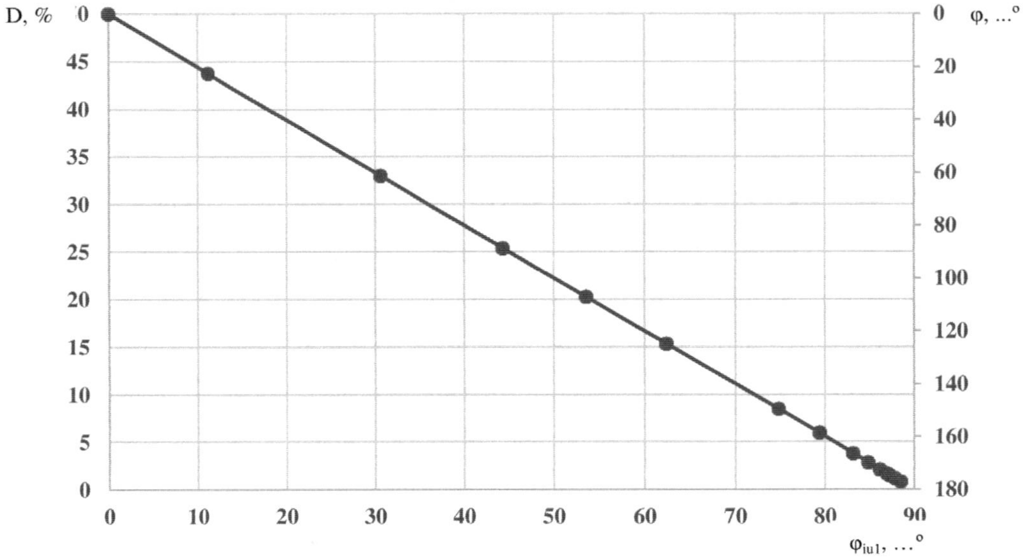

Figure 2. The dependence for minimum permissible relative pulse duration D with pulse and latitude regulation and the maximum permissible phase shift ϕ with phase and pulse regulation depending on phase ϕ between output current and output voltage on the first harmonic of the inverter

With the set misalignment of the inverter on the first harmonic ϕiu1 = 44.289 ° - the limiting minimum permissible relative pulse duration of the control pulses of the regulated part of the inverter, based on the dependence shown in Figure 2, is: D = 25.5%. However, because of the introduced dead time between the control pulses of opposite diagonals equal to 100 nS, which is more than enough for a confident switching of the opposite diagonals of the bridge on field-effect transistors in the absence of through currents through the inverter half-bridges (arms) - the limiting minimum relative pulse duration at which the soft switching mode is maintained slightly differs from that determined based on the dependence in Figure 2 and is equal to: D = 26.5%. Figure 4 shows timing diagrams explaining the operation of the inverter (Fig. 1) with a relative pulse duration of the control pulses of the regulated arm of the inverter D = 26.5% and a phase deviation on the first harmonic of the output current and voltage of the inverter ϕiu1 = 44.289 °: Figure 4a - control pulses with an unregulated relative pulse duration of transistor VT2 (Vg2) and transistor VT4 (Vg4) with an adjustable relative pulse duration corresponding to curves 1 and 2; Figure 4b – transistor current VT2 (ID(VT2)); Figure 4c – transistor current VT4 (ID(VT4)); Figure 4d - output current (I (Ru)) and voltage (Vout) of the inverter corresponding to curves 2 and 1. It should be noted that the pattern of the inverter operation, shown in Figure 4, will remain with the phase and frequency regulation, except for the type of control pulses, which will be deviated by ϕ = 88.5 °.

Comparing the operating modes of the inverter according as per the timing diagrams shown in Figures 3 and 4, the following points can be identified: the output current of the inverter dropped from 67.4 A to 51.4 A; the current switched by transistors from 48 A changed to 1 A for the pulse and latitude unregulated arm and up to 51.3 A for the regulated one; based on a linear approximation, the average current (area) of bypass transistor diodes during the conduction time changed from 51.4 μA to 11 nA for the pulse and latitude unregulated transistor and to 115.2 μA for an adjustable one; the average current (area) of the forward conductance of the transistor during the open state changed from 356.3 μA to 309.6 μA for a pulse-width unregulated and to 167.1 μA for the regulated one.

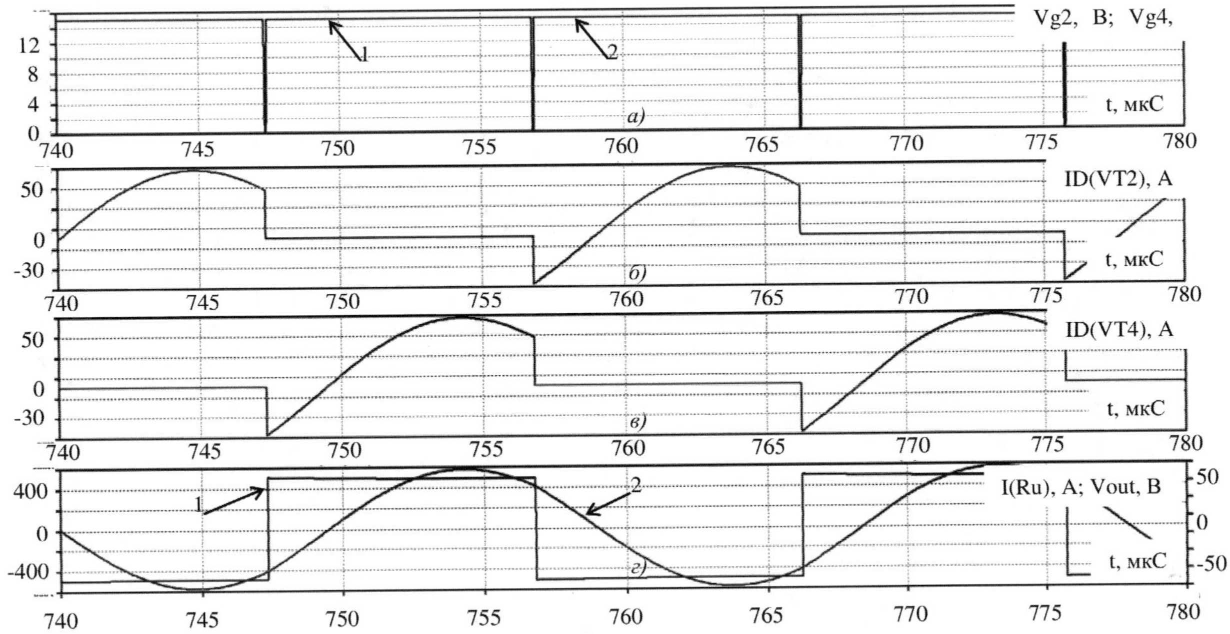

Figure 3. The timing diagrams explaining the operation of the inverter at the specified operating point with the set parameters of the load circuit

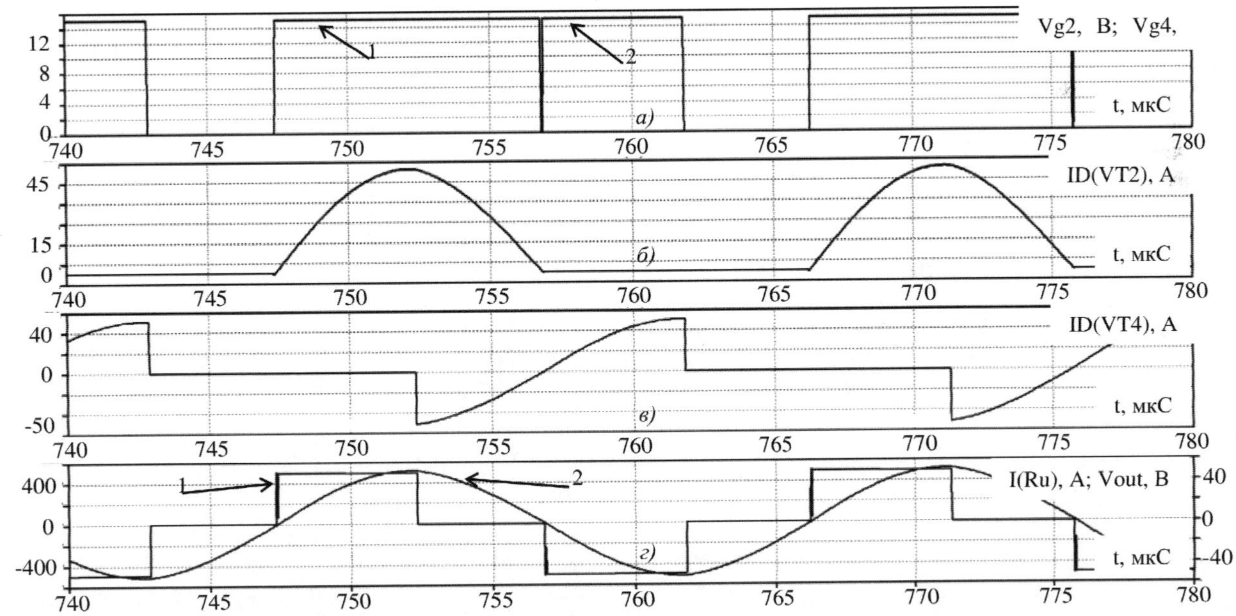

Figure 4. The timing diagrams explaining the operation of the inverter with a relative pulse duration of the control pulses of the regulated arm of the inverter D = 26.5% and a phase deviation on the first harmonic of the output current and voltage of the inverter ϕiu1 = 44.289 °

It may be said that when an additional regulation channel is used together with a frequency (phase or latitude) channel, the switching losses of the additionally regulated transistors practically do not change – of those unregulated are practically excluded, forward conduction losses for all transistors of the bridge are reduced, the conduction losses of the bypass diode increase for regulated transistors - unregulated ones are practically excluded, however, the total bypass conduction losses for the entire inverter remain the same as when using frequency regulation only.

Thus, we can conclude that the use of an additional regulation channel integrally reduces losses in the entire inverter. Thus, the possibility is confirmed and the limits of the mixed regulation of the output power of the resonance source voltage inverter are demonstrated, at which soft switching mode of the semiconductor power keys of the inverter is necessarily provided. The purpose of using the mixed regulation method is to reduce the frequency range of the regulating converter output power. Figure 5 shows the dependences of rated output power Pn on the operating rated frequency of the inverter relative to resonance circuit fn when using only a particular regulation method (Figure 5a) and mixed pulse-frequency-latitude or frequency-phase method (Figure 5b) ensuring the soft switching mode of power keys with different Q-factor values of the load circuit. Figure 5 shows that using the mixed method compared to the frequency method exclusively, the ration of the maximum achievable narrowing of the operating frequency range is 1.41 times.

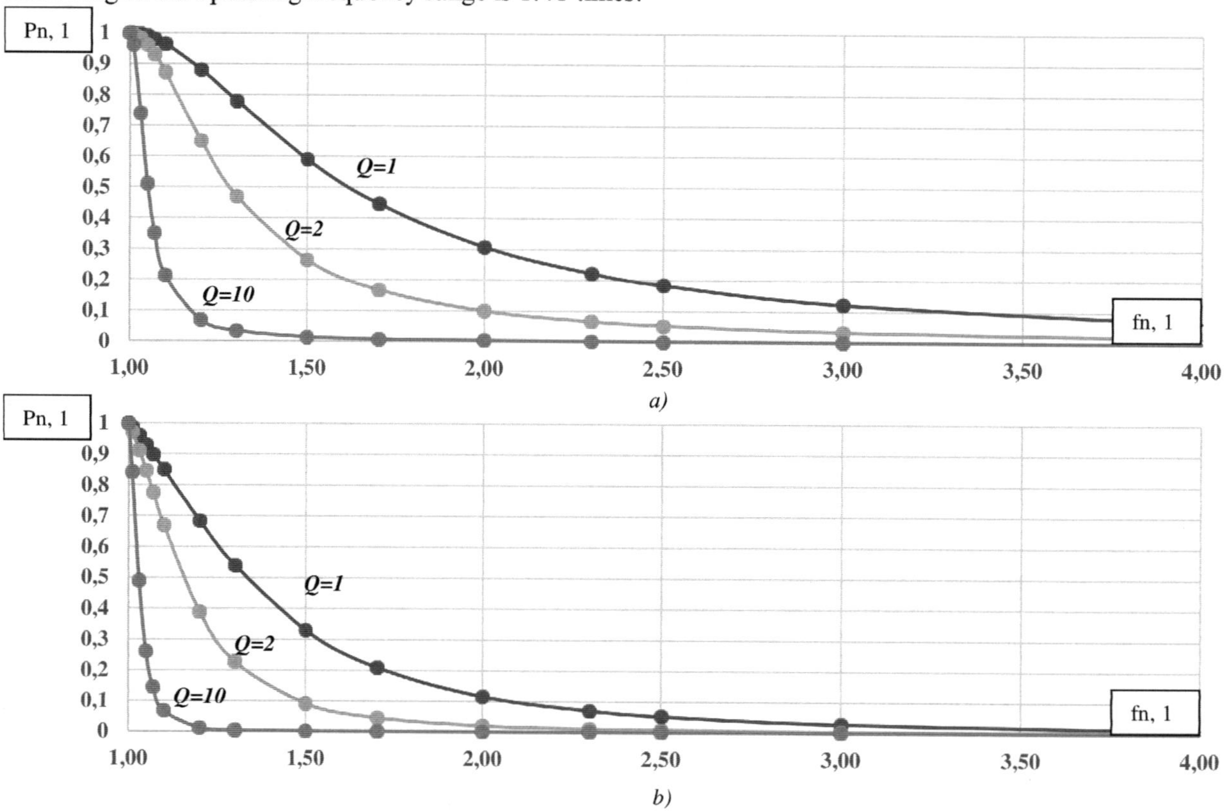

Figure 5. The dependences of rated output power Pn on the operating rated frequency of the inverter relative to resonance circuit fn

Thus, the article demonstrates the possibility of implementing a mixed regulation of the output power based on pulse latitude-frequency and phase-frequency methods. At the same time, implementation of these methods is based on the linear dependence of the maximum permissible relative pulse duration or phase deviation of control pulses from the output phase of the inverter on the first harmonic, at which the soft switching mode of the power keys of the inverter is maintained.

Using the output phase of the inverter as the main parameter is convenient because the dependence on the load parameters of the inverter, which also includes an inductor, is excluded, and the parameters of the inductor, in turn, can change by ten folds [7], and the mixed regulation based on the established linear relationship become possible to be implemented. Linear dependence can be implemented both programmatically on the process controller, and through feedback with a selected amplification factor based on the data of the output phase of the inverter.

3. Conclusions

Using an additional control channel together with a frequency channel allows to narrow the control frequency range by 1.41 times in the limit at which the soft switching mode of the power keys of the inverter with minimum output capacitances is still maintained, compared to exclusively frequency regulation. Narrowing of the frequency range provides the following advantages: expanding the regulation capabilities of converters with a limited operating frequency range with a low Q-factor value of the load circuit (or heating stage, when the Q-factor value of the inductor is minimal); reducing the cross impact of induction systems with magnetically coupled inductors and individual power supply by increasing the difference between the operating frequency of the inverter and the resonance frequency of the magnetically coupled inductor circuit [8]; complying with the special requirements of the induction heat treatment technology in terms of the frequency of currents induced in the part.

In the case of magnetically coupled induction systems with individual power supply - somewhat reducing their cross impact by increasing the difference between the operating frequency of the inverter and the resonance frequency of the magnetically coupled inductor circuit; also complying with the special requirements of induction processing technology.

References

[1] Perevalov, Yu.Yu. Modeling and development of an automated induction hardening plant for large rolls of rolling mills: author. dis. Cand. tech. Sciences: 05.09.10 / SPb: SPbGETU, 2017.- 16 p.

[2] Dzliev, SV Characteristics of a resonant transistor voltage inverter with phase and frequency control // Actual problems of the theory and practice of induction heating (APIH-2005). - Materials of the international scientific and technical conference .- St. Petersburg, 2005.- P. 363-369.

[3] Hamamura, S. Combined PWM and PFM Control for Universal Line Voltage of a Piezoelectric Transformer Off-Line Converter / S. Hamamura, T. Ninomiya, M. Yamamoto, M. Katsuno // IEEE TRANSACTIONS ON POWER ELECTRONICS. - 2003. - Vol. 18, no. 1. - pp. 270-277.

[4] Zeman, S.K. Pulse-code method for regulating the technological parameter of the frequency converter of the induction heating installation // Bulletin of the Tomsk Polytechnic University. - 2007. - T.310, No. 1. - S. 191-196.

[5] Afanasyev, A.M. Synchronous time-pulse method of controlling a resonant inverter with sequential capacitive load compensation / A.M. Afanasyev, Yu.M. Golembiovsky, A.V. Kulikov // Reports of TUSUR. - 2018. - T.21, No. 1. - S. 95-99.

[6] Vasiliev AS Power sources for high-frequency electrothermal installations: monograph. T. 4 / A.S. Vasiliev, G. Konrad, S.V.Dzliev / Novosibirsk: Publishing house of NSTU, 2006.426 p.

[7] Influence of Magnetic Steel Induction Heating Power Density on Inductor Resistance Range / S. Dzliev, K. Pishchalev // Proceedings of the 2019 IEEE Conference of Russian Young Researchers in Electrical and Electronic Engineering (ElConRus), Spb, 2019.- pp ... 956-959.

[8] Abdulkhakov, I.Yu. Increasing the noise immunity of transistor generators with voltage inverters with a strong magnetic coupling of induction systems / I.Yu. Abdulkhakov // Izvestiya SPbGETU "LETI". - 2018. - No. 1. - S. 55-62.

[9] Demidovich V. Andrushkevich V, Perevalov Yu, Induction heat treatment of large diameter pipes and coils // 9th International Symposium on Electromagnetic Processing of Materials (EPM2018): IOP Conf. Series, 2018. - p.1-4. doi:10.1088/1757-899X/424/1/012064

IOP Conf. Series: Materials Science and Engineering **950** (2020) 012024 doi:10.1088/1757-899X/950/1/012024

Model of power quality system using electric meter suppressor of fire and electric damage and hazards in the residential sector

N V Korneev[1,2,4], Y V Prus[1], A V Perikov[1], N M Smolenskaya[3], V V Smolenskii[3]

[1] Gubkin Russian State University of Oil and Gas, Moscow, Leninsky Prospect, 65, Russian Federation

[2] Financial University under the Government of the Russian Federation, Moscow, Leningradsky Prospekt, 49, Russian Federation

[3] Togliatti State University, Togliatti, Belorusskaya st. 14, Russian Federation

[5] Email: niccyper@mail.ru

Abstract. The analysis, detailing, theoretical and technological justification of the concept of fire and electric damage. It is proposed to use the electric meter suppressor of fire and electric damage and fire hazards in the residential sector. The cognitive model taking into account possibilities of influence of the electric meter of the suppressor on decrease in damage for the electrical reasons and the functional model of electric power quality system are constructed. The cognitive model is considered as one of the components describing the external impact in the neurograph as a model to support control over the comprehensive objects safety for reducing fire risks in the residential sector associated with electrical causes of fires.

1. Introduction

Fires in the residential sector lead to numerous human losses, cause significant material damage. According to statistics, the residential sector accounts for up to 80% of the total number of fires, occurring annually in Russia. About 90% of dead in fires are also in the residential sector, the main causes of death are combustion products (up to 76% of deaths) and high temperature (up to 19%).

There is a correlation between the amount of fire damage in the residential sector with the corresponding quantity of electricity consumption. A significant number of fires in the residential sector are caused by various electrical reasons (Figure 1). Detailed statistical data detailing on sources and damage from fires in the residential sector allows to identify the main sources, among which the most common are various electrical devices and electrical networks equipment (Figure 2).

From the above it follows that in order to reduce the damage caused by fires in the residential sector, it is necessary to reduce the risk of fire break-out for the electrical reasons.

According to probabilistic-physical models of reliability, quality and safety [1-8], a significant risk reduction of fires for the electrical reasons requires the implementation of various measures related to two technological areas which ensure:

1. Reducing the fire hazard of electrical appliances.

IOP Conf. Series: Materials Science and Engineering **950** (2020) 012024 doi:10.1088/1757-899X/950/1/012024

2. Improving the quality of the consumed electricity.

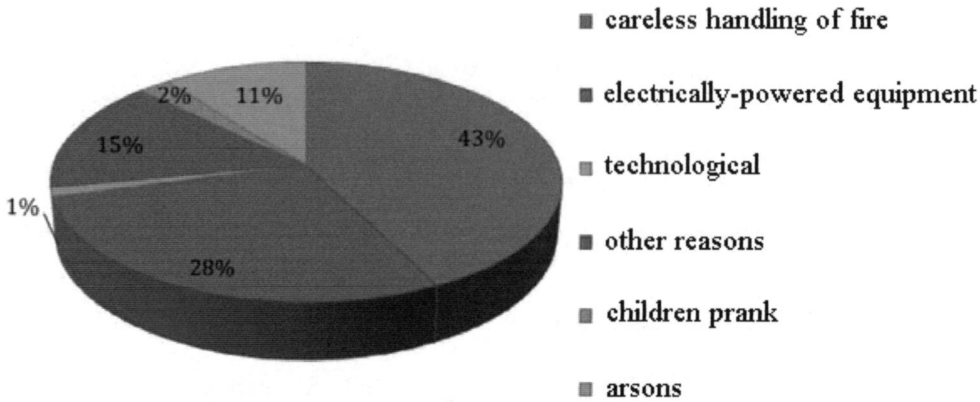

Figure 1. The main causes of fires in the residential sector

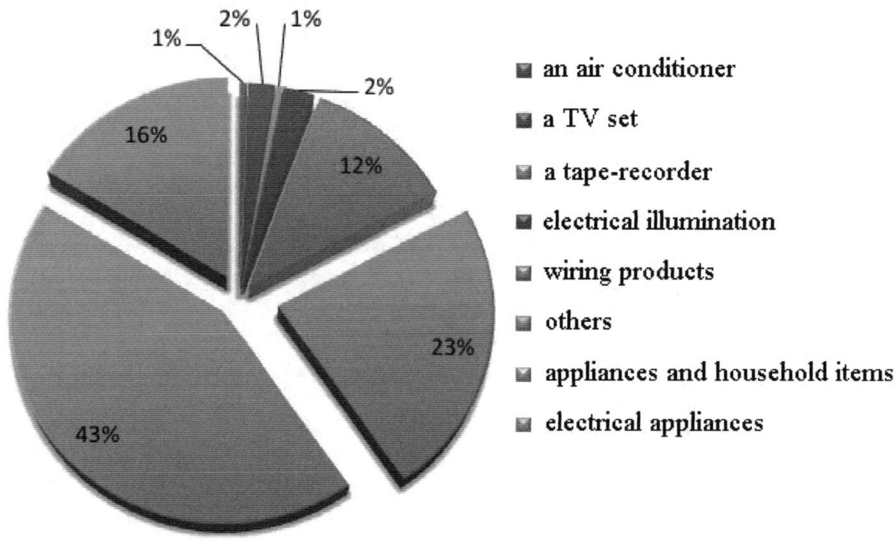

Figure 2. Sources of fires for the electrical reasons

Nowadays, the development of risk-oriented methods and models of security management in sociotechnical systems are well-spread [8-10]. When applying the risk-oriented method to decrease the fire damage for the electrical reasons we should use the concept of fire and electric damage (FED) based on the complex factors study that affect the level of fire hazard while using electrical appliances [1].

2. Results and Discussion

In [1] FED is considered as a phenomenon, caused by a negative change in the quality of electrical appliances and wiring products at the facility for the conversion/transmission of high and low-quality electricity. As a result of the development of physical degradation processes in materials and components, the period of fireproof recourse of the electrical equipment is reduced that leads to the probability of fire for the electrical reasons occurrence.

Further development of ideas about the nature of the FED phenomenon led to the original interpretation change. In [1] investigated the connection of fire hazard of electric equipment with quality indicators of consumed electricity. As the electricity consumption with invalid quality indexes leads to a significant increase in the probability of fires for the electrical reasons. It is proposed to differentiate consumed electricity by quality indicators using electricity meters modified with additional functions.

Meanwhile, in terms of "low-quality electricity identification" [1] it is proposed to use a "statistical method" to determine the power consumption, in which for a specified time interval the deviations established by GOST 13109-97 values are defined (when measuring and digitizing alternating voltage curve). The consumed electricity is differentiated according to certain algorithms:

$$W = W_a + W_u \quad , \qquad (1)$$

where W – the total amount of electricity distributed to a consumer during the time T; W_a – the amount of consumed electricity with values within acceptable limits; W_u – the amount of consumed electricity with deviation parameters outside the acceptable limits.

According to [1], FED can be identified while measuring the amount of consumed electricity with its differentiation into acceptable and unacceptable components of quality assurance established by GOST 13109-97. Taking into account various probabilities of fires for the electrical reasons during the time T, FED is represented by the sum of fractions of consumed electricity components:

$$\text{FED} = P_a \times W_a + P_u \times W_u \quad , \qquad (2)$$

where P_a – a probability of fire for the electrical reasons with acceptable deviation parameters of electricity; P_u – a probability of fire for the electrical reasons with unacceptable deviation parameters of electricity.

According to [1], W_a and W_u are registered in the process of continuous monitoring of the consumed electricity quality, and the probabilities P_a and P_u can be identified based on the statistic data analysis of fires for the electrical reasons.

In this interpretation, the cause of the fire is electrical equipment, due to the electricity consumption with both: valid and invalid quality parameters. In this case, the components of FED have a dimension (kW) and are proportional to the corresponding amount of consumed electricity.

Accepting the undoubted usefulness of the proposed approach to the study of the fire hazard nature of electrical appliances operation and the significance of the concept of FED introduced in [1] , it is necessary to emphasize, that the above interpretation of FED does not provide the possibility of its practical use for solving problems of fire risks for the electrical reasons. This is due to the fact, that authors [1] they did not offer a practical method for constants determination P_a and P_u from (2). Moreover, the dimension of the defined in accordance with (2) FED is in the logical contradiction with the original meaning of this rational concept that reflects the level of fire risk and is inextricably connected with the expected material losses from fires for the electrical reasons. It should be noted, that in the considered interpretation FED is a damage, directly related to electrical equipment, it isn't determined as a damage to the facility where it is operated.

However, it follows from the above discussion that so far FED indicator has a contradictory and blurred character for practical use when controlling fire risks for the electrical reasons. For this purpose, the identification and numerical evaluation of FED require clarification, detalization, theoretical and technological justification.

Based on the meaning of the notion of harm as a synonym for damage to a facility, FED parameters must be related to the extent of the expected material damage from fires for the electrical reasons within a certain time interval. At the same time, the damage must be measured in monetary terms.

For FED's representation in a monetary format it is necessary to move from the probabilistic characteristics of the fire for the electrical reasons within a certain time interval P_a and P_u to indexes, related to the specific amount of expected material damage. A specific damage expected from fires for the electrical reasons is assessed by the value of material damage expressed in a monetary term per unit of consumed energy; meanwhile, the consumed energy can be differentiated by quality characteristics. Two options can be specified for introducing such indicators.

The first option is: FED identification when setting the average fire damage indexes $\langle U \rangle$ and converting (2) to:

$$\text{FED}=\left(W_a \times P_a^s + W_u \times P_u^s\right) \times \langle U \rangle \quad , \tag{3}$$

Where P_a^s – the specific probability of fire in the electrical reasons with acceptable deviations of electric power parameters (1/kW); P_u^s – the specific probability of fire in the electrical reasons with unacceptable deviations of electric power parameters (1/kW).

It is possible to offer more convenient interpretation (3), highlighting the works of medium damage from a fire at the specific probability and considering them as specific indicators K_a and K_u, characterizing the corresponding values of expected damage per unit of consumed high-quality and low-quality energy:

$$K_a = P_a^s \times \langle U \rangle, \;\; K_u = P_u^s \times \langle U \rangle \quad , \tag{4}$$

We introduce these specific rates in (3), and then FED is represented as:

$$\text{FED}=W_a \times K_a + W_u \times K_u \quad . \tag{5}$$

As can be seen in (5), FED represents the sum of fire and electrical damage components, connected with electrical appliances exploitation with acceptable and unacceptable characteristics of the consumed electricity quality:

$$\text{FED}_a = W_a \times K_a, \;\; \text{FED}_u = W_u \times K_u \quad . \tag{6}$$

The usability of (4) specific indicators is obvious, however, it is necessary to assess an average fire damage to set them, as well as determine the specific probabilities of (3) fire for the electrical reasons.

The second option is: direct determination of specific indicator quantity of FED K_a and K_u, characterizing the extent of expected damage, accounted for one unit of consumed high and low-quality energy, based on the summary statistics of consumed electricity volume with differentiation by quality characteristics and statistics of fire damage for the electrical reasons.

Each of the specific indicators of FED is identified by the damage change ΔU from fires for the electrical reasons, caused by the corresponding increase in electricity consumed:

$$K_a = \frac{\Delta U}{\Delta W_a}, \;\; K_u = \frac{\Delta U}{\Delta W_u} \quad , \tag{7}$$

by $\Delta W_a \to 0$ and $\Delta W_u \to 0$.

Further development of the above approach involves such measures as monitoring of consumed electricity quality, also, statistics data accumulation of fire damage for the electrical reasons.

IOP Conf. Series: Materials Science and Engineering **950** (2020) 012024 doi:10.1088/1757-899X/950/1/012024

When implementing these measures, from the analysis of fire statistics and electricity quality, it is possible to directly assess local fire and electric damage. While operating a certain set of electrical equipment, if electricity consumption data is available at various points in time as well as damages from fires for the electrical reasons from the system:

$$FED=W_a \times K_a + W_u \times K_u ;¿¿¿¿ \qquad (8)$$

it is easy to determine specific indicators of FED:

$$K_a=\frac{FE\{D' \times W_{a'} - FED \times W_a}{W_{a'} \times W_{u'} - W_a \times W_u} ¿;¿¿¿¿ \qquad (9)$$

Monitoring the quality of consumed electricity through the use of electric meters of a special type with the functions of suppressing fire electric damage and fire hazards (ESSFED and FH) creates a technological and organizational base for the electric power quality management system formation (EPQMS) in the residential sector.

This article proposes the definition of local EPQMS as sets of organizational structure, responsibility, procedures, processes, and resources ensuring the implementation of general quality management of the consumed electricity in a certain area.

The interaction of factors affecting the occurrence and the change of FED can be analyzed using the cognitive modeling tools (Figure 3).

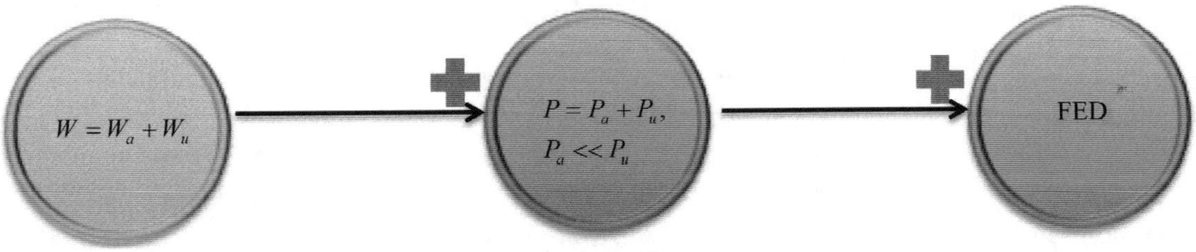

a)

IOP Conf. Series: Materials Science and Engineering **950** (2020) 012024 doi:10.1088/1757-899X/950/1/012024

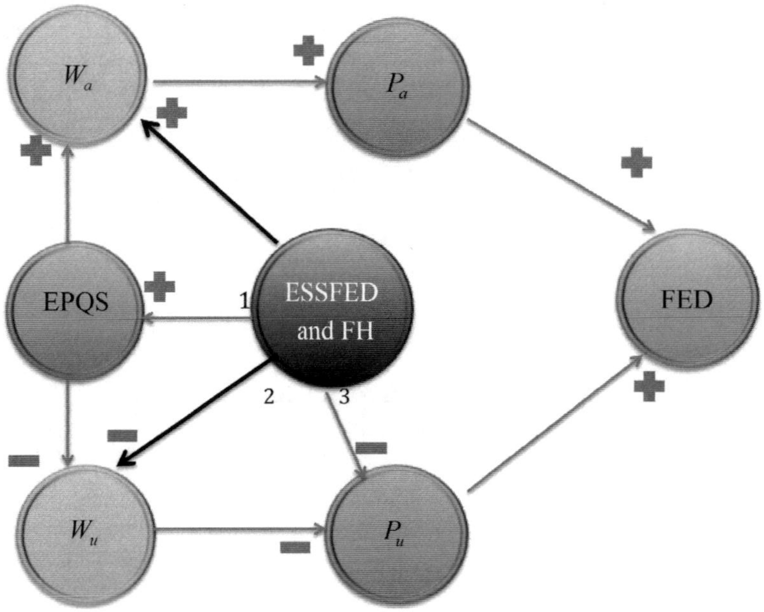

b)

Figure 3. A cognitive model, that includes the possibility of influence of ESSFED and FH on decrease of fire damage for electrical reasons: (a) if there is no accounting for the amount of consumed electricity with valid and invalid quality indicators; (b) when monitoring the consumed electricity quality through the application of ESSFED and FH

If there is no accounting for the amount of consumed electricity with valid W_a and invalid quality indicators W_u, FED it can only be connected with the total electricity consumption and the probability of fires for the electrical reasons (Figure 3a).

When monitoring the consumed electricity quality based on the use of ESSFED and FH there is a number of possibilities for influencing FED, that's shown by cognitive modeling (Figure 3b).

The most significant possibility is a differentiation of the total amount of consumed energy W by quality characteristics, that is the basis for a number of measures aimed at improving the quality of supplied electricity to consumers.

As shown by the corresponding arcs of the cognitive graph (Figure 3b), ESSFED and FH, in addition to monitoring the consumed electricity quality that is mentioned above (function 1) can fulfill the additional function of suppressing FED [1] using the built-in adjustment mechanisms in the device. There can be some improvement in the quality of consumed electricity as a result of the function 2 ESSFED and FH compensation of deviations from acceptable values of power quality. If fire hazard conditions occur and the compensation is not possible, ESSFED and FH fulfill the function 3 of emergency disconnection of electrical equipment and de-energizing the power supply network.

For the power quality management system formation (PQMS), along with technical tools introduction, it is necessary to integrate them into a single information system for electric power quality monitoring (ISEPQM). Further development of appropriate legal, economic, and organizational mechanisms based on local PQMS forms a united state-wide Electric Power Quality System (EPQS), which includes (Figure 4) functional subsystems:

1. Power quality standardization system (PQSS).
2. Power quality monitoring system (PQMS).

3. Fire and electric damage suppression system (FEDSS) and local automated microsystem diagnostics and protection of the residential sector (LAMDP) and information diagnostics, suppression and notification (IDSN).
4. Energy quality management system (EQMS).
5. Legal and regulatory system for ensuring electric quality (LRSEEQ).

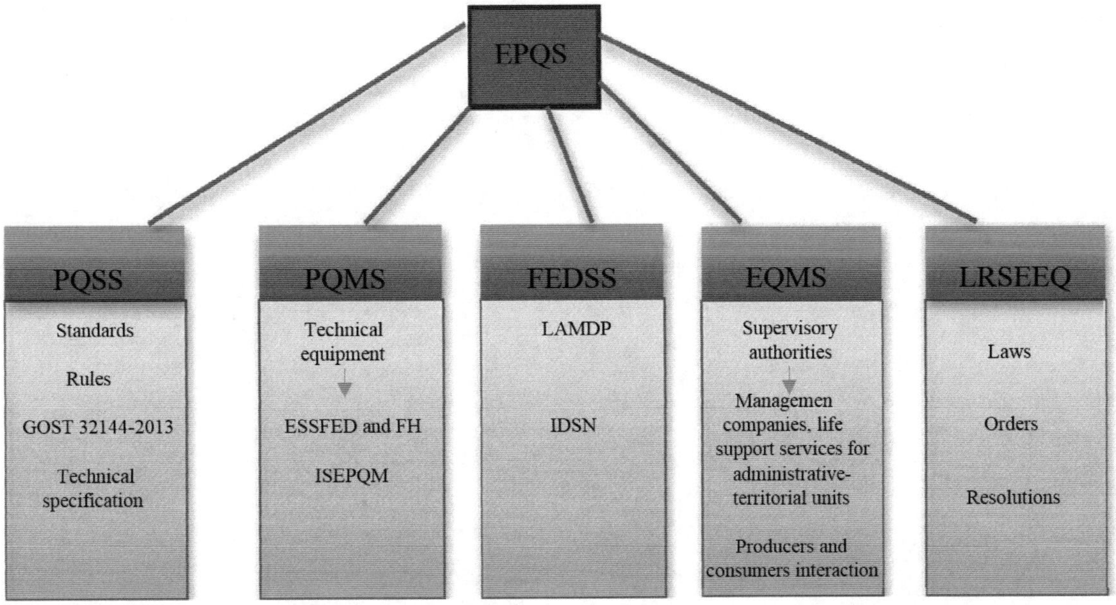

Figure 4. Functional structure of the united power quality system

As can be seen from the proposed model EPQS, its main functions are related to the impact on management, power supply and energy inspection companies with the help of appropriate economic and legal mechanisms. As a result of this influence, energy retail and energy inspection companies will be interested in ensuring the proper quality of electricity supplied to consumers.

3. Conclusion

ESSFED and FH integration in the residential sector with the further development of EPQS, representing the total sets of organizational structure, responsibilities, procedures, processes, and resources, ensuring the implementation of general management of the consumed electricity quality, will allow to create effective technological, economic and legal mechanisms to reduce one of the main components of fire risks for the electrical reasons in the residential sector. For further scientific research, the proposed cognitive model is considered as one of the components describing the external impact in the neurograph as a model to support control over the comprehensive objects safety [8] to reduce fire risks in the residential sector associated with electrical causes of fires.

References

[1] Belozerov V V 2018 *Electronics and Machinery* **3** 17–50
[2] Ashu G and Paul S 2004 *Proc Winter Simulation Conference* (Catonsville: INFORMS) pp 866-873
[3] Volkan U and Jeffrey S 2008 *Proc Winter Simulation Conference* (Catonsville: INFORMS) pp 879–885
[4] Garcia M L 2008 *Design and Evaluation of Physical Protection Systems* (Oxford: Butterworth-Heinemann) p 370

[5] Fennelly L 2016 *Effective Physical Security* (Oxford: Butterworth-Heinemann) p 458
[6] Thompson P and Marchant E 1995 *Fire Safety Journal* **24(2)** 131–148
[7] Owen M, Galea E R and Lawrence P J 1996 *Journal of Fire Protection Engineering* **8(2)** 65–84
[8] Korneev N V 2019 *IOP Conf. Ser.: Earth Environ Sci.* **224** 012021
[9] Biagini C and Ottobri P 2018 *Proc 18th International Conference on Geometry and Graphics* (Milan: Springer) pp 2163–2166
[10] Brushlinskiy N N, Sokolov S V and Grigoryeva M P 2017 *Fire and Explosion Safety* **26(2)** 72–80.

Numerical simulation and investigation of induction system for hardening of internal combustion engine camshaft

T Kozulina, I Fomin, S Galunin and M Kudryash

St. Petersburg Electrotechnical University,
Prof. Popova St. 5, 197373 St. Petersburg, Russia

E-mail: tpkozulina@mail.ru

Abstract. This paper describes the numerical simulation and investigation of the electrothermal processes in the induction system for hardening of cams and journals of camshaft. Hardening of the camshaft elements is carried out by one simple inductor of round shape due to shift of the camshaft axis inside it. During the process of the hardening of the cams, the camshaft is stationary, during the process of the hardening of the journals it rotates. The results of the investigations are presented in the form of the temperature distribution in the cam and the journal. The optimum parameters of the induction system are also given, at which the best temperature distribution in the elements of the camshaft is achieved.

1. The camshaft hardening problem

A camshaft, which consists from cams and journals, is very important detail of any internal combustion engine. The camshaft is used to operate engine valves, which give fuel mix or let out the fulfilled. The number of camshaft cams is equivalent to number of engine valves. As a result there is an individual cam for each valve, which opens it, when runs on a pusher lever of valve. When the cam "runs" from the lever, the valve closes because of force of returnable spring.

The journals of a camshaft rotate in bearings, and are exposed to destruction as well as cams, that's why a surface of these camshaft parts should have a high hardness. This property can be achieved by the way of induction surface hardening, which consists in preheating the workpiece to a temperature above the Curie point and its following cooling to temperatures of martensitic transformations. Induction surface hardening has a number of advantages over other methods of hardening workpieces with complex geometry [1].

The main problem of induction hardening of camshafts is uneven distribution of hardness on a surface of its elements and defect of cams after cooling.

The existing standard hardening methods have their disadvantages. The most broadcast method for hardening camshafts is when the shaft is inside the inductor, their axes coincide, and the shaft rotates. All shaft elements are hardened with one inductor. With this heating method surface of cam unevenly heats up in zones of a nape and a nose, and journal surface badly gets warm because of a big gap between a workpiece and inductor. As a result we get the low depth of the hardened layer, big thermal losses and low efficiency of process.

2. Numerical simulation of induction system for hardening a camshaft

Within the framework of this work, a solution to the problem of hardening cams and bearing journals of camshaft is presented using numerical simulation in ANSYS® Software.

IOP Conf. Series: Materials Science and Engineering **950** (2020) 012025 doi:10.1088/1757-899X/950/1/012025

This solution consists in:

- development of a 3D electrothermal model of the inductor-workpiece system;
- determination of the optimal position of the shaft axis about the inductor axis when various shaft elements hardened;
- optimization of the inductor geometry.

The developed induction system provides an uniform distribution of the hardened layer depth and hardness on the surface of the cams and the bearing journals of the camshaft.

Heating during hardening of the cams and bearing journals is carried out using one simple inductor by shifting the shaft axis inside the inductor. When the cams are heating, the shaft is stationary; when the bearing journals are heating, the shaft rotates.

The use of numerical simulation to solve a similar problem is shown on the example of a camshaft with three pairs of cams, which are located at an angle of 120 degrees relative to each other.

Figure 1 shows the parameters of 1/3 part of the camshaft, because the shaft consists of three symmetrical parts that differ only in the angle of the cams rotation.

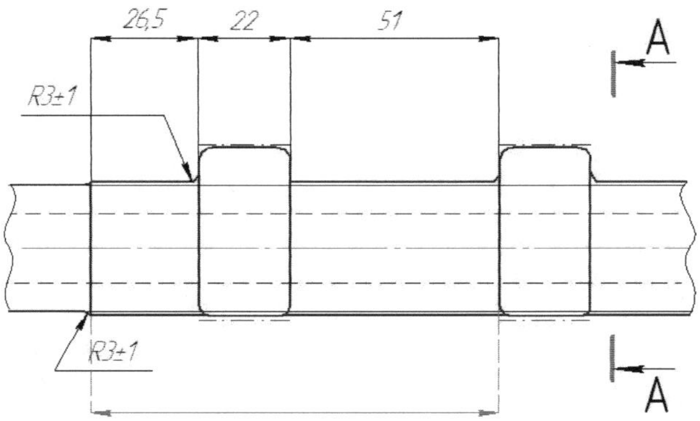

Figure 1. The parameters of 1/3 part of the camshaft

The position of the cams pairs relative to each other and the geometric parameters of the cam are shown in Figures 2 and 3, respectively.

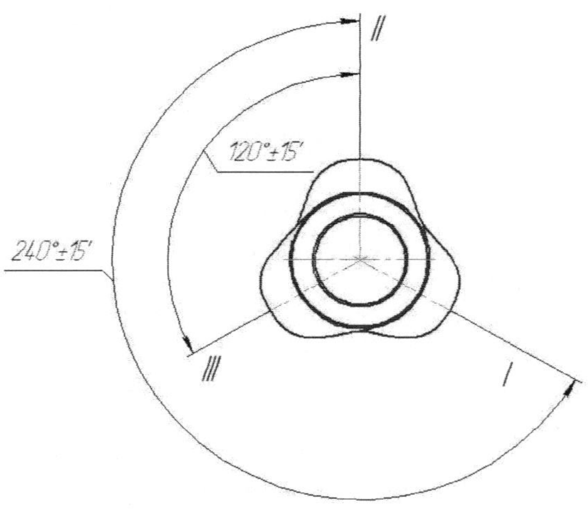

Figure 2. The position of the cams pairs relative to each other

IOP Conf. Series: Materials Science and Engineering **950** (2020) 012025 doi:10.1088/1757-899X/950/1/012025

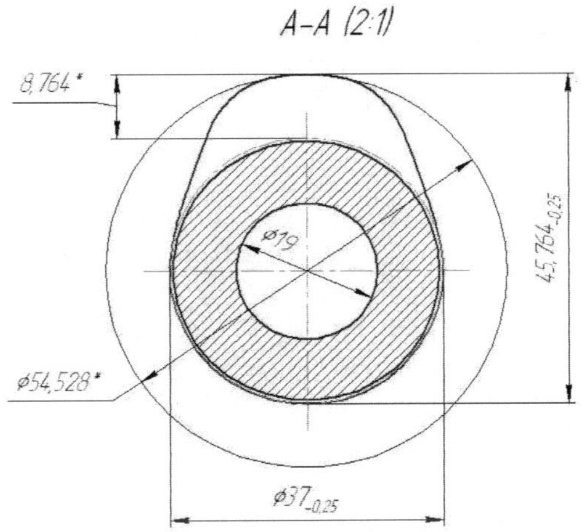

Figure 3. The parameters of the cam

2.1. Numerical simulation of induction heating of the camshaft cam

Using the approaches described in the literature [2, 3], a 3D model of the cam-inductor system was created in the ANSYS® Software, and then a numerical model of this system was developed on its basis. The cam is located inside the inductor, the inductor protrudes 6.5 mm from both sides of the cam along the Z axis. The 3D model of the cam-inductor system is shown in Figure 4.

Figure 4. The 3D model of the cam-inductor system

A single-turn round inductor has the following initial parameters:

- copper tube of rectangular profile 35×10 mm with a wall thickness of 2 mm;
- inner radius of the inductor – 85 mm.

Figure 5 shows a numerical model of an induction system.

IOP Conf. Series: Materials Science and Engineering **950** (2020) 012025 doi:10.1088/1757-899X/950/1/012025

Figure 5. The numerical model of the cam-inductor system

Electrothermal calculations were carried out for several options for the location of the cam inside the inductor at a frequency of 2.4 kHz and an initial inductor current of 13 kA. In the course of research, the cam was displaced along the X and Y axes inside the inductor and changed the angle of rotation about the Z axis. Also, such system parameters as the height of the inductor and the current in the inductor was changing.

Based on the results obtained, presented in the form of the temperature distribution in the cam, it was found that the optimal heating of the cam occurs with the following parameters of the induction system:

- the shaft axis is displaced relative to the inductor axis by 2 mm in negative directions along the X and Y axes;
- the cam is rotated 90 degrees relative to the inductor current leads;
- inductor height – 33 mm;
- current in the inductor – 11.6 kA;
- heating time – 5 sec.

The results of numerical simulation of induction surface heating process of the cam at the optimal parameters of the induction system are shown in Figure 6 as a temperature distribution.

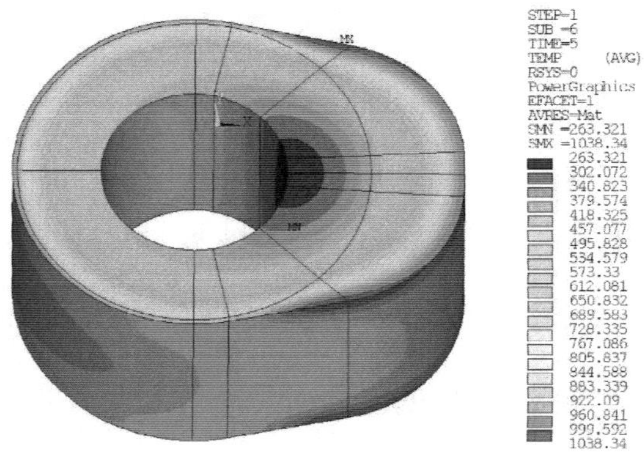

Figure 6. Optimum temperature distribution in the cam

The temperature difference between the regions of the back and the nose of the cam is about 40 degrees, and the overheating of the nose corners is insignificant, which together is the best result among all the calculations performed.

IOP Conf. Series: Materials Science and Engineering **950** (2020) 012025 doi:10.1088/1757-899X/950/1/012025

2.2. Numerical simulation of induction heating of the camshaft bearing journal

Further, a numerical simulation and investigation of induction heating process of the bearing journal of this camshaft in a copper inductor was carried out, the optimal geometry of which was calculated earlier. The purpose of these investigations was to determine the optimal parameters of the induction system at which the bearing journal is uniformly heated to a temperature of 850-880 °C up to 2 mm in depth.

The inner diameter of the bearing journal is 19 mm, the outer diameter – 34 mm, and the height – 30 mm. A single-turn inductor of a round shape is made of a rectangular profile of 33×10 mm with a wall thickness of 2 mm, the outer diameter of the inductor is 85 mm.

The bearing journal is located inside the inductor, the inductor protrudes 1.5 mm from both sides of the bearing journal along the Z axis. Figures 7 and 8 show a 3D model and a numerical model of the bearing journal-inductor induction system.

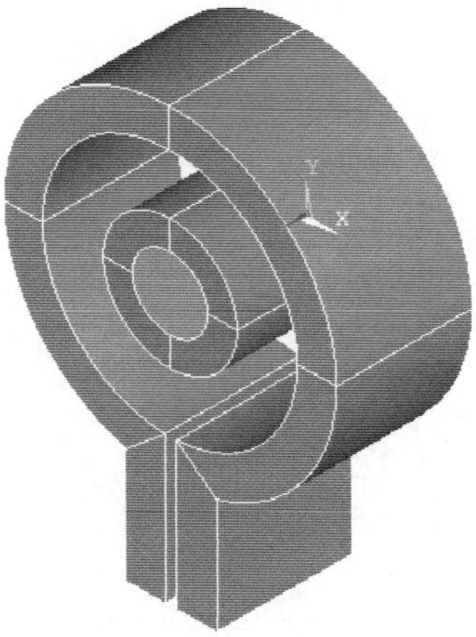

Figure 7. The 3D model of the bearing journal-inductor system

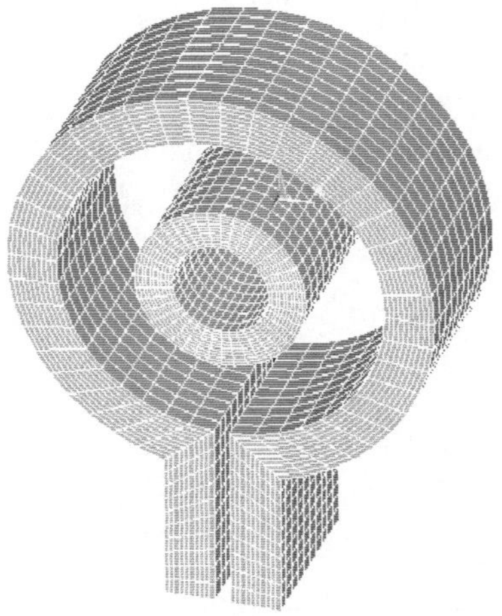

Figure 8. The numerical model of the bearing journal-inductor system

Electrothermal calculations were carried out for several options for the location of the bearing journal inside the inductor. During the research, the bearing journal was displacing relative to the X and Y axes inside the inductor, while the effect of the shaft rotation about the Z axis was simulated.

According to the research results obtained in the form of the temperature distribution in the bearing journal, it was found that the camshaft bearing journal heats up evenly to a depth of 2 mm to a temperature of 850-880 °C at the following parameters of the induction system:

- displacement of the shaft axis inside the inductor by 2 mm to the current leads;
- current strength – 11.2 kA;
- current frequency – 2.4 kHz;
- heating time – 9 sec.

This heating mode provides a uniform depth of the hardened layer and a uniform distribution of hardness on the surface of the bearing journal due to its rotation and shift of the shaft axis inside the inductor.

The results of numerical simulation of the induction surface heating process of the camshaft bearing journal at the optimal parameters of the induction system are presented in Figure 9 in the form of a temperature distribution.

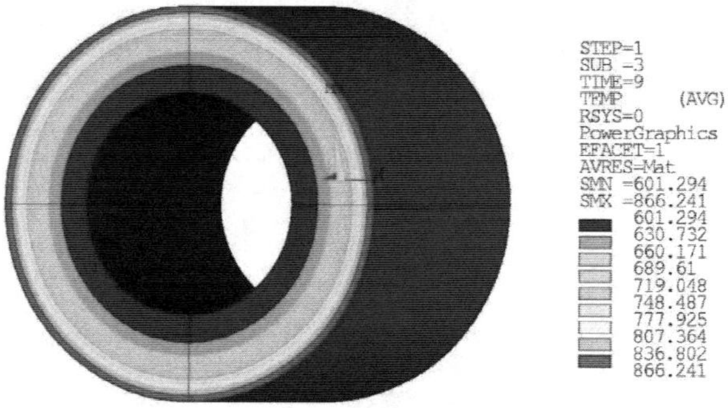

Figure 9. Optimum temperature distribution in the bearing journal

3. Conclusions and outlooks

Numerical simulation significantly reduces the time for developing real induction installations, as well as for investigations and debugging of the induction heating and hardening processes of workpieces with complex geometry.

The obtained results of numerical simulation of induction heating of camshaft elements can be used as a basis for investigations and developing of induction installations for hardening camshafts of various configurations.

References

[1] Kozulina T, Galunin S, Nikanorov A and Nacke B 2017 Numerical simulation and optimization of induction surface hardening systems with rotating workpiece *Proc. of XVIII International UIE-Congress Electrotechnologies for Material Processing (Hannover)* pp 514–519

[2] Schlesselmannn D, Nacke B, Nikanorov A and Galunin S 2015 Coupled numerical multiphysics simulation methods in induction surface hardening *6th International conference on computational methods for coupled problems in science and engineering COUPLED PROBLEMS (Venice)* pp 392–403

[3] Galunin S, Kudryash M, Blinov K and Kozulina T 2015 Numerical simulation and optimization of heater for induction hardening of leading wheel gear *2015 IEEE North West Russia section young researchers in electrical and electronic engineering conference ELCONRUSNW (St. Petersburg)* pp 192–195

IOP Conf. Series: Materials Science and Engineering **950** (2020) 012026 doi:10.1088/1757-899X/950/1/012026

Numerical calculation of large-sized objects induction heating

I. Smolyanov, F. Tarasov, E. Shvydkiy

Ural Federal University, Yekaterinburg, Russia

E-mail: i.a.smolianov@urfu.ru, F.E.Tarasov@urfu.ru, e.l.shvydky@urfu.ru

Abstract.
The paper describes a numerical procedure for modeling induction heating of large objects. A coupled calculation of the electromagnetic and thermal fields is performed by the finite element method in a three-dimensional formulation. The setting of the computational mesh is described. The results of numerical calculations are demonstrated.

1. Introduction

In many industrial processes, there is a need for uniform heating. For example, during the hot stamping light-alloy parts for the aircraft industry, it is necessary to preheat the stamping tool [1]. This tool is a steel part weighing 10-20 tons. Previously, this problem was solved by simple heating in resistance furnaces. However, this method takes a very long time and is accompanied by large heat losses.

It was proposed to use induction heating for this technological operation [2, 3]. Induction heating, compared to resistance heating, is characterized by a high heating rate and efficiency, since energy is released directly into the object [4].

The uniformity of heating in this case can be achieved by selecting the inductor geometry and the power supply parameters (power, frequency). Coupled electromagnetic and thermal analysis for the selection of these parameters is performed either analytically or numerically. However, induction heating provokes the skin effect. This effect is described by the fact that the electromagnetic field and Joule heat realizing are concentrated in a surface layer called the penetration depth. In the case of heating large-sized objects, the scale of the object and the depth of penetration can differ dramatically [5, 6] and can be a problem for numerical calculations. This work describes the features of the numerical procedure for calculating induction heating of large scale objects.

2. Experimental setup

Considering installation is shown in the Fig 1. The unit consist of the stamp tool (orange color), the steel plate (magnet color) and the base plate (light green color). The general part of the unit is the stamp tool. Temperature of the part should be acquired up to 450-550 degree Celsius, depending on the type of heated product, and maintained to the value during whole working day (about 21600 second). Particularly important is the upper part of the die, as it is in contact with the workpiece. It can be finded more information in works [2]. To maintain the set temperature

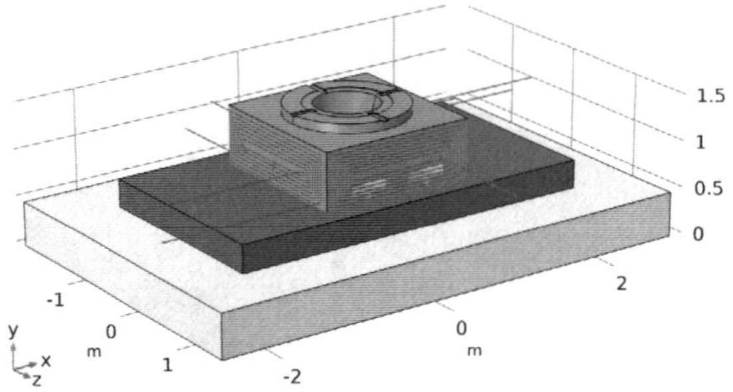

Figure 1. Geometry of stamp induction heating installation

level, induction heating is carried out using four inductors (dark green color) on each side of the stamping tool. The maximum power for each pair of inductors is 30 kW and depends on the source.

The inductor is a specially shaped water-cooled coil made of copper tube (electrical conductivity $6e7$ S/m). The steel (magnet) and base (light green) plate and stamp tool (orange) made of difference kind of steel, but it can be neglected because of weak influence on temperature field. The properties of the parts have following values, the specific heat capacity is 440 J/(kg·K), the mass density is 7870 kg/m^3, the thermal conductivity is 76.2 W/(k·m), the electrical conductivity is $6.25e6$ S/m. The thermal insulation maded of kaolin is set between stamp tool (orange) and coils (dark green). The properties of the insulation following: thermal conductivity is 0.13 W/(K·m), the mass density is 300 kg/m^3, the specific heat capacity is 1000 and the thickness of the insulation is 20 mm.

3. Numerical model

The problem is multyphysics and have difference values of time derivativeties for each task. The term task here means the solution of the field equation in partial derivatives. The problem consist of magnetic and temperature fields. The magnetic field serves for to produce energy suggested for transforming to heat. The field changes its value over time, but the size of calculating domains is much less than wavelength of the field. Therefore, the equation (1) can be rewriten as (2). This approach will make it possible to rationally use the computing resources of a computer, since the estimated heating time is about 21600 seconds, and the frequency of the magnetic field is about 10 kHz. For the correct resolution of this equation, assuming that the source of the magnetic field changes according to the harmonic law in time, it will be necessary to recalculate the equation at least 10 times per period. From this it follows that the minimum step of the magnetic field equation solver is 10 μsec. This makes it practically impossible to use the dynamic equations of the magnetic field. Therefore, it is advisable to use in this study the harmonic equation of the magnetic field, where the time derivative is replaced with the help of the complex Laplace term.

$$curl\left(\frac{1}{\mu}curl\mathbf{A}\right)+\gamma\left(\frac{\partial\mathbf{A}}{\partial t}-\mathbf{v}\times curl\mathbf{A}\right)=\mathbf{J_{ext}} \tag{1}$$

$$curl\left(curl\underline{\mathbf{A}}\right)=j\mu\omega\gamma\underline{\mathbf{A}}=\mu\underline{\mathbf{J}}_{\mathbf{ext}} \tag{2}$$

Here, μ denotes the magnetic permability, γ is the electric conductivity, $\mathbf{v}$ is the vector of velocity, $\mathbf{J_{ext}}$ represents the current density, $\mathbf{A}=rot\mathbf{B}$ is the vector magnetic potential.

The temperature distribution is calculated only in the area of the instrument. Heat removal to the environment is modeled using the heat transfer coefficients obtained experimentally.The temperature field in the heated stamp tool due to magnetic field is described by the heat transfer equation in the general form

$$div(\lambda gradT) = \rho C_p \left(\frac{\partial T}{\partial t} + \mathbf{v} \cdot gradT \right) - w \tag{3}$$

Here, (λ denotes the thermal conductivity, ρ denotes the mass denesty, c_p is the specific heat at constant pressure and w stands the volumetric internal sources of heat in the stamp tool. The volumetric heat sources w are represented by Joule losses w_j and losses arising from the phenomenon of hysteresis w_h. The latter of them can be neglected in this study because of their insignificant significance. Joule heat can be calculated using the well-known formula

$$w_j = \frac{|\mathbf{J_{ind}}|^2}{\gamma}, where \mathbf{J_{ind}} = j \cdot \omega \gamma \underline{\mathbf{A}} \tag{4}$$

Here the induced current density is defined as $\mathbf{J_{ind}}$. The Joule heat is determined based on calculations obtained in a magnetic field and comparing them with experiment. It should be noted that the investigated magnetic field is not inert with respect to the temperature field, in other words, changes in the temperature distribution in the investigated region do not play a significant role in the values of the magnetic field. Therefore, in this study, the magnetic problem is calculated once from which the calculation of the thermal energy arising in the heated part is made. This value is considered the maximum value of the heat energy released in the load. During the calculation, the value of thermal energy in the die tool is varied using a non-linear coefficient, which depends on the temperature of the workpiece at a certain point and the maximum temperature in the workpiece. This control mechanism is similar to the principle of energy control with a P-controller. All these equations are solved using the finite element method in a three-dimensional formulation in the Comsol Multiphysics numerical package.

One of the most difficult tasks in solving this problem is the correct grid setup. The dimensions of the structure are meters, and the penetration depth of the magnetic field is millimeters. This disproportion should be solved in an optimal way, so that the accuracy of the calculations corresponded to the engineering concept, and the computing power was acceptable for an average computer. In part, this was solved in this work, the basic representation of the grid is shown in the figures 2 and 3. For clarity of representation of the computational grid, the studied regions of the inductor and air were suppressed in Figure 2. The inductor mesh is shown separately in Figure 3.

4. Results

The main goal of this work is to conduct a preliminary evaluation of the main physical phenomena occurring in this installation, which affect the efficiency of the die press. Also evaluate the correctness of the used mathematical models and assumptions for the analysis of the results. One of the first points for research is the influence of the dimensions of the installation, how to take them into account correctly, as well as the construction heterogeneity on the distribution of the temperature and magnetic fields.

In the process of designing an installation for heating a die tool, more than 20 types of inductor were considered, some of them are considered in [2, 3]. In this work, one of them was chosen at random to assess the distribution of fields in zones with complex geometry. To do this, we plot the distribution of magnetic induction on the surfaces of the installation (Figure 4). It should be especially noted that almost the entire field is concentrated in the area opposite the inductor. This leads to the fact that the central part of the stamping tool has a significant

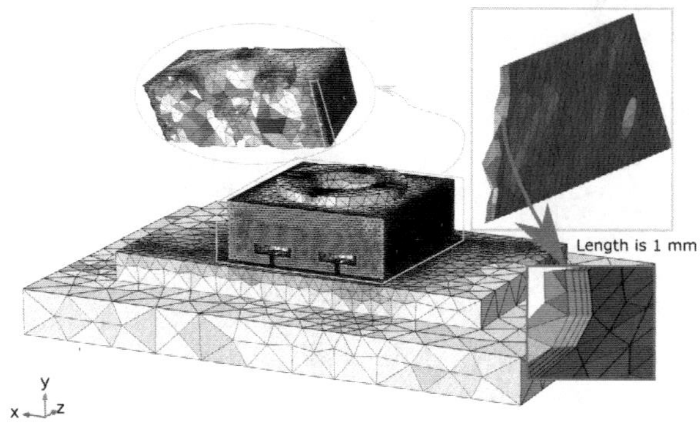

Figure 2. The computational mesh of the stamp installation.

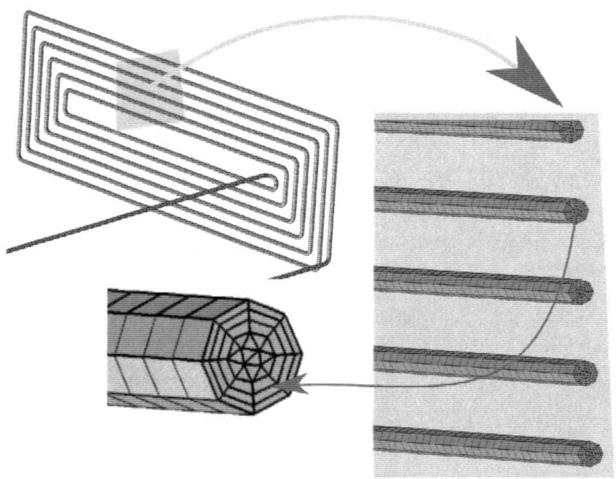

Figure 3. Numerical mesh of one of four inductors

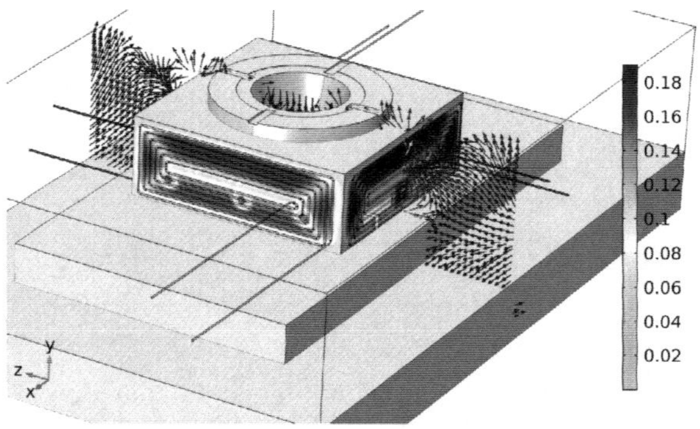

Figure 4. Normal Magnetic Flux Density on the surfaces of the unit

subsidence in relation to the value of the magnetic induction. Also, this phenomenon can be observed in places with uneven geometry (notches and gaps).

Therefore, it is worth evaluating the distribution profile of the released heat power in the die tool (Figure 5). Lowest heat dissipation in areas with low induction. These designs help to

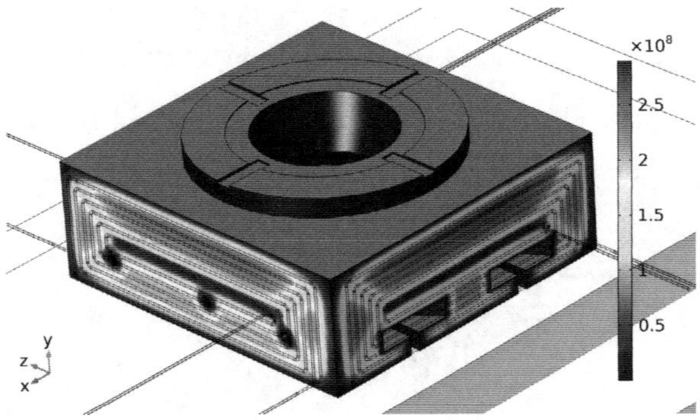

Figure 5. Volumetric Joule heat power, W/^{3} loss density

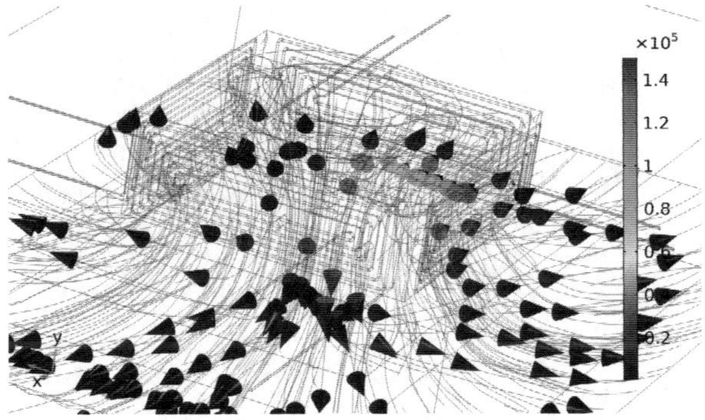

Figure 6. Streamlines of heat flow through the study area

optimize and redesign the inductor design to improve the efficiency of the die heating system.

In addition to the effect of the magnetic field on the efficiency of the test facility, the thermal component also makes a significant contribution. Namely, it is necessary to look through which parts of the study area the maximum outflow of thermal energy occurs. This distribution is shown in Figure 6 using color-coded streamlines. It can be seen that the maximum energy outflow occurs at the junction between the die tool and the steel plate, as well as into the plate itself. The temperature gradient is significantly greater between the environment and the punching tool than between the steel plate and the punching tool. This is due to the presence of a layer of kaolin wool between the inductor and the stamping tools, which significantly reduces heat losses through this surface. It should be noted in the future that a layer of thermal insulation can be added at the junction between the stamping tool and the steel plate in the corner part of it, which should lead to an increase in thermal efficiency.

A decrease in efficiency does not always lead to a high-quality implementation of the planned technological process. For example, in this case, the presence of a significant value of the temperature gradient leads to an increase in the phenomenon of heat transfer due to the thermal conductivity of the medium. This allows in some cases to level the uneven distribution of the magnetic field, heat sources due to the complexity of the design and to equalize the temperatures due to the phenomenon of thermal conductivity. This can be seen in the temperature profile in general, and in its various projections (Figure 7)

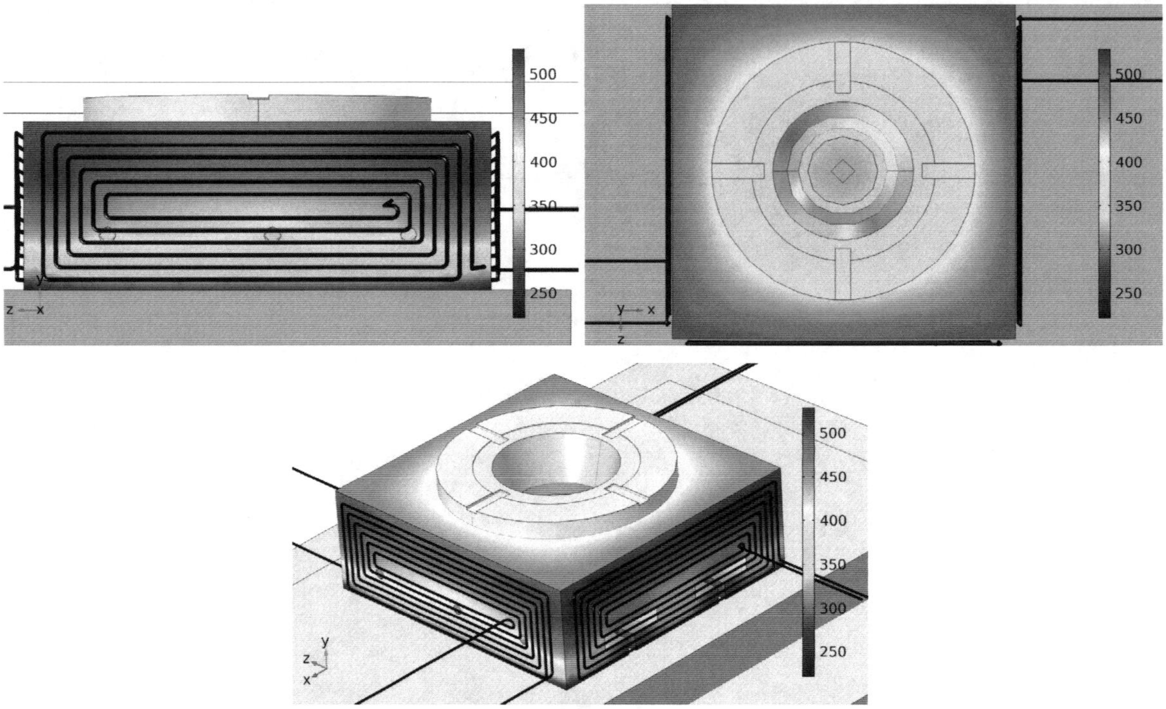

Figure 7. Temperature in the stamping punch

5. Conclusions

In this paper, the basics of modeling large-sized installations using numerical analysis, namely the finite element method, were considered. The main problem solved in this work was to qualitatively evaluate the main features of heating a die tool of large dimensions and complex shape using induction heating. In the work, the main points related to the unevenness of the magnetic field due to the unevenness of the space of the geometry of the die installation and the design of the inductor were considered. The main disadvantages and advantages of this heating approach and plant design were also discussed. Conclusions are drawn about the effect of the temperature gradient on the alignment of the temperature profile in the entire volume of the workpiece. In subsequent works, it is planned to consider systems for controlling the heating of a die tool, design optimization, as well as a number of numerical methods that will reduce the load on computing resources.

Acknowledgments

This work was financially supported by the Ministry of Science and Higher Education of the Russian Federation, project No. 075-11-2019-028.

6. References

[1] Karbasian H and Tekkaya A 2010 *Journal of Materials Processing Technology* **210** 2103 – 2118 ISSN 0924-0136 URL `http://www.sciencedirect.com/science/article/pii/S092401361000213X`

[2] Tarasov F E, Marinkova E O, Bychkov S A and Frizen V E 2017 Calculation of inductor for stamp tools heating systems *2017 IEEE Conference of Russian Young Researchers in Electrical and Electronic Engineering (EIConRus)* pp 1594–1599

[3] Bychkov S A, Tarasov F E, Frizen V E and Smolyanov I A 2017 Stamp tool induction heating system *2017 18th International Conference on Computational Problems of Electrical Engineering (CPEE)* pp 1–4

[4] Lupi S, Forzan M and Aliferov A 2015 *Induction and direct resistance heating: Theory and numerical modeling*

[5] Demidovich V and Perevalov Y 2019 *Metallurgist* **63** 617–626
[6] Demidovich V and Perevalov Y 2019 *Russian Electrical Engineering* **90** 781–785

IOP Conf. Series: Materials Science and Engineering **950** (2020) 012027 doi:10.1088/1757-899X/950/1/012027

Research of a dual-frequency power supply system for induction crucible furnaces for melting ferrous and non-ferrous metals

D.A. Kamaev, A.S. Koptyakov, V.I. Lusgin, V.E. Frizen, E.V. Krivtsova

Ural Federal University, Ekaterinburg, Russian Federation

kamaev_dima@mail.ru

Abstract. The article describes the power supply system of induction furnace based on a semiconductor converter, in which it is possible to work simultaneously in the exciting mode of low-frequency currents and frequencies three times higher when switching the compensation circuit of the furnace inductor at different stages of metal melting. Calculations of electromagnetic processes have shown the advisability of operating the power supply system in the mode of tripling the current frequency at the stage of heating and melting of the metal, but in the mode of low frequency - at the stage of technological processing of the melt. The ratio between the parameters of the furnace power supply in various operating modes are established.

1. Introduction

Induction melting of metals at medium frequencies is widely used in many industries. The high efficiency of induction crucible furnaces is achieved by the optimal choice of design parameters, operating frequency and specific power, which helps to reduce the energy consumption of molten metal per ton. When melting ferrous and non-ferrous metals in induction furnaces, it is necessary to solve a number of technological problems associated with the implementation of metallurgical processes. During technological processes of alloying or carburization, it is advisable to actively mix the melt, however, the electromagnetic field force decreases with increasing frequency of the inductor current, and the intensity of the magnetohydrodynamic processes of metal mixing decreases. To increase the efficiency of electromagnetic effect on the metal melt, it is necessary to supply the furnace with low-frequency currents at the stage of melting metal, and the power supply system must ensure the formation of high and low frequency currents [1, 2].

2. Induction furnace supply system

The power supply systems of modern induction smelting installations for metals are design on the basis of semiconductor frequency converters, usually consisting of a controlled rectifier and a self-commutated inverter. Functional diagram of the power supply system is shown on the Figure 1. It consists of a controlled rectifier (Rectifier) with a filter choke L_d in the DC circuit of a bridge current inverter (Inverter), a block of compensating capacitors (Capacitor banks) connected in parallel with the induction crucible furnace (ICF).

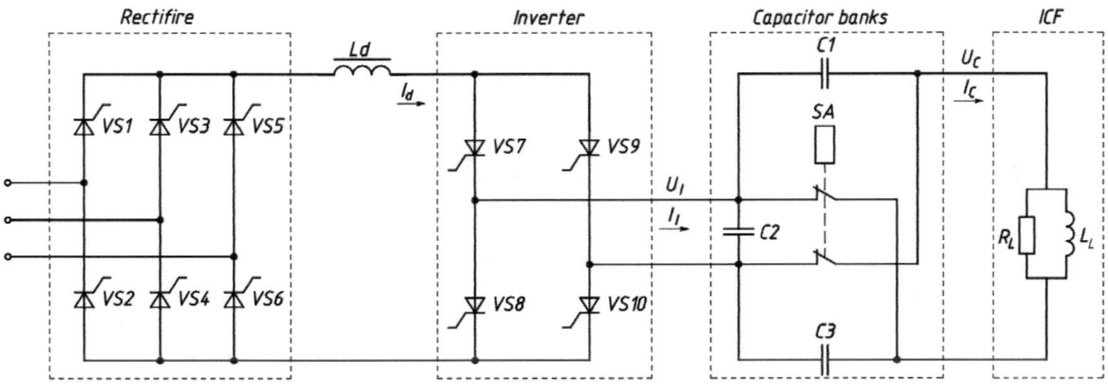

Figure 1. Functional diagram of the induction furnace power supply system

The capacitor block contains three series-connected groups of capacitors *C1*, *C2*, *C3* and a switch *SA*. When *SA* in the closed position, all groups of capacitors are connected in parallel with the furnace inductor. The parallel load circuit formed by capacitors and inductor has its own phase change rate:

$$\omega_0 = 1/\sqrt{L_{Ind} \cdot C_C} = 1/\sqrt{3L_L \cdot C} \quad (1)$$

L_L – inductance of the furnace inductor;

$C_C = C1 + C2 + C3 = 3C$ – capacity of the capacitor banks.

With the synchronized control of the self-commutated current inverter valves by the voltage at the inverter output, its value can be determined from the ratio:

$$U_I = \frac{U_d}{0,9 \cdot \cos\beta} \quad (2)$$

U_d – average voltage of the rectifier;

β – angle of advancing the supply of control pulses to the inverter valves.

A rectangular form current with an amplitude equal to the average value of the rectifier current Id and frequency f is formed at the inverter output, the fundamental harmonic of which can be determined by expanding it in a Fourier series.

$$I_I = 0,9 \cdot I_d \quad (3)$$

Active power at the inverter output is determined by the main harmonics of the output current and voltage:

$$P_I = I_I \cdot U_I \cos\beta \quad (4)$$

Based on the equality of active powers at the input and output of the inverter, assuming there is no loss in its elements $P_d = P_l$, the ratio between the active component of the load resistance R_L and the reduced resistance at DC inverter terminals R_d can be determined:

$$R_d = 0{,}81 \cdot R_L \cos^2 \beta \tag{5}$$

Therefore, the input current and input power are determined as:

$$I_d = U_d / R_d \tag{6}$$

$$P_d = U_d \cdot I_d \tag{7}$$

The switching stability of the inverter operation is ensured during the capacitive reaction of the load circuit, when the operating frequency of the inverter is greater than the resonant frequency of the load circuit $\omega > \omega_0$. In this case, a reverse voltage will be applied to the inverter valves after they are turned off for a time interval expressed in angular measure $\theta_r = \beta$. The time provided for recovery the inverter thyristor controllability can be set by tuning the inverter operating frequency from the resonant frequency of the load circuit:

$$\theta_r = \beta = arctg\left[Q_L\left(q - \frac{1}{q}\right)\right] \tag{8}$$

$Q_L = \dfrac{R_L}{\sqrt{L_L / C_L}}$ – q-factor of the load circuit;

$q = \omega / \omega_0$ – coefficient of the inverter operating frequency detuning from the load circuit resonant frequency.

The condition for the switching stability of the inverter during the capacitive reaction of the load circuit (q>1) is the following inequality:

$$\theta_r = \omega \cdot t_q \tag{9}$$

t_q – turn-off time of the thyristor.

From the above equations, the ratio of the average value of the inverter input current I_d and the loop current I_c can be established:

$$I_c = 0{,}9 \cdot I_d \cdot \sqrt{1 + Q_d^2} \cos \beta \approx 0{,}9 \cdot I_d \cdot Q_L \cdot \cos \beta \tag{10}$$

Thus, we can conclude that the external characteristic of the current inverter in the self-excitation mode (β = const) is equivalent to the characteristic of the voltage source, in which when the active component of the inductor resistance R_L changes over a wide range, the output current I_I changes, and the output voltage U_I remains constant.

3. Characteristics of the power system

Let's consider the characteristics of the furnace power system in the *SA off* position. In this case, the groups of compensating capacitors are connected in series with respect to the furnace inductor and the total capacity of the compensating capacitor decreases and becomes equal to $C1 = C2 = C3 = C$.

$$C_{C.off} = \frac{1}{3}C \tag{11}$$

In this case, a nine-fold decrease in the compensating capacitor capacity occurs, and the resonant frequency of the load circuit increases:

$$\omega_{0.off} = 1/\sqrt{L_L \cdot C_C} = \sqrt{3}/\sqrt{L_L \cdot C} \tag{12}$$

and it becomes three times higher than in the *SA on* position with a slightly changing inductance of the furnace inductor ($L_L \approx const$). Frequency ratio coefficient is the following:

$$K_f = \frac{\omega_{0.off}}{\omega_0} = \frac{\sqrt{3}/\sqrt{L_L \cdot C}}{1/\sqrt{3L \cdot C}} = 3 \tag{13}$$

When changing the current frequency, the equivalent resistance of the furnace inductor $R_{L.off}$ can be determined from the relation:

$$R_{L.off} = R_L \cdot \left(\frac{\omega_{0.off}}{\omega_0}\right)^{n_1} \tag{14}$$

R_L – equivalent inductor resistance at low frequency;

ω_0, $\omega_{0.off}$ – resonant frequencies of the load circuit in the first and second *SA* position;

$n_1 = 1,35 \div 1,4$ – index for crucible furnaces [1].

If $n_1 = 1,35$, then $R_{L.off} = 4,41 \cdot R_L$.

The inductance of the inductor insignificantly depends on the current frequency and the reactance of the inductor increases and is determined as:

$$X_{L.off} = X_L \cdot \left(\frac{\omega_{0.off}}{\omega_0}\right)^{n_2} \tag{15}$$

$n_1 = 0,8 \div 0,95$ – index for melting furnaces [1].

If $n_1 = 0,9$, then $X_{L.off} = 2,7 \cdot X_L$.

The q-factor of the furnace inductor is determined by the ratio of active and reactive resistance:

$$Q_L = \frac{R_L}{X_L} \tag{16}$$

When switching from one operating mode to another, the q-factor of the furnace inductor and load circuit change. The multiplicity of increasing the q-factor of the load is defined as:

$$Q_{L.off} = \frac{R_{L.off}}{X_{L.off}} = 1,63 \cdot Q_L \tag{17}$$

The compensating capacitor consists of three series-connected capacitors, one of which ($C2$) is connected to the inverter output. Such kind of connection of the inverter capacitors and the load represents a capacitive voltage divider, the transmission coefficient of which at a sufficiently high q-factor of the load circuit is determined as:

$$K_U = \frac{U_{C.off}}{U_I} = 1 + \frac{C2}{C_S} = 1 + \frac{C}{C/2} = 3 \tag{18}$$

$C_S = \frac{C1 \cdot C3}{C1 + C2} = \frac{C}{2}$ – capacity of the serial leg of the voltage divider;

$C2 = C$ – capacity of the voltage divider parallel leg.

When switching SA, if the voltage at the inverter output U_I remains unchanged ($\beta = const$), then the voltage at the inductor changes three times ($K_U = 3$), therefore, the change in the active power of the furnace can be determined by the coefficient of increase in active power:

$$K_P = \frac{P_{L.off}}{P_L} = \frac{U_{C.off}^2 / R_{L.off}}{U_C^2 / R_L} = \frac{(3U_C)^2 / (4,41R_L)}{U_C^2 / R_L} = 2 \tag{19}$$

$U_{C.off} = K_U \cdot U_L$ – voltage at the inductor;

$U_C = U_I$ – voltage at the output of inverter.

The magnitude of the current in the inductor is determined as:

$$I_{C.off} = \frac{U_{C.off}}{Z_{L.off}} \tag{20}$$

$U_{C.off} = K_U \cdot U_I$ – voltage at the inductor at the operating frequency $\omega_{off} = 3\omega$;

$Z_{L.off}$ – resistance of the load circuit at a frequency ω_{s}.

Based on the above relations determine $Z_{L.off}$:

$$Z_{L.off} = \frac{R_{L.off}}{\sqrt{1+Q_{L.off}^{2}}} = 2,7\frac{R_{L}}{Q_{L}} = 3,33\frac{R_{d}}{Q_{L}\cdot\cos^{2}\beta} \tag{21}$$

Then $I_{C.off} = 0,333\cdot K_{U}\cdot Q_{L}\cdot I_{d}\cdot\cos\beta = I_{d}\cdot Q_{L}\cdot\cos\beta$.

Determine the rate of change of current when switching *SA*:

$$K_{I} = \frac{I_{C.off}}{I_{C}} = \frac{I_{d}\cdot Q_{L}\cdot\cos\beta}{0,9\cdot I_{d}\cdot Q_{L}\cdot\cos\beta} = 1,11 \tag{22}$$

Thus, when switching the power supply mode of the furnace, the magnitude of the inductor current changes insignificantly, while its frequency changes in 3 times. However, active power at a low frequency is halved.

4. Calculation of power system parameters

Consider the calculation of the dual-frequency power system parameters for an induction crucible for melting cast iron with a capacity of $V = 400$ kg. Calculation of the furnace parameters was carried out according to the method of computer simulation given in [3]. With a supply voltage of $U = 380$ V and an operating frequency of the furnace inductor current in the metal melting mode (*SA* - turn-off) $f_{H} = 1000$ Hz, the furnace parameters and the electrical load characteristics of the power supply system are given in table 1. The power supply system parameters in the mixing mode of the metal at a low operating frequency of $f_{L} = 325$ Hz (*SA* - turn-on) are also given there. According to the calculation data, the capacitance of the compensating device capacitors of $C_{c} = 3C;\ C = \frac{C_{c}}{3} = \frac{1223}{3} = 408$ µF can be determined and their type selected.

Table 1. Electrical parameters of the power supply system and load.

Electric parameters of the furnace inductor		
The operating frequency of the inductor current, Hz	1000	325
The turns number of the furnace inductor	30	
The reduced resistance of the inductor, Ohm	13	2,15
Inductor reactance, Ohm	1,2	0,419
Power factor inductor, $\cos\varphi_{L}$	0,093	0,148
Inductor impedance Z_{L}, Ohm	1,206	0,424
Electric efficiency of inductor	0,799	0,795
Power system parameters		
Inverter control angle β	30	25

Inverter output voltage U_L, V	700	650
Inverter output voltage U_C, V	2000	650
Active power of the furnace P_L, kW	310	148
Reactive power of capacitors Q_C, kVAr	3445	1041
Inductor current I_C, A	1660	1535
Switch current $I_{S.i}$, A	-	1013
Compensating Capacitance C_C, µF	136	1223
Capacitor Voltage	700	650

According to the calculation data, it is seen that when switching from the metal melting mode to the mixing mode, the power consumed by the load decreases, and the current in the inductor changes slightly.

When switching the operating mode, the inductor parameters change in approximately the same proportions that were determined by the formulas (14, 15):

$$\frac{R_{L.off}}{R_L}=\frac{13}{2,85}=4,56;\ \frac{X_{L.off}}{X_L}=\frac{1,2}{0,419}=2,86$$

Thus, the proposed method for calculating the parameters of a dual-frequency power supply system for induction crucible furnaces can be used with sufficient accuracy in engineering calculations.

5. Conclusion

According to the results of the study, it can be concluded that the proposed power supply system is advisable to use in high-frequency operation mode for forced heating and metal melting, and the low-frequency operation mode is necessary during technological operations when efficient metal mixing is ensured.

Also, authors in researches [4, 5, 6,7] proved that dual-frequency power supply provide an increase in the rate of the molten metal motion in the case of melting, and also allows the processing of parts with a complex geometric shape of the surface in case of heating.

References

[1] Gitgarts D.A., Ioffe Yu.S. 1972 New power supplies and automation of induction installation for heating and melting (Moscow, Energy)

[2] Trauzedal D. 2003 Realization of special technological and metallurgical tasks in optimized medium frequency furnaces *J. Russian Foundrymen* 5 pp 20-23

[3] LuzginV.I., Sarapulov S.F., F.N. Sarapulov 2005 Melting complexes based on induction crucible furnaces and their mathematical modeling: a training manual, (Ekaterinburg).

[4] Frizen V E, Luzgin V I, Koptyakov A C, Bolotin K E, Tarchutkin N V and Mironov S E 2017 Induction crucible furnace with dual frequency power supply *Proc. Int. Conf. on Electrical Machines, Drives and Power Systems, 15th* (Bulgaria: Technical University of SofiaSofia) pp 423-426

[5] V. Esteve, J. Pardo, J. Jordan, E. Dede, E. Sanchis-Kilders, E. Maset, 2005 High power resonant inverter with simultaneous dual-frequency output, *Power Electronics Specialists Conference*, pp. 1278-1281.

[6] Luzgin V. I., Koptyakov A. S., Petrov A. U., Zinovev K. A, Kamaev D. A., and Yakovenko S. R. Power supplies for dual–frequency induction melting of metals *IOP Conf. Series: Materials Science and Engineering 313 (2018) 012008,* APEET 2017

[7] Sergio Lucia, Héctor Sarnago, Denis Navarro, Oscar Lucia, 2019 Improved Multi-Load Resonant Power Conversion Using Model Predictive Control, *Industrial Electronics Society IECON 2019 - 45th Annual Conference of the IEEE*, vol. 1, pp. 5075-5080

IOP Conf. Series: Materials Science and Engineering **950** (2020) 012028 doi:10.1088/1757-899X/950/1/012028

Research of options for maintaining the operability of the traction switched reluctance motors in emergencies

A Petrushin[1], V Smachney[1], D Petrushin[1]

[1] Rostov State Transport University, 2, Rostovskogo Strelkovogo Polka Narodnogo Opolcheniya sq., Rostov-on-Don, 344038, Russia

E-mail: alex331685@yandex.ru; smachney87@mail.ru; empanger@yandex.ru

Abstract.

The switched reluctance machines (SRM) in most cases of their application do not have a galvanic connection between the phase windings. It is widely known when one phase fails, SRM as a short circuit in the coil of one of the phases, the others continue to work. This gives you the ability to retain partial performance of the engine. This is especially important for transport systems: an electric car can drive to a depot on its own, and an electric car can drive to a service station. And for some transport systems, such as manned underwater vehicles, maintaining partial performance in emergency situations is a priority requirement-the life support of the crew depends on it.

For using a widely applied semi-bridge power supply scheme, the SRM with two transistors for each of the phases of the electric motor, it is not possible to maintain partial performance of the SRM if the only power source fails. To ensure the performance of the type, in the case of a short circuit in the source, a new power supply scheme for the SRM is proposed, containing one transistor for each of the phases of the engine (patent No. 2704494). Each phase is electrified charged from a separate source. For transport systems that have batteries in their composition, the total battery is divided into components by the number of phases. At the same time, the entire electrical circuit of the SRM phase, including the source, is backed up.

A simulation of the proposed new power supply circuit was performed in order to confirm its performance and evaluate its functionality. The analysis of the existing Converter circuits used for SRM is carried out, which allows to evaluate the advantages and disadvantages of the proposed circuit solution.

1. Introduction

The traction electric drives are used in a wide class of transport systems of the railway, automobile, surface and underwater transports. Electric drives are also used for transport robots.

The requirements for reliability and survivability of transport systems are usually high. Nevertheless, special requirements are placed on reliability and survivability. This means that movement in ultra-low or ultra-high temperatures in remote areas. This category also includes autonomous inhabited vehicles for servicing underwater communications, carrying out rescue and research work at various depths. To create such traction electric drives, it is advisable to use switched-reluctance electric machines (SRM) which have a simple design and high energy performance.

SRM cannot work independently, such as asynchronous motors, but only as part of an electric drive. The study of options for maintaining the operability of traction switched-reluctance drives (SRD) in emergency situations was carried out by analyzing the reliability of the functioning of the components of the entire electric drive. SRD consists of the following components: SRM; converter's, which in turn consists of a power part and a control system including a rotor position sensor; power source. An emergency here means a complete failure without the possibility of further use of the SRD. The possibility of an automatic (without human participation) conversion of a complete failure to a partial one, in which a SRD can be partially used with incomplete power, is being investigated. It does not address the obvious cases in which a faulty unit can be repaired on site or replaced with a new one.

SRM possesses design features, such as a winding-free rotor and concentrated stator windings without intersecting frontal parts, which makes it more reliable among the closest competitors.

As a power source in traction electric drives of autonomous vehicles, batteries or supercapacitors are often used, which are low-voltage elements connected in series until the required voltage level and energy supply are achieved. For example, a Tesla S battery has 2100x1500x150mm dimensions and a weight of 540kg, which is about 25% of the total weight of the car and consists of 16 blocks, each of which in turn consists of 74 elements. A large number of elements with an equal probability of failure, connected in sequence, form a system whose reliability may not correspond to what is desired. For this reason, for traction drives with increased requirements for reliability and survivability, the power source is duplicated or divided into several independent parts and is used in circuits that allow to maintain the partial operability of the traction drive in the event of failure of one of the parts of the power source.

2. The objectives of the investigation

The power part of the converter and the SRD control system have a fairly high degree of reliability, taking into account the current level of quality of semiconductor devices and microprocessor technology. However, there are still reserves for improving circuitry solutions to improve the reliability and survivability of SRD in emergency situations.

Currently, the most common circuit of the power part of the converter for powering the SRD phases is an asymmetric half-bridge circuit, which contains a source, two transistors and two reverse diodes, a capacitor that acts as a filter and an energy storage device for exchanging it with the winding through reverse diodes (fig. 1) [1-9].

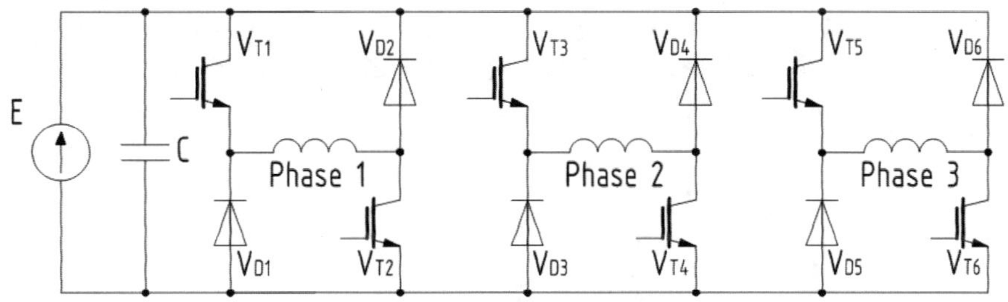

Figure 1. The power circuit diagram of an asymmetric converter in 3-phase design

The main properties of the converter:

1. Provides greater flexibility in management, as the converter can realize the following circuits: positive - connecting the full voltage of the source to the phase winding, zero - gradually reducing the current and dissipating the energy stored in the phase on the internal resistance of the circuit elements, negative - applying the phase voltage with negative polarity to the motor winding and the energy

conservation circuit — the energy conservation of the residual current of the phase winding in the capacitor or source for its further use at the next switching cycle [6].

2. All phases are connected to the source independently, and it is possible to disconnect the damaged phase without affecting the operation of the others.
3. The converter can be performed on any number of phases.
4. The converter can implement the motor and generator operating modes of the SRM.

The power part of the converter has two semiconductor switches for one phase SRM. To reduce the cost of the power part and increase the reliability of the SRD, the search for circuitry solutions such that the number of power semiconductor switches for one phase of the SRM will be less than two, while maintaining the quality of operation, does not stop. By the number of switches per phase, the existing power part of the converters are divided into several groups [1-5]:

- single-switches converters: a converter with a power dissipating resistor [2-4, 6-9]; converter with bifilar winding [1-3, 6, 8-9]; converter with separation of the DC link [1-4, 6, 8-9]; converters with one switch and two diodes per phase with adjacent phases [2, 10] and with independent phases [2]. These power parts of the converters have a lower cost due to the use of a single switch, but do not satisfy several basic requirements for the effective operation of the SRD such as: independent operation of the phases, it is not possible to form all 4 operating modes of the circuits (positive, zero, negative circuits, energy conservation circuit), have low energy efficiency, and also some of them have a limit on the number of phases.
- converters with (q + 1) switches for q phases: a converter with double demagnetization [2, 6]; a converter with a common switch for 2 phases [2, 4, 6, 8]; converter with storage capacitor - classic [1-5, 7-9, 11], modified [2, 4, 7, 12] (with storage capacitor and zero circuit), energy-efficient [12]; Miller Converter [1, 4-6]; a converter with a minimum number of switches and an increased input voltage [2]; a converter with a minimum number of switches and an increased pulse input voltage [2]; resonant converter (developed on the basis of a converter with a storage capacitor) [9]. These converters have one or several drawbacks compared to the circuit (fig 1): they do not provide independent operation of the phases, they do not implement one or several operating modes, they have voltages higher than the source voltage, they have a complex and not flexible control system, high torque ripples, long time dispersion and, consequently, low switching speeds, have generally lower efficiency and, often, quite high cost (due to the use of additional diodes, inductors, capacitors).
- converter with 1.5q switches for q phases and independent phase current control [2, 4, 13], provides independent control and allows you to implement all operation algorithms, but is suitable only for SRD with an even number of phases, some of the switches work for two phases, which increases the load on them and reduces the reliability of the drive.

According to the power section of the converter (fig. 1), the SRM phases receive power from a single source. This does not ensure the survivability of the entire electric drive system in case of failure of the power source (fig. 2*a*).

IOP Conf. Series: Materials Science and Engineering **950** (2020) 012028 doi:10.1088/1757-899X/950/1/012028

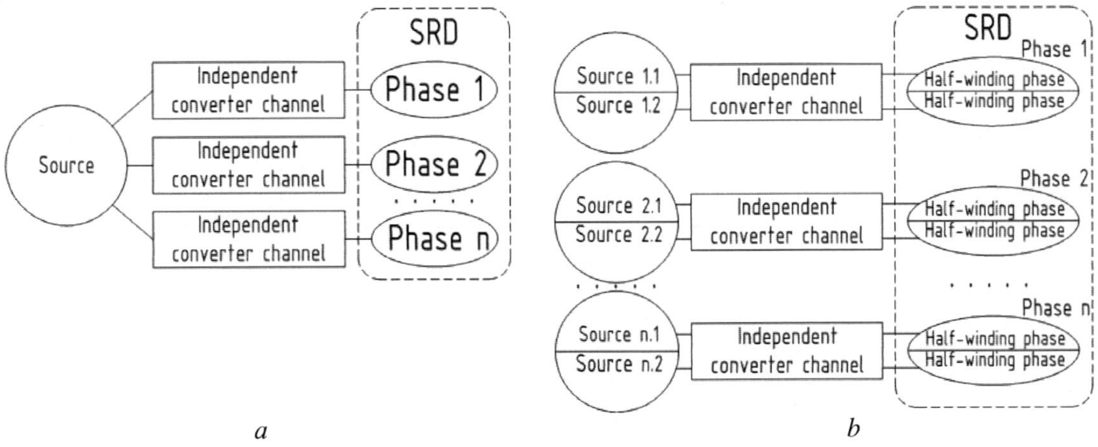

Figure 2 SRD phase power scheme:
a – without the implementation of a full reservation;
b – using the proposed single-switch power part scheme and the implementation of full redundancy.

3. The solution of the problem

It is proposed to use SRD with a converter containing a smaller number of power switches and allowing full backup of all components of each phase, including power sources, as a traction drive for autonomous vehicles [14-15]. Full redundancy involves the creation of independent channels in phases, including the source. The batteries used in autonomous transport batteries can be divided into groups, each of which feeds a separate phase. Failure of one group of batteries, for example, with an internal circuit or other cause of loss of operability, will deenergize only one phase, and the rest will remain in operation. In the proposed technical solution, the operation of the SRD power supply circuit with one power switch (transistor) for switching each phase is illustrated in fig. 2*b*.

The single-switch circuit for powering the phase of the SRD [14-15] contains two sources E_1 and E_2, two capacitors C_1 and C_2, a reverse diode V_D, a semiconductor switch V_T and a phase winding divided into two half-windings L_1 and L_2 located on diametrically opposite teeth stator (fig. 3*a*). This single-switch scheme works when the parameters of the sources E_1 and E_2 are equal. If the sources E_1 and E_2 have unequal parameters, then diodes V_{D1}, V_{D2} are used to exclude equalizing currents (fig. 3*b*). Using the proposed single-switch SRD phase power circuit, multi-phase SRD converters can be implemented, as shown in fig. 3*c* [15].

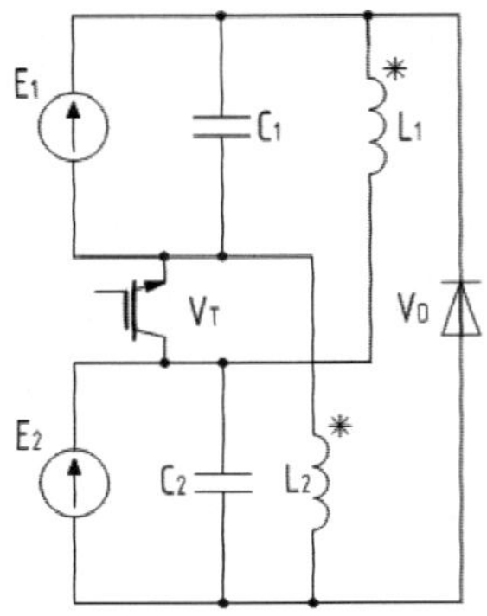

a

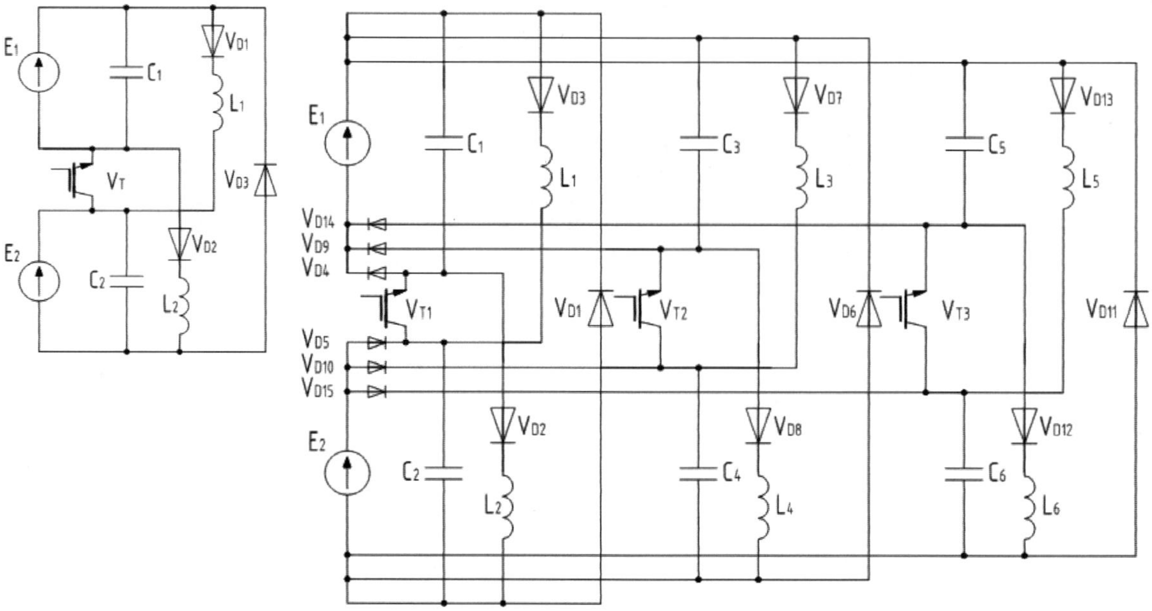

b c

Figure 3 Single-switch scheme of the SRM converter:
a – with voltage sources having the same parameters;
b – with voltage sources having different parameters; c – three-phase design

The scheme (fig. 3*b*) allows the SRD to work according to the following algorithm. In the initial state, the V_T switch is closed, the current in the half-windings L_1 and L_2 is zero, the capacitors C_1 and C_2 are charged. The V_T switch open field through the half-windings L_1 and L_2 flows current – circuits $E_1 - V_{D1} - L_1 - V_T - E_1$, $C_1 - V_{D1} - L_1 - V_T - C_1$, $E_2 - V_T - V_{D2} - L_2 - E_2$, $C_2 - V_T - V_{D2} - L_2 - C_2$. In this case, the electromagnetic energy is converted into mechanical energy in the SRD (the positive voltage circuits work). After closing the V_T switch, the energy accumulated in the half-windings L_1 and L_2 charges the capacitors C_1 and C_2 – circuits $L_1 - C_2 - V_{D3} - V_{D1} - L_1$, $L_2 - V_{D3} - C_1 - V_{D2} - L_2$ (energy conservation circuits work). Next, the cycle repeats. When the V_T switch is opened, the V_D diode opens and a negative voltage is applied to the half windings of the phase (negative voltage circuits work).

The calculation of the operation parameters of the single-switch electric drive circuit was made using the example of the operation of one phase in different modes. For this, substitution schemes for the moments of opening (fig. 4a) and closing (fig. 4b) of the power semiconductor switch have been compiled.

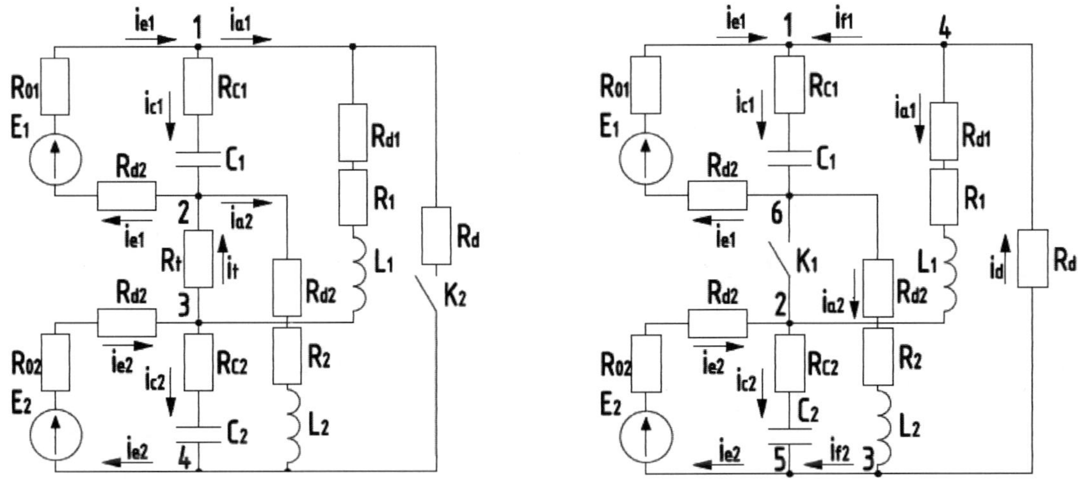

a *b*

Figure 4 SRM replacement Circuit with a single-switch converter circuit:
a – the semiconductor switch is open; b – the semiconductor switch is closed

According to Kirchhoff's laws, the equations for 1-4 nodes (fig. 4*a*) and contours $E_1 - R_{01} - R_{C1} - C_1 - R_{d2} - E_1$; $E_1 - R_{01} - R_{d1} - R_1 - L_1 - R_t - R_{d2} - E_1$; $E_2 - R_{02} - R_{d2} - R_{C2} - C_2 - E_2$; $E_2 - R_{02} - R_{d2} - R_t - R_{d2} - R_2 - L_2 - E_2$. After converting the obtained equations and reducing them to the Cauchy form, a system of equations is obtained for calculating the electromagnetic and electromechanical processes of the SRD when the transistor is opened. An expression for the electromagnetic moment has been added to this system of equations.

$$\begin{cases} \dfrac{dU_{c1}}{dt} = \dfrac{E_1 - U_{c1} - i_{a1}R_{01} - i_{a1}R_{d2}}{C_1\left(R_{01} + R_{d2} + R_{c1}\right)} \\ \dfrac{dU_{c2}}{dt} = \dfrac{E_2 - U_{c2} - i_{a2}R_{02} - i_{a2}R_{d2}}{C_2\left(R_{02} + R_{d2} + R_{c2}\right)} \\ \dfrac{di_{a1}}{dt} = \dfrac{1}{L_g}\left(E_1 - i_{a1}\left(R_{01} + R_{d1} + R_1 + R_t + R_{d2}\right) - C_1\dfrac{dU_{c1}}{dt}\left(R_{01} + R_{d2}\right) - \omega\dfrac{\partial\psi\left(i_{a1},\theta\right)}{\partial\theta} - i_{a2}R_t\right) \\ \dfrac{di_{a2}}{dt} = \dfrac{1}{L_{g2}}\left(E_2 - i_{a2}\left(R_{02} + R_{d2} + R_t + R_{d2} + R_2\right) - C_2\dfrac{dU_{c2}}{dt}\left(R_{02} + R_{d2}\right) - \omega\dfrac{\partial\psi\left(i_{a2},\theta\right)}{\partial\theta} - i_{a1}R_t\right) \\ \dfrac{d\omega}{dt} = \dfrac{1}{J}\left(M_\partial - M_н - B_t\omega\right) \\ M_\partial = \dfrac{\partial\int_0^i \psi\left(i_{a1},\theta\right)di}{\partial\theta} + \dfrac{\partial\int_0^i \psi\left(i_{a2},\theta\right)di}{\partial\theta} \end{cases}$$

Similar transformations are performed for the transistor closing mode. According to Kirchhoff's laws, the equations for 1-6 nodes (fig. 4*b*), and for contours $E_1 - R_{01} - R_{C1} - C_1 - R_{d2} - E_1$; $L_1 - R_{C2} - C_2 - R_d - R_{d1} - R_1 - L_1$; $E_2 - R_{02} - R_{d2} - R_{C2} - C_2 - E_2$; $L_2 - R_d - R_{C1} - C_1 - R_{d2} - R_2 - L_2$. After the transformations, a system of equations is obtained:

$$
\begin{cases}
\dfrac{dU_{c1}}{dt}=\dfrac{E_1-U_{c1}+i_{a2}R_{01}+i_{a2}R_{d2}}{C_1(R_{01}+R_{d2}+R_{c1})}\\
\dfrac{dU_{c2}}{dt}=\dfrac{E_2-U_{c2}+i_{a1}R_{02}+i_{a1}R_{d2}}{C_2(R_{02}+R_{d2}+R_{c2})}\\
\dfrac{di_{a2}}{dt}=\dfrac{-1}{L_{g2}}\left(U_{c1}+C_1\dfrac{dU_{c1}}{dt}R_{c1}+i_{a2}(R_d+R_2+R_{d2})+\omega\dfrac{\partial\psi(i_{a2},\theta)}{\partial\theta}+i_{a1}R_d\right)\\
\dfrac{di_{a1}}{dt}=\dfrac{-1}{L_{g1}}\left(U_{c2}+C_2\dfrac{dU_{c2}}{dt}R_{c2}+\omega\dfrac{\partial\psi(i_{a1},\theta)}{\partial\theta}+i_{a1}(R_1+R_{d1}+R_d)+i_{a2}R_d\right)\\
\dfrac{d\omega}{dt}=\dfrac{1}{J}(M_\partial-M_н-B_t\omega)\\
M_\partial=\dfrac{\partial\int_0^i\psi(i_{a1},\theta)di}{\partial\theta}+\dfrac{\partial\int_0^i\psi(i_{a2},\theta)di}{\partial\theta}
\end{cases}
$$

To test the circuit's performance (fig. 3b), a simulated computer simulation of the SRD was performed using the MATLAB (Simulink) program for a 3kW electric motor. From the OrCad v17 program, the IGBT models of the IGW40N60DTP_L2 transistor and the Infineon IR2110 driver were imported into Simulink PSpice, the remaining components of the circuit were selected from the standard Simulink library. As a result of the simulation, the characteristics of the SRD are obtained, similar to the scheme of the power part of the asymmetric Converter (fig. 1). Thus, while maintaining the basic functionality of the most common circuit of the power part of the SRD Converter (fig. 1), the proposed scheme (fig. 3*b*) can be used for electric drives with full redundancy of each of the phases of the SRD, including the power source. At the same time, the number of semiconductor switches in the power part of the SRD will be reduced by half.

4. Conclusions

As a result of consideration of the various circuitry of the power part of the SRD and save health in emergency situations for cases where the reliability and survivability of SRD is the defining requirement, we can recommend the single-switch circuit power Converter circuit (fig. 3b). For this option, each phase of the SRM is powered from a separate power source, and the number of power semiconductor elements per phase is minimal. In this version of the circuit solution, the SRD phase combines only a common magnetic core and each of the phases can be disabled in an emergency without compromising the performance of the others.

References

[1] **Miller, T.J.E.** Switched Reluctance motor and their Control. – Magna Physics Publishing and Clarendon Oxford Press, 1993. – 203 p.
[2] **Krishnan, R.** Switched reluctance motor drives. Modeling, simulation, analysis, de-sign, and applications. – Virginia: the Bradley Department of Electrical and Computer Engineering, 2001. – 416 p.
[3] **Jin-Woo Ahn**, Switched reluctance motor / M. T. Lamchich (Ed.), Torque Control, InTech, 2011, Ch. 8, pp. 201 – 252. doi:10.5772/10520.
[4] **Mehta, N.D.** A review of switched reluctance motor converter topologies for automotive applications /N.D. Mehta, A.M. Haque, A.P. Patel// International Journal of Engineering

Technology, Management and Applied Sciences, February 2018, Volume 6, Issue 2, ISSN 2349-4476, p. 7-17.

[5] **Torsten Wichert, M.Sc**. Design and construction modifications of switched reluc-tance machines. Ph.D. Thesis. / Warsaw University of Technology, Institute of Elec-trical Machines // Warsaw, 2008.

[6] **Cunningham J.D.** Switched reluctance motor drive circuit evaluation criteria for vehicle efficiency and responsiveness. Ph.D. Thesis. The University of Texas at Austin, 2011. 234 p.

[7] Samia M. Mahmoud. Studying different types of power converters fed switched reluctance motor / Samia M. Mahmoud, Mohsen Z. El-Sherif, and Emad S. Abdel-Aliem / International Journal of Electronics and Electrical Engineering, 2013, Vol. 1, no. 4.

[8] **Suppharangsan, W.** A new switching technique for minimization of dc-link capacitance in switched reluctance machine drives / A thesis submitted for the degree of PhD in the department of Electronic and Electrical Engineering, The University of Sheffield, August, 2012.

[9] **Hoang Le-Huy.** A current-controlled quasi-resonant converter for switched-reluctance motor / H/ Le-Huy, K. Slimani, P. Viarouge / IEEE Transactions on industrial electronics, Vol. 38, №5, pp 355-362, October 1991.

[10] **Pollock, C.** A unipolar converter for a switched reluctance motor / C. Pollock, B. W. Williams / IEEE Transactions on Industry Applications, Vol26, №2, pp. 44-49, 1990.

[11] **Archana G. Thosar**. C-Dump converter for switched reluctance motor / Archana G. Thosar, Rakesh D. Zalke // IEEE International Conference On Recent Trends in Electronics Information & Communication Technology (RTEICT), India, 2017.

[12] **Yong-Ho Yoon**. Control of c-dump converters fed from switched reluctance motors on an automotive application / Yuen-Chung Kim, Sang-Hoon Song, Chung-Yuen Won // Journal of Power Electronics, Vol. 5, No. 2, April 2005, pp 120 – 128.

[13] **Jin-Woo Ahn**. Design and Drive Characteristics of Low Voltage 8/6 SRM for Fan Application / The Transactions of the Korean Institute of Electrical Engineers. Vol. 63, No. 10, pp. 1371-1376, 2014.

[14] **Patent 2459341,** IPC H02K 29/00, H02R 6/00, H02R 25/08, H02K 19/06. Single-switch electric drive / A.D. Petrushin, R. M. Devlikamov, V. Yu. Smachny. Application no. 2011119730/07; application no. 16.05.2011.

[15] **Patent 2704494,** H02K29/00, H02P 25/086. Single-switch multiphase electric drive / A.D. Petrushin, V. Yu. Smachny. Date of registration: 29.10.2019. Application no. 2019102994; application no. 04.02.2019 G.

IOP Conf. Series: Materials Science and Engineering **950** (2020) 012029 doi:10.1088/1757-899X/950/1/012029

Research of various inductors configurations for heating titanium slabs

Evgeniy Shmakov, Vasiliy Frizen and Dmitriy Nikolaev

Department of Electrical Engineering and Electrotehnlogy Systems, Ural Federal University, Ekaterinburg, Russia

E-mail: shmakov.evgeny@urfu.ru

Abstract. In this paper various inductors configurations for heating a titanium slab, as well as the influence of frequency changes on efficiency of inductors are considered. The calculations are carried out by means of finite element method. In this research a three-dimensional time-harmonic model in A-formulation of Maxwell's equations is used. In the framework of this work, only electromagnetic problem was calculated, since the main aim of this work is to assess the effectiveness of inductor configurations in terms of electromagnetic problem. Efficiency of inductor and the uniformity coefficient introduced herein have been chosen as the comparison parameters. Ultimately, distribution of electromagnetic power density in the workpiece is shown, also conclusions are drawn for each of inductor configurations.

1. Introduction

The use of titanium and its alloys is a promising area in modern industry. Titanium has optimal parameters: the mechanical strength approximately equals to steel, but density is less (7.8 $\frac{g}{cm^3}$ is the density of steel; 4.54 $\frac{g}{cm^3}$ is the density of titanium; 2.7 $\frac{g}{cm^3}$ is the density of aluminium), due to these physical properties it is one of the most important material in aerospace engineering. These industry requires high quality of fabrication operation, that increases the level of production knowledge intensity.

Titanium and its alloys go to production in the shape of finished blanks (slabs). A slab is a thick blank of rectangular cross section with a large ratio of width to height. To process the workpieces, it is necessary to heat them for further plastic deformation. The main requirement for heating is minimum temperature difference between center and surface of workpiece, as well as uniform heating. One way to meet this requirement is application of electric resistance furnaces. But the productivity of such furnaces is low, since heating is due to radiation from heaters. An alternative solution is application of induction heating, which provides heating of workpiece by induced current in it, thereby it is possible to carry out rapid heating by increasing power [1, 2, 3, 4].

Induction heating is a complex multiphysics problem (coupled thermal and electromagnetic problem) from the simulation point of view. Therefore, in modern realities, in the most cases the simulation of induction heating is performed by finite element method, which allows solving complex systems of partial differential equations by solving systems of linear algebraic equations [1, 2].

2. Formulation of the problem

The main aim of this research is to consider various configurations of inductors. The comparison is taken place according to the efficiency of the inductor and by analyzing the uniformity of electromagnetic power density distribution.

The object of heating is a slab with dimensions 4.1 m x 1.12 m x 0.28 m. The configurations of inductors are presented in fig. 1. The properties of titanium, electrical steel and copper which are necessary for the electromagnetic problem, are summarized in tab. 1.

Table 1: Material Properties

Material	Property	Value
Titan	Electrical conductivity, MSm/m	1.798
	Relative magnetic permeability	1
Electrical steel	Electrical conductivity, Sm/m	10
	Relative magnetic permeability	500
Copper	Electrical conductivity, MSm/m	60
	Relative magnetic permeability	1

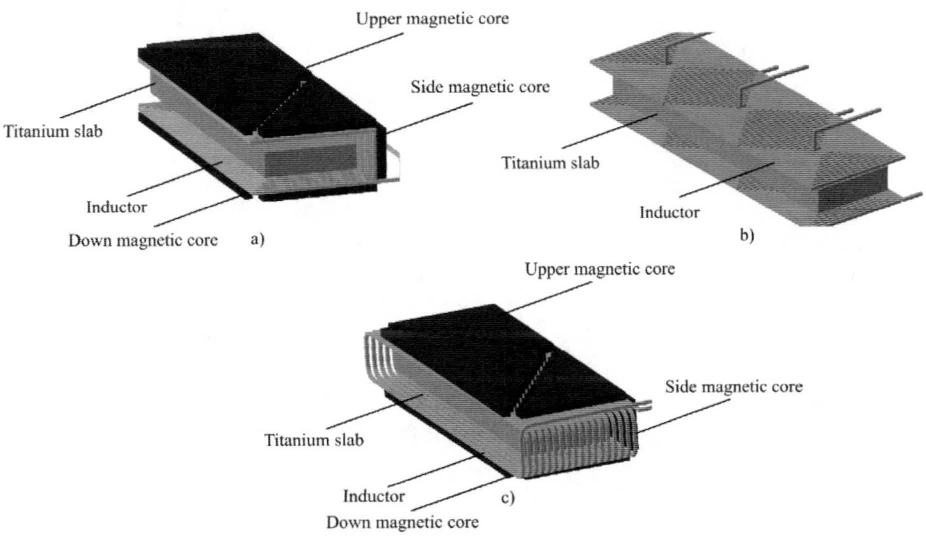

Figure 1. Inductor configurations: a) C-type inductor, b) six flat-type inductors, c) solenoid-type inductor

The cross section of the inductors is made so that the skin effect is clearly expressed, therefore, the calculation of the active resistance of the inductor is performed according to eq. (5).

3. Mathematical description

In this paper, the magnetic field is calculated based on the Maxwell's equations in A-formulation:

$$\Delta \mathbf{A} - j\mu\sigma\omega\mathbf{A} = -\mu\mathbf{j}_{ext}, \tag{1}$$

where **A** is the value of the magnetic vector potential, σ is the electrical conductivity of the material, μ is the magnetic permeability of the material, $\mathbf{j}_{ext}$ is the current density vector.

The induced current density in workpiece is calculated as follow

$$\sigma \mathbf{E} = \mathbf{j}, \tag{2}$$

where **E** is the induced electric field strength.

Electromagnetic power in workpiece is calculated by eq. (3):

$$Q = \iiint \frac{1}{2} \mathbf{E} \cdot \bar{\mathbf{j}} dV, \tag{3}$$

where Q is the electromagnetic power in workpiece.

The active resistance of the inductor is calculated taking into account the fact that there is a clearly narrowed skin effect in it, therefore, the calculation is performed according to the following formulas

$$\delta = \sqrt{\frac{2}{\sigma \cdot \omega \cdot \mu}}, \tag{4}$$

$$R = \frac{l}{\sigma \cdot \delta \cdot a}, \tag{5}$$

$$\eta = \frac{Q}{Q + 0.5 \cdot R \cdot I^2}, \tag{6}$$

where δ is the penetration depth of the electromagnetic wave into the material, ω is the angular frequency, R is the active resistance of the inductor, l is the length of the inductor, a is the width of the cross section, I is the amplitude of the current in the conductor of the inductor.

4. Results

All calculations were performed at a constant value of the current in inductor. Due to the symmetry of the arrangement, only one part of geometry is used, thereby reducing the required computation power [3]. Due to the skin effect of the electromagnetic field, the surface mesh of titanium billet is finer than that of the center [3].

Also, to assess the uniformity of electromagnetic power distribution in workpiece, we introduce a uniformity coefficient according to the following formula

$$k_u = \frac{Q_{savg}}{Q_{smax}}, \tag{7}$$

where k_u is the uniformity coefficient, Q_{savg} is the average density of electromagnetic energy on the surface facing the inductor, Q_{smax} is the maximum density of electromagnetic energy on the surface facing the inductor

4.1. C-type inductor

Calculation of electromagnetic field for this configuration is performed only for a quarter of the model.

4.1.1. Configurations with different arrangement of magnetic cores

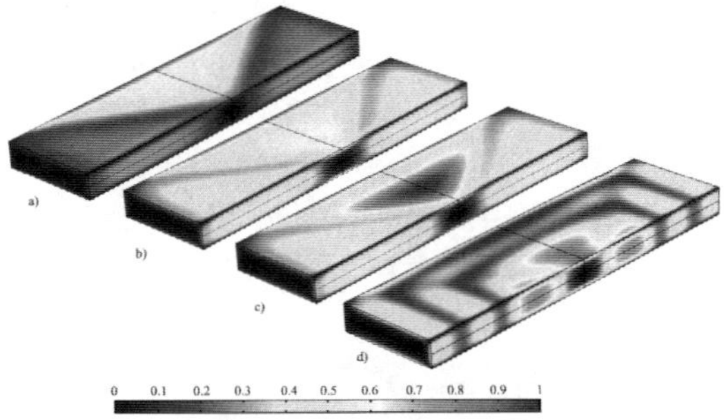

Figure 2. The distribution of electromagnetic power in the workpiece: a) all magnetic cores, b) there is no middle magnetic core, c) half of the middle magnetic core, d) all magnetic core and three-phase power supply of the inductor

Table 2: Efficiency values for various cases

Parameter	a)	b)	c)	d)
η	0.63	0.55	0.57	0.434
k_u	0.337	0.45	0.505	0.308

4.1.2. Different inductor supply frequencies

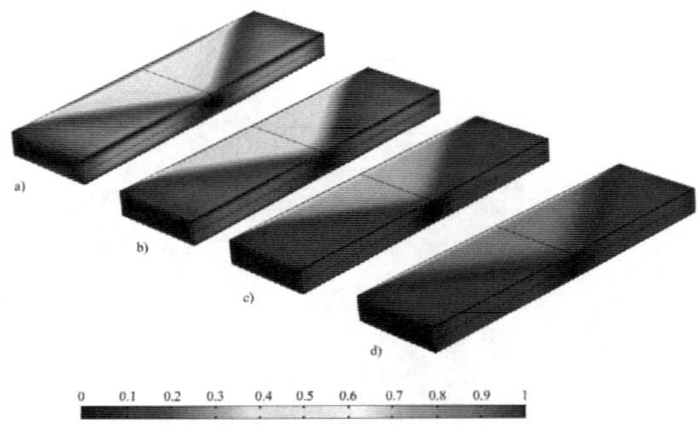

Figure 3. The distribution of electromagnetic power in the workpiece: a) 50 Hz, b) 100 Hz, c) 250 Hz, d) 500 Hz

Table 3: Efficiency values for various cases

Parameter	a)	b)	c)	d)
η	0.63	0.64	0.646	0.65
k_u	0.337	0.273	0.203	0.163

IOP Conf. Series: Materials Science and Engineering **950** (2020) 012029 doi:10.1088/1757-899X/950/1/012029

4.2. Solenoid-type inductor

Calculation of electromagnetic field for this configuration was also performed only for a quarter of the model.

4.2.1. **Different inductor supply frequencies**

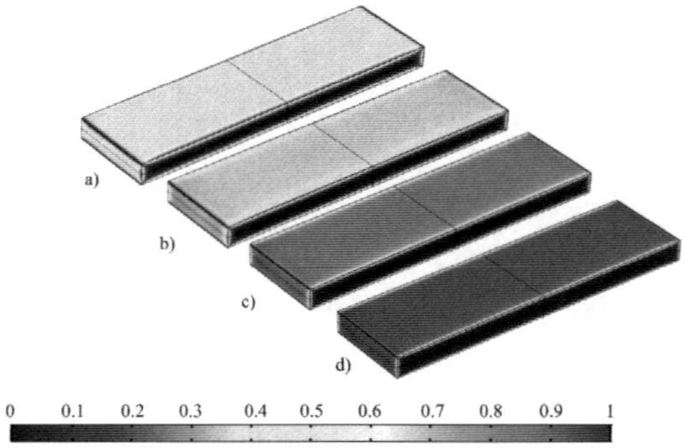

Figure 4. The distribution of electromagnetic power in the workpiece: a) 50 Hz, b) 100 Hz, c) 250 Hz, d) 500 Hz

Table 4: Efficiency values for various cases

Parameter	a)	b)	c)	d)
η	0.607	0.613	0.62	0.623
k_u	0.485	0.391	0.29	0.23

4.2.2. **Without side magnetic core with different inductor supply frequencies**

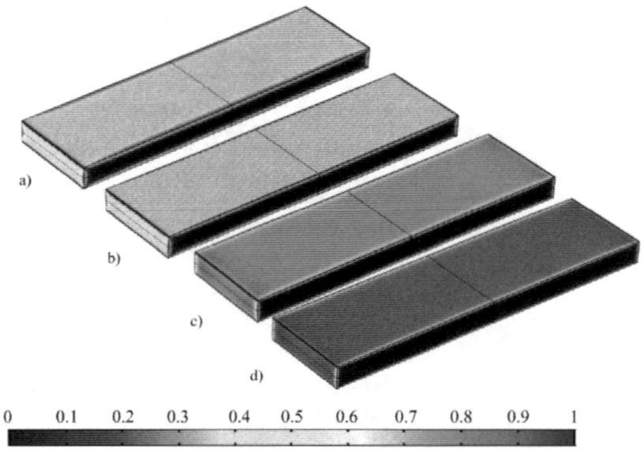

Figure 5. The distribution of electromagnetic power in the workpiece: a) 50 Hz, b) 100 Hz, c) 250 Hz, d) 500 Hz

IOP Conf. Series: Materials Science and Engineering **950** (2020) 012029 doi:10.1088/1757-899X/950/1/012029

Table 5: Efficiency values for various cases

Parameter	a)	b)	c)	d)
η	0.6	0.606	0.612	0.615
k_u	0.568	0.457	0.338	0.268

4.3. Six flat-type inductors

Calculation of electromagnetic field for this configuration was performed only for a half of the model.

For this configuration of six flat-type inductors, various schemes of connecting individual inductor are possible, these schemes are shown in fig. 6– 8 .

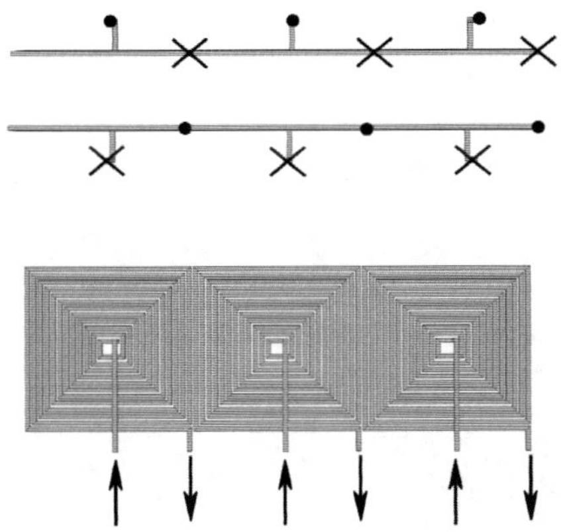

Figure 6. With accordant connection of adjacent coils and opposite connection of opposite coils

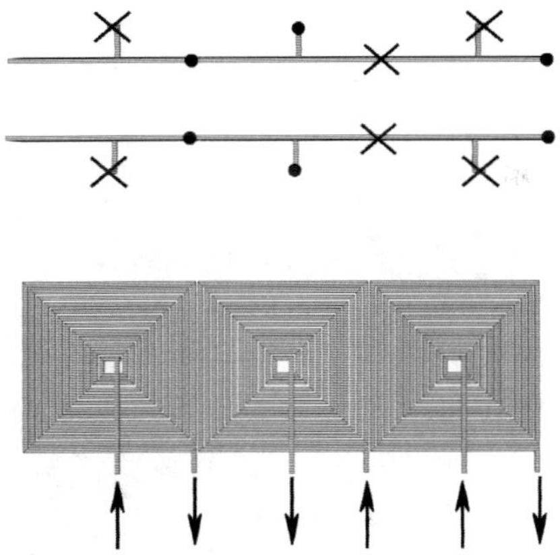

Figure 7. With opposite connection of adjacent coils and opposite connection of opposite coils

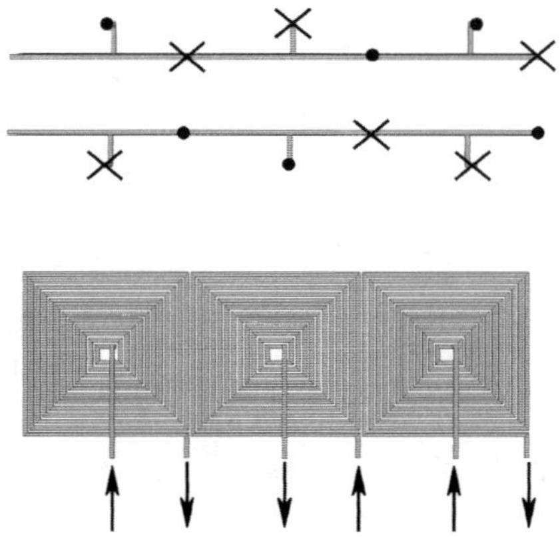

Figure 8. With opposite connection of adjacent coils and accordant connection of opposite coils

4.3.1. With accordant connection of adjacent coils and opposite connection of opposite coils

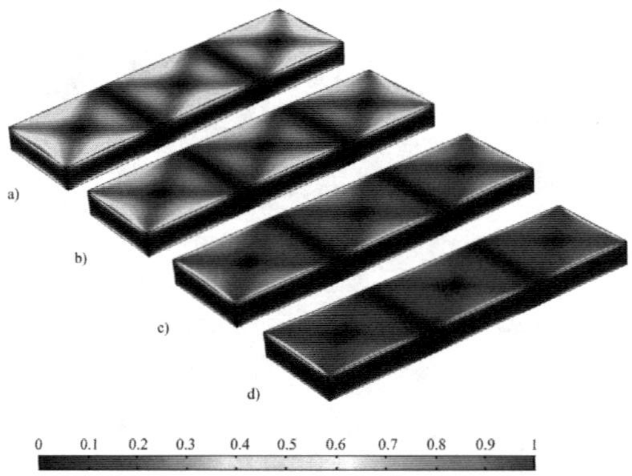

Figure 9. The distribution of electromagnetic power in the workpiece: a) 50 Hz, b) 100 Hz, c) 250 Hz, d) 500 Hz

Table 6: Efficiency values for various cases

Parameter	a)	b)	c)	d)
η	0.396	0.416	0.433	0.443
k_u	0.27	0.224	0.171	0.138

4.3.2. With opposite connection of adjacent coils and opposite connection of opposite coils

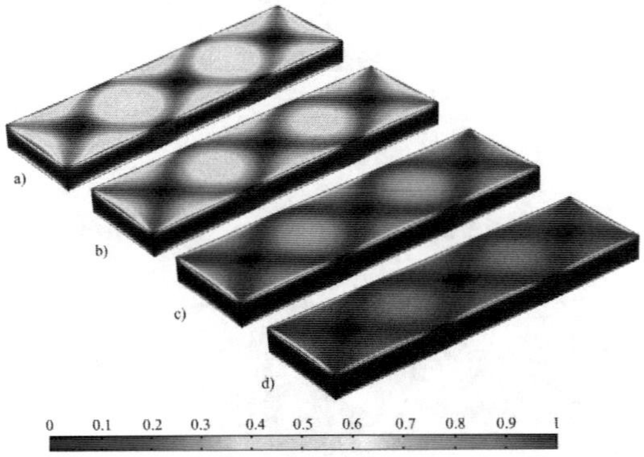

Figure 10. The distribution of electromagnetic power in the workpiece: a) 50 Hz, b) 100 Hz, c) 250 Hz, d) 500 Hz

Table 7: Efficiency values for various cases

Parameter	a)	b)	c)	d)
η	0.462	0.479	0.494	0.502
k_u	0.357	0.292	0.221	0.178

4.3.3. With opposite connection of adjacent coils and accordant connection of opposite coils

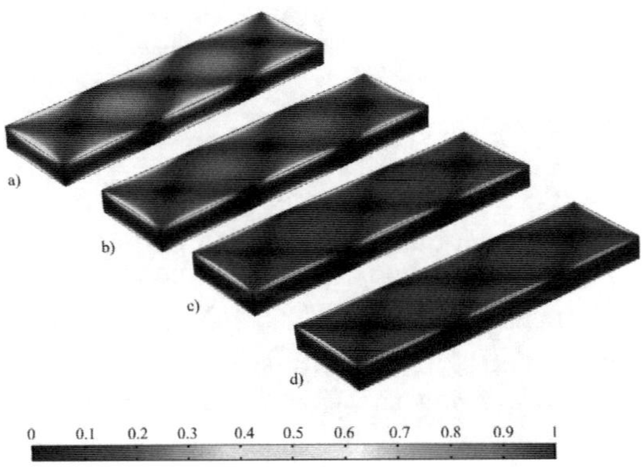

Figure 11. The distribution of electromagnetic power in the workpiece: a) 50 Hz, b) 100 Hz, c) 250 Hz, d) 500 Hz

Table 8: Efficiency values for various cases

Parameter	a)	b)	c)	d)
η	0.51	0.522	0.533	0.54
k_u	0.205	0.173	0.134	0.109

4.3.4. With three-phase power supply and accordant connection of opposite coils

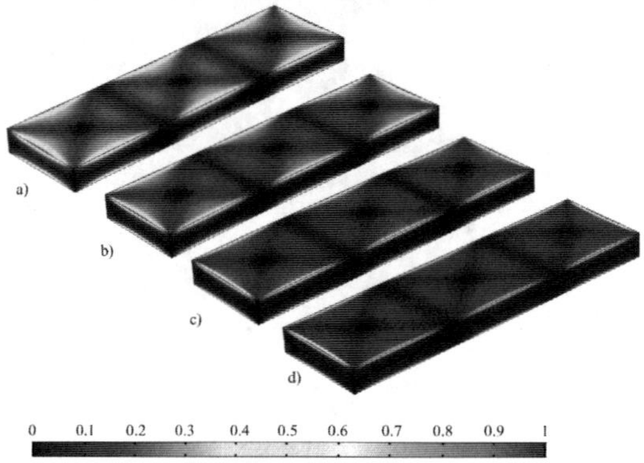

Figure 12. The distribution of electromagnetic power in the workpiece: a) 50 Hz, b) 100 Hz, c) 250 Hz, d) 500 Hz

Table 9: Efficiency values for various cases

Parameter	a)	b)	c)	d)
η	0.458	0.471	0.484	0.491
k_u	0.212	0.178	0.138	0.113

5. Conclusions

Based on fig. 2– 5, fig. 9– 12, as well as tab. 2– 9, we can conclude that the best variant in terms of efficiency and uniformity of electromagnetic power density is the solenoidal-type inductor without a side magnetic core (although its presence increases the efficiency, but the uniformity of the power distribution is worsened). This variant of the inductor is the most inconvenient for the operating procedure, because loading the workpiece into the inductor is complicated.

C-type inductor with all magnetic cores also has a good efficiency equals to 0.63 for 50 Hz (see tab. 2), but with this configuration, there is a significant nonuniformity in electromagnetic power density distribution (fig. 2). A compromise in this case can be a configuration with half of middle magnetic core, which provides efficiency equals to 0.57 and uniformity coefficient equals to 0.505.

In the case of six flat-type inductors, the best variant for optimal efficiency and uniformity would be a configuration with with opposite connection of adjacent coils and opposite connection of opposite coils (fig. 7).

The use of a three-phase power supply system for inductors of the C-type and six flat-type inductors, although it provides load symmetry, does not show satisfactory results in terms of efficiency and power distribution in the workpiece.

An increase in the supply frequency shows an increase in efficiency, but the distribution of power density becomes less uniform, the maximum power is on the edges of the workpiece, which can cause overheating in these places.

References

[1] Rudnev V, Loveless D and Cook R 2017 *Handbook of Induction Heating* Manufacturing Engineering and Materials Processing (CRC Press) ISBN 9781351643764
[2] Lupi S, Forzan M and Aliferov A 2014 *Induction and Direct Resistance Heating: Theory and Numerical Modeling* EBL-Schweitzer (Springer International Publishing) ISBN 9783319034799
[3] Liu H P, Wang X H, Si L Y and Gong J 2020 *Journal of Iron and Steel Research International*
[4] Mei R, Li C, Han B and Liu X 2008 *Materials Science Forum* **575-578 PART 1** 282–287

IOP Conf. Series: Materials Science and Engineering **950** (2020) 012030 doi:10.1088/1757-899X/950/1/012030

Technology and automation of low-temperature distributed induction heating in the oil and gas industry

Yu. A. Nikitin

Ufa, associate Professor of Ufa state aviation technical University, scientific adviser of LCC "Geoburservis"

nikyu@yandex.ru

Abstract. Energy efficient electrotechnologies are one of the most promising areas of development of any industry. To improve the energy efficiency of the processes of keeping up the temperatures and heating technological facilities in the oil and gas industry by reducing electricity consumption, the development and introduction of modern technologies and equipment is needed of electroheating. To such means of electroheating includes low-temperature induction heating, which provides a guaranteed increase in quantitative and qualitative production indicators, which can significantly reduce the cost of production, which is important to save energy resources especially with increased tariffs. Various aspects of the development of such induction heating systems are discussed, with a particular focus on ensuring the safety and automation of installations used in the oil and gas industry. The issues of using distributed induction heating in the oil industry are considered using the example of the induction method for oil and gas producing wells with the aim of facilitating the production of highly viscous oil, eliminating tar-tar deposits and preventing the formation of hydrate-paraffin plugs. The results of modelling the heating of the inductor - tubing system at various positions of the load-bearing and non-load-carrying cable-inductor are presented. During the simulation, data were obtained on the temperature of the oil product were obtained under various oil production conditions.

1. Introduction

Maintaining technological temperatures or heating the product in various technological devices, pipelines, tanks and containers, heat exchangers, when heating valves, water pipes and other equipment is widely used for technological purposes at the enterprises of the oil and gas industry. At the same time, such technological operations as compensation of heat losses to the environment, and directly heating the product in accordance with a given technology are characteristic. To increase the energy efficiency of these operations, the latest achievements in the field of heating electrical technology should be used.

In the case of extended pipelines, tanks or large-volume containers, usually the product heating temperature does not exceed the permissible temperature for the transported or stored product, i.e.

such heating systems belong to the class of low-temperature heating (for oil products, the surface of a heated object usually does not exceed 2500 ° C). Therefore, it is advisable in this case to use heating systems distributed over the surface of the heated object. Such heating systems, in contrast to concentrated (for example, heating elements) heating systems, have a large heating area (tending to the full surface of the heated object in the limit), and, therefore, better heat transfer and a significantly lower temperature gradient for transferring thermal energy from the heater to the product.

At present, indirect electric heating using resistive [1] or self-regulating heating cables [2] is increasingly used for distributed heating systems, which replaced steam heating, which has low efficiency, but is widely used in the petrochemical industry. However, a significant disadvantage of this method of electric heating is that even with good contact of the heating cable with the heated surface of the object, it is not possible to provide low thermal resistance between the heating cable and the surface of the heated object. As a result, the heating cable in order to transfer heat to the heated object must be heated to a high temperature significantly exceeding the required surface temperature of the heated object. This imposes a restriction on the use of the heating cable in explosive areas. In addition, when using self-regulating heating cable, local overheating of the cable is in loose contact with a heated surface will lead (at the expense of self-regulation) to reduce the temperature of cable (to decrease its specific capacity), and, consequently, to reduce heat transfer from the cable to the heated object, which ultimately will lead to the inability to provide specific technological regimes in certain heating zones.

An alternative to the heating cable is induction distributed heating, which, in fact, is a direct method of heating the metal surface.

2. Description of the method of distributed low-temperature induction heating of technological objects

For the first time, induction distributed heating is mentioned in [3], where the principal possibility of heating extended pipelines to maintain process temperatures is considered. It is noted that for heating pipelines, cylindrical inductors can be used from a wire with heat-resistant insulation or a single-phase cable with a large winding step. As the winding step increases in the limit, we get a two-wire line laid along the pipeline. The pipe itself can be used as a return pipe. The power is determined from the calculation of heat losses, taking into account the adopted insulation and the temperature of the pipe and the environment. Typically, the average power is a fraction of a watt per square centimeter of pipe.

Thus, in a distributed induction heating on the metal surface of the heated object (technological apparatus, pipeline, container, etc.) is laid (linearly or along any given contour), a conventional mounting copper stranded cable in a shell that can withstand a given surface heating temperature (for example, in a fluoroplastic shell or a shell made of organosilicon rubber, the cable temperature can reach 180 ° C). In contrast to the heating cable, a conventional mounting cable has a significantly lower cost. This cable is connected to an AC power source. Induction heating occurs as a result of the penetration of electromagnetic waves into the metal volume. Since the conducting circuit or coil through which the alternating current flows is located near the metal object of heating (through a layer of heat and electrical insulation), as a result of the penetration of the electromagnetic field into the metal, eddy currents occur in it (due to the phenomenon of electromagnetic induction), which heat the metal (Joule heat is released). Thus, heating is carried out without direct contact of the induction cable with the metal, which provides a particularly low thermal inertia and high efficiency of converting electrical energy into heat.

At an average and high operating frequency, the depth of the heated layer is small and the alignment of the thermal field in the heating object is provided only by heat conduction processes. At lower operating frequencies, the depth of penetration of the electromagnetic field increases, but the achievement of uniform heating is also achieved through the mechanisms of thermal conductivity.

Lowering the operating frequency allows you to maximize the volume of the heated area, but in some applications it is more effective to use the increased frequency of the electromagnetic field. It should be noted that the heating rate is strongly dependent on the magnetic characteristics of the heated metal. The frequency range and shape of the alternating current is selected individually in each case and can reach wide limits (from the industrial frequency-50 Hz to the ultrasonic frequency range-tens of kHz). For use in the oil industry, the frequency selection method consists not only in the correct choice of the metal heating depth, but also in some cases, the need for vibration or electromagnetic action on the petroleum product in order to meet the specified technological requirements, taking into account the necessary heat transfer in individual elements of the heating system. Therefore, the development and implementation of induction technologies in the oil and gas industry is currently an urgent and promising direction and requires appropriate development in the field of scientific and engineering practice.

3. Advantages of distributed low-temperature induction heating

A current research show [4, 5] that induction heating distributed over the surface has the following advantages compared to other types of electric heating:

- Uniform heating throughout the volume and low thermal inertia. In installations with induction heating, rapid cooling and heating of the product is possible.
- Since the thermal resistance of the walls, which must be taken into account, for example, for indirect heating, is absent, the heating temperature decreases, and the total heating efficiency increases.
- It is easy to reproduce the heating conditions of the product in accordance with the specified technological process, so the spread of product parameters is insignificant; its quality is usually stable.
- The temperature limits of heating are wide, temperature control is possible within 0-500 °C.
- Long service life and low cost of the inductive cable.
- For the operation does not require special maintenance personnel.
- Inductors and control cabinets can be explosion-proof.
- Wide possibilities for heat exchange intensification by controlling the geometric shape of the heated layer.

In addition, with induction heating, all the advantages of electric heating are preserved: good sanitary and hygienic operating conditions, the absence of special maintenance personnel (except the apparatchik himself), lower capital investment, the ability to regulate and automate the process, and save scarce primary energy resources.

As you know [5], the choice of a particular type of heating equipment for conducting a particular process is due to a number of requirements:

- The resource of work, which determines to a large extent the cost of production, the duration of operation.
- The quality of heating, on which the quality of the processed product depends (the ability to control heating, the accuracy of regulation or stabilization of temperature).
- Profitability due to the minimum consumption of energy and materials.
- Electrical, fire and explosion safety.
- Constructive perfection of the equipment: minimum weight and dimensions, manufacturability of the design, ease of use, ease of installation and dismantling, ease of maintenance, etc. This requirement is relevant, since devices with heating contain technological and electrical elements, i.e. obviously quite complicated.
- The minimum cost of equipment.
- Good sanitary and hygienic working conditions and compliance with environmental requirements.

Induction heating satisfies all the above requirements.

In contrast to the considered distributed induction heating, the method of concentrated induction heating using an inductor coil is very widely used in mechanical engineering in induction furnaces for melting metal, when heating work pieces for forging, quenching [3]. A special feature and main advantage of concentrated induction heating is that it is heated to high temperatures (local heating). Currently, semiconductor frequency converters are increasingly used as power sources for inductors (generators) operating at frequencies higher than industrial frequency. The efficiency of using a particular class of converters for local induction heating, used in the engineering industry, is a well-studied issue and there are appropriate engineering methods for choosing a particular type of converters [6]. The circuitry of the power part and the control circuit of these converters is based mainly on ensuring safe switching of power semiconductors using an oscillating circuit that is the load of the Converter and an additional resonant circuit formed by the inductor and a compensating capacitance connected in parallel or in series with the inductor. This type of induction heater is characterized by heating the metal above the Curie point, which consequently leads to a sharp change in the magnetic properties of the latter (the magnetic permeability drops and becomes in some cases at the level of the magnetic permeability of the air). The consequence of such a sharp change in the physical properties of the metal is a significant detuning of the resonant circuit, which is the load of the semiconductor converter. Therefore, controlling the operation of such a Converter is quite a complex task and requires the mandatory use of phase-locked frequency in control systems.

With distributed induction heating, the temperature usually does not reach the Curie point, so there is no detuning of the resonant circuit and the requirements for Converter control systems are significantly reduced. As a result of the research and design work [4, 7], simple technical solutions were proposed for implementing algorithms and circuit solutions for controlling the power part of induction heaters using frequency-pulse modulation. It is shown that energy-efficient (high efficiency, ensuring the safe operation of power switches and drivers in a wide range of power control in the load and other indicators of the efficiency of the power supply) is the use of quasi-resonant circuits in of the power part of the converters [7], which reduces the cost of induction distributed heating systems in general.

4. Various aspects of development of distributed low-temperature induction heating systems

At the present stage of development of distributed induction heating systems due to only optimization of the design of converters and induction cables, it is quite difficult to substantially improve the technical characteristics of the heating system. Further research is required to identify the regularities of induction distributed heating systems in order to ensure their more economical, reliable, and safe operation (especially in high explosive conditions at oil and gas industry facilities), as well as ensuring coordination of their interaction with the work of existing automation systems for technological processes and production, which requires the use of more advanced in terms of integration of electronic control and management systems. Most often, various safety systems are connected to distributed heating systems, which (in automatic or semi-automatic mode) monitor the performance of devices, prevent the occurrence of emergency situations, protect converter equipment from damage, which leads to the need to develop:

- alarm devices that indicate an unauthorized stop of the heating system or the operation of the Converter in emergency mode;
- devices for protecting heating systems from short-circuiting the inducing cable to the metal surface of the heating object (short-circuiting the inductor to the ground in case of accidental mechanical damage to the latter);
- inducing cable protection devices that disable the Converter when the temperature rises above a certain value.

Also, devices that automatically turn on or turn off the heating system in a particular situation, for example, various sensors and actuators, can be connected to the system [8]:

- relays that turn the heating system on and off at certain times of the day;
- devices that automatically turn on the heating system after interruptions in the operation of the power supply network;
- devices that control several heating objects, including one or more generators and, for example, switch inducing cables at certain intervals;
- pressure and temperature sensors;
- interfaces allowing to connect external control and monitoring tools to heating systems;
- other electronic devices.

When choosing automation, it is necessary to take into account the power, type and number of semiconductor converters of the induction distributed heating system, the number and type of the induction cable used, on which the heating system can work. Since, in heating systems, the operation of the generator and induction cable is often tightly connected with the operation of the pump, moreover, both the heating system and the pump are controlled by the automation of the heating system.

Often, various control and monitoring systems of induction distributed heating systems are combined into sufficiently powerful multi-channel programmable electronic units that control not only the heating system directly, pumps, but also the corresponding shut-off valves.

Of great importance in an induction distributed heating system is the estimation of the cost of the "generator - induction cable" system, which is quite difficult to implement, since it depends on a variety of parameters and the mode of operation of the system. At the same time, despite the relatively high cost of powerful industrial converters, they pay off quickly. Moreover, since converters of induction distributed heating systems are necessarily equipped with soft start and soft stop devices, their use avoids sudden changes in the power consumed from the network, which in turn increases the service life of both the converter itself and induction cable. In contrast to the use of traditional resistive heating cables, the overall efficiency of the heating system increases, since the heating cable is characterized by the presence of significant starting currents until the cable and the heating object itself are warmed to the nominal temperature. With the help of modern semiconductor converters, can not only control the operation of the inducing cable and the heating system as a whole, but also monitor their main parameters in real time. The most advanced models can capture and transmit up to several dozen operating parameters of the heating system in real time. Processing this information allows you to perform deep diagnostics of equipment and prevent the occurrence of emergency situations.

5. Application of distributed induction heating in the oil industry as an example of induction effects on oil and gas wells

As an example of the use of distributed induction heating in the oil industry, we consider the method of induction impact on oil and gas wells in order to facilitate the production of high-viscosity oil, eliminate asphaltene, wax deposits and prevent the formation of hydrate and paraffin plugs [9, 10]. The method of induction impact on the well includes:

- Induction heating of the geophysical cable in metal armor, which is immersed either directly into the pump-tubing of an oil and gas well through a lubricator, or in the tube space between the pump-tubing and the casing of the well, and maintaining the temperatures in the wells, as well as additional heating by the well of the liquid from the heated geophysical cable.
- Ultrasonic vibration effect on downhole fluid and oil plates.
- Additional electromagnetic effects on the borehole fluid.

In the case of using geophysical cable with an outer shell of polypropylene or fluoropolymer, while passing high frequency current through the veins of the geophysical cable thermal, vibrational, acoustic and electromagnetic field is formed directly in a conducting body [11] (production string and tubing hanger) and on bottom-hole formation zone and the downhole fluid, respectively, without heating the cable itself (the cable should have proper insulation and mechanical strength to hold it well).

Method of induction impact on the well [10] (Fig. 1) ensures the safety of work, reducing the time spent on the elimination of asphaltene and paraffin deposits in the inter-pipe and internal space of the tubing and, accordingly, the time of preventive treatment without stopping the well operation, high efficiency and, as a result, low operating costs. Work is carried out without involving repair teams. During preventive work, complete cleaning of the inner column space from paraffin is provided, the interprophilactic period of well treatment is significantly increased, and the power consumption during preventive work is usually no more than 25 kW.

For fig.2 shows the appearance of the control station for induction processing, which contains a high-voltage high-frequency transformer and a power converter on IGBT transistors, made according to the bridge scheme, and also shows how the geophysical cable is immersed in a well with a rod pump. The geophysical cable is located on the winch and is smoothly fed through the lubricator into the well into the inter-tube space between the casing and the tubing.

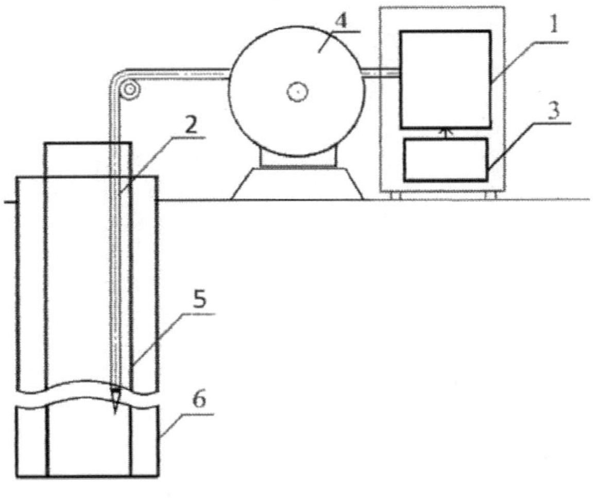

Fig. 1 Composition of equipment for induction effects on the well: 1 - high-frequency high-voltage transformer; 2 - geophysical cable; 3 - control station; 4 - a drum with a geophysical cable; 5 - tubing (a pump-compressor pipe, pump-tubing); 6 - casing string.

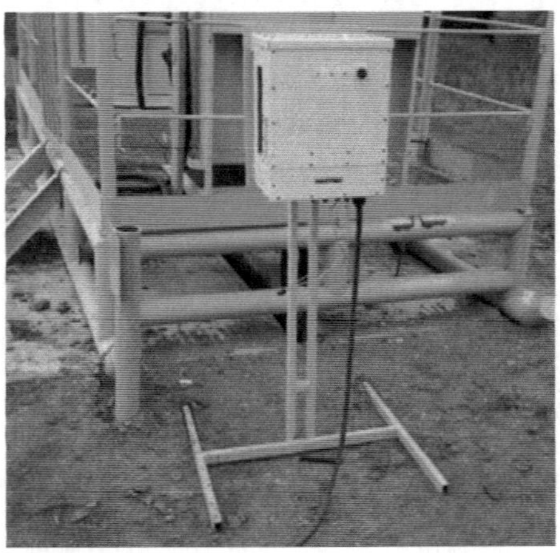

Fig. 2 Control station (left) and immersion of the geophysical cable into the well (right).

In all types of wells, the cable-inductor can be installed by hanging the cable at the mouth: in wells with an electric pump through the lubricator, in wells with a rod or screw deep pump through the gland device into the inter-tube space (the cable is lowered and lifted by a standard geophysical winch without stopping the well pump).

To model the phenomena occurring during the induction action on the tubing, consider a heating system consisting of an inductor – 2 and a pump-compressor pipe – 5 (Fig.1) [12]. The heating system is placed in the casing 6. The pump power cable is located in the inter-tube space. The length of the inductor cable must be greater than the zone of formation of asphaltene and paraffin deposits in the tubing (usually no more than 1000 m).

Modeling of electromagnetic and thermal processes in the inductor-tubing system was carried out in the COMSOL Multiphysics application package [13]. In the model creation wizard, we step by step define the model's design area, enter the model's geometric parameters, select materials for each geometry object, select AC/DC (Magnetic Fields) and Heat Transfer (Heat Transfer in Solids) physics, the Frequency-domain solver, and the required type of study, such as Temperature. The considered model of tubing with an inductor placed in it is shown in Fig.3.

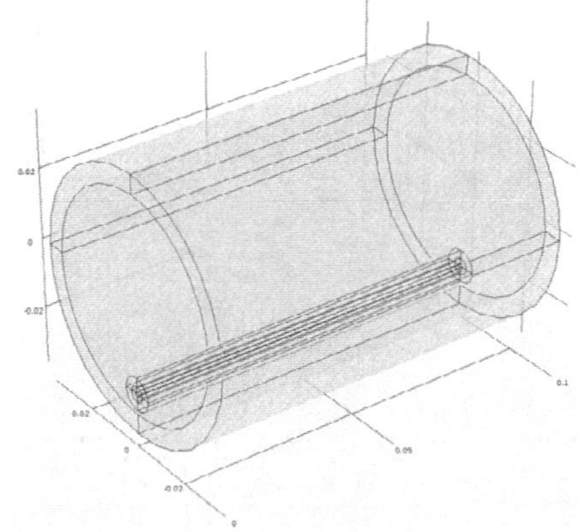

Fig. 3 Tubing model in COMSOL Multiphysics

The simulation was performed under the condition that the cable-inductor is either adjacent to the inner surface of the tubing, or is located in the center of the tubing. During the simulation, tubing with an average nominal diameter of 73 mm was selected. Single-core load-carrying and non-load-carrying cables were considered as an inductor. The following simplifications are introduced into the model: the current-conducting cores of the inductor are represented as a single conductor with an equivalent cross-section area, the inner and outer layer of the load-carrying cable's armor is represented as a single shell made of galvanized steel. Physical characteristics of the model are as follows: inductor length – 0.3 m, conductor diameter – 11.38 mm for load-carrying cable, 5.64 mm for non-load-carrying cable, cable insulation thickness – 1.6 mm, armor thickness of the load-carrying cable – 2.4 mm, tubing diameter-73 mm, tubing wall thickness-5.5 mm, frequency of current passed through the veins of the cable-inductor -20 kHz, cable-inductor current-10 A, materials-iron, copper, fluoroplast, oil.

When oil production 1-20 m^3/day flow rate of oil in the tubing will be respectively equal to 0.23 m/ min 4.6 m/min. The length of the inductor is equal to 1000 m choose To simplify the construction of models and presentation of modeling of processes of heating wells, we consider only the part of the inductor, which is many times smaller than the heated portion of the tubing (Fig. 3). This assumption is correct, since the induction heating of the tubing has a uniform increase in the temperature of the tubing surface under the condition of mixing the oil in the tubing. Based on this, knowing the speed of the oil flow in the tubing, we can determine the time during which the oil will rise to the surface from a depth of 1000 m. All this time, the oil will be heated inside the tubing under the action of the heated inductor shell. The lifting time of oil in the tubing, respectively, is from 217.4 minutes to 4347.8 minutes.

When using a non-load-carrying inductor, the tubing in well is heated by eddy currents. In turn, the tubing surface in contact with the oil heats it and, accordingly, the tubing surface is cleaned of asphaltene and paraffin deposits. Some results of modeling of oil heating under induction influence on the well by load-bearing and non-load-bearing cables with production of 1 m^3/day are presented in Fig.4 with oil temperature indicators.

When modeling oil heating, when the non-load-carrying inductor was adjacent to the tubing surface, the required temperature was reached after 1050 minutes (17.5 hours) of inductor operation. When the required temperature is reached, the device turns off the current supply to the inductor, and the temperature increase stops. The current supply will resume when the oil temperature set for a particular well is reached. In the oil heating model with a non-load-carrying inductor located in the center of the tubing, the required temperature was reached after 2300 minutes (38.3 hours) of inductor operation.

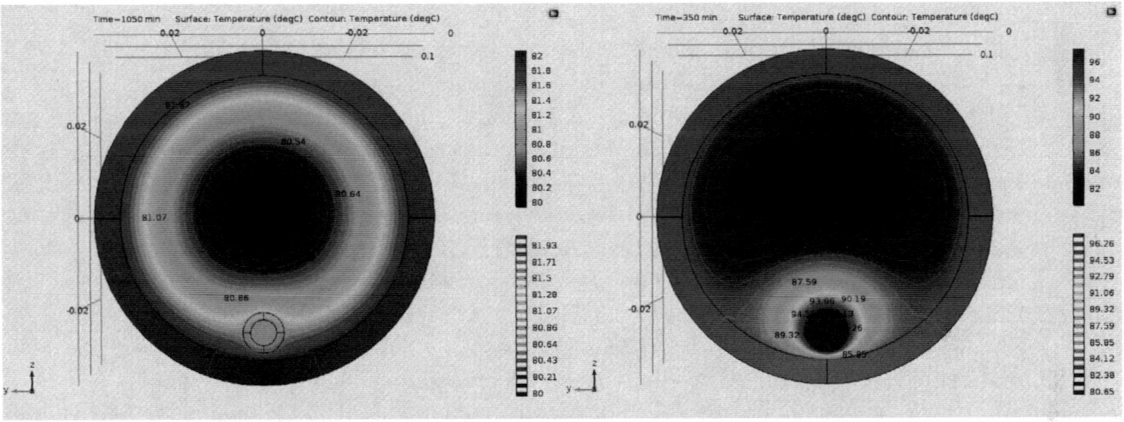

Fig. 4 Section of a model with a non-carrying cable (left) and with a carrying cable (right).

When using a load-carrying inductor, oil is heated from the surface of the armor heated by eddy currents. It took 350 minutes (5.8 hours) for the inductor to heat to the required temperature. With the central position of the inductor, the required temperature was reached after 250 minutes (4.2 hours) of operation of the inductor.

Similarly, the heating of oil was considered under the induction effect on a well with a production of 20 m^3/day. During the time required to rise to the mouth well, the oil was heated only to a temperature of 47–48 °C. When simulating heating with a non-load-carrying cable in the center of the tubing during production of 20 m^3/day, the average oil temperature was 43 °C. As a result of the simulation, during the time necessary for the ascent to the mouth, the oil was heated mainly to a temperature of 65–67 °C. When simulating heating with a load- carrying cable in the center of the tubing during production of 20 m^3/day, the average oil temperature was 75–85 °C.

Thus, using the developed models, information was obtained on the nature of heating the tubing with a non-load- carrying and load- carrying geophysical cable used as an inductor at their various positions inside the tubing. When modeling electromagnetic and thermal processes in a well, oil temperatures were obtained under various production conditions.

6. Conclusions

The use of low-temperature distributed induction heating in the oil and gas industry helps to reduce the cost of production and increase the quantitative and qualitative indicators of production. Information is provided on the development of such induction heating systems that ensure the safety and automation of installations.

With distributed induction heating, the temperature of the metal surface of the heating object is significantly lower than the Curie point, so there is no mismatch in the load loop of power converters and the requirements for their control systems are significantly reduced.

Modeling of electromagnetic and thermal processes using COMSOL Multiphysics shows that the downhole temperature regime for distributed induction heating of well pipelines satisfies the conditions under which asphaltene and paraffin deposits in the well are removed, which facilitates the production of high-viscosity oil.

References

[1] Fonarev Z. I. Electric Heating of pipelines, reservoirs and process equipment in the oil industry. – Leningrad: Nedra, 1984. – 148.

[2] Strupinsky M. L. Design and operation of electric heating systems in the oil industry: reference book / M. L. Strupinsky, N. N. Khrenkov, A. B. Kuvaldin. – Moscow: Infra-Engineering, 2015. – 272 p.

[3] Induction heating installations: A textbook for universities/ A. E. Slushotsky, V. S. Nemkov, N. A. Bamuner; edited by A. E. Slushotsky – L.:, Energoizdat, 1981. – 224 p.

[4] Makulov I. A., Nikitin Yu. A. Equipment and features of application of induction heating in the oil and gas industry. - Moscow: Industrial electric heating and electric heating. 2014, No. 3. Pp. 50-53.

[5] Nikitin Yu. A. Technology and automation of low-temperature induction distributed heating. - Automation and management of technological and production processes: Materials of the all-Russian scientific and practical conference (Ufa state aviation technical University. – Ufa: USATU, 2011. – Pp. 226-231.

[6] Development and design of thyristor frequency converters /A. K. Belkin, S. A. Gorbatkov, Yu. m. Gusev, I. I. Parfenov, A. A. Shulyak-M.: Energoatomizdat, 1994. – 272 p.

[7] Nikitin Yu. a. Transistor converters of constant voltage of class E. / / Kand. thesis. - MAI, 1990. - 225 p.

[8] Nikitin Yu. a. Technical means of automation and quality management systems: A textbook for students of higher educational institutions studying in the direction of training "Automated technologies and production" / Yu. a. Nikitin; Federal Agency for education, State educational institution of higher education. Prof. of education Ufa state aviation technical University-T. Ufa, 2008. – 223 p.

[9] Makulov I. A., Nikitin Yu. A., Nikitin A. Yu.., Makulov R. I. Method of destruction and prevention of formation of deposits and traffic jams in oil and gas wells and device for its implementation. - Patent for invention RU no. 2503797. Registered in the State register of inventions of the Russian Federation. - Byul. No. 14. Published on 10.01.2014.

[10] Nikitin Yu. a., Osipov V. V., Nikitin A. Yu. Method of elimination, prevention of formation of deposits and intensification of oil production in oil and gas wells and device for its implementation. - Patent for invention RU No. 2630018. Registered in the State register of inventions of the Russian Federation. - Byul. No. 25. Published on: 05.09.2017.

[11] Makulov I. A., Nikitin Yu. a., Nikitin A. Yu., Makulov R. I. Method of transportation and discharge of high-viscosity fluids. - Patent for invention RU No. 2568084. Registered in the State register of inventions of the Russian Federation. - Byul. No. 31. Published on: 10.11.2015.

[12] Nikitin Yu. a. Modeling and calculation of the inductor-pump-compressor pipe system. – In the collection: machine tool Construction and innovative engineering. problems and points of growth: Materials of the all-Russian scientific and technical conference. Ufa, 2018. Pp. 170-175.

[13] Pracht V. A., Dmitrievsky V. A., Sarapulov F. N. Modeling of thermal and electromagnetic processes in electrical installations. Comsol program: Textbook / Edited by F. N. Sarapulov- M.: Sputnik + publishing House, 2011. - 158 p.

IOP Conf. Series: Materials Science and Engineering **950** (2020) 012031 doi:10.1088/1757-899X/950/1/012031

The influence of electric current application configuration on the electro-vortex flow structure of conductive medium in cylindrical cell

S Mandrykin, I Kolesnichenko

Institute of Continuous Media Mechanics, Perm, Russia

E-mail: msd@icmm.ru

Abstract. A numerical study of the liquid metal electro-vortex flows (EVF) in a closed cylindrical cell, the radius of which is equal to its height, is performed. A direct current of 1000 A is applied through the electrodes at the bottom cylinder face, and is collected at the entire surface of the cylinder top end face. Various configurations of electrode location are considered. Namely, from one to five electrodes are placed on the cylinder bottom end face. The three-dimensional fields of the conducting medium flow velocity in the cell are obtained as a result of numerical simulations. The EVF is non-stationary in all considered regimes. In the one-electrode case, the flow is poloidal and is represented by one large-scale vortex. In the multielectrode case, the flow consists of multiple small-scale vortices, the size and quantity of which depend on the configuration of the electric current application. At the same electric current value of 1000 A, the mean flow velocity and characteristic frequency of the process are higher for the localized current application — 6.4 and 2.5 times respectively.

1. Introduction

Electro-vortex flows (EVF) are at the root of many modern technological applications for liquid metals, such, for example, as arc furnaces [1, 2], aluminum reduction cells [3], as well as actively developing liquid metal batteries [4–6]. EVF arise as a result of the interaction between an electric current in a conducting medium and its own magnetic field, provided that the current distribution is nonuniform [7].

The structure and characteristics of EVF significantly depend on the configuration of the problem. There are well-studied canonical EVF configurations, such as hemispheres [8] and flat layers [9–11] (including the multielectrode electric current application [9]). Albeit it is shown that in some cases the formation of stationary EVF is feasible, in the real-world industrial applications they are usually non-stationary. The non-stationary nature of EVF can have a positive effect on the product of the technological process, for example, it can enhance the mixing of the molten metal, which inevitably improves the quality of the ingots produced.

However, in liquid metal batteries the EVF have both positive and negative effect. On the one hand, they introduce a positive effect of mixing the melt, thereby decreasing formation of the intermetallic compounds, which can affect corrosion rate and reduce the device lifespan. Moreover, this mixing can help equalizing temperature fluctuations of the melt [12]. At the same time, the expected current values during operation of such devices are thousands of amperes, which corresponds to EVF with high intensity. Thus, EVF can lead to deformation of the electrolyte layer [13], previously observed in aluminum reduction cells [14], and its excessive thinning, causing a subsequent short-circuiting and failure of the device [15–17].

The simplest way to decrease the EVF intensity, while maintaining the electric current value, is to increase the electrode cross section, that is, to reduce the non-uniformity of the electric current distribution. Another approach is to increase the number of electrodes used. Obviously, the latter option is more preferable from both the economic and a technical points of view.

In this work, the EVF of a conducting medium in a closed cylindrical cell is studied numerically for various electric current application configurations. A direct current of 1000 A is applied through electrodes at the cell bottom end face and is collected over the entire surface of the cell top end face. Problem configurations with electrodes amount from 1 to 5 are considered.

2. Methods

The problem is solved numerically using the Ansys software. Namely, the Ansys Emag is used for simulating the electromagnetic part of the problem, and the Ansys Fluent — for the hydrodynamic part. The computational domain is a cylinder with a height of $H = 0.100$ m and a diameter of $D = 0.200$ m, as shown in Fig. 1. The electrodes with a radius of 0.020 m are are located at the lower end 1 of the cylinder. The electric current is collected at the entire top surface 2 of the cell. The Cartesian coordinate system is used in the simulations and in the presentation of results. Its origin is located in the center of the cylinder bottom end face, and the Oz axis is collinear to the central symmetry axis of the cell. An unstructured mesh with tetrahedral elements is used. The linear size of its elements does not exceed 3 mm. For the velocity, the no-slip conditions are applied at all boundaries. The problem is considered isothermal. All computations begin from a state of equilibrium.

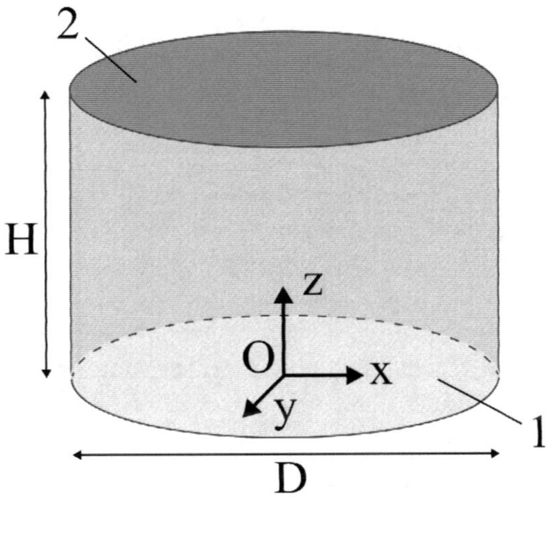

Figure 1. Computational domain scheme

The following configurations of the distributed current application are considered, as shown in Fig. 2: 5 electrodes, one of which is located in the center, and the other four are equidistant from it with the inter-space 0.070 m; 4 electrodes, also equidistant from the center; 3 electrodes are lined up, so that two electrodes are equidistant from the third one, located in the center; 2 electrodes are equidistant from the center. The results are also presented for the case of localized

IOP Conf. Series: Materials Science and Engineering **950** (2020) 012031 doi:10.1088/1757-899X/950/1/012031

current application (the case when a single electrode is located at the center) to compare it with that of the distributed electric current application. For the sake of simplicity, in the following, each of the above configurations will be designated by the number of electrodes it includes.

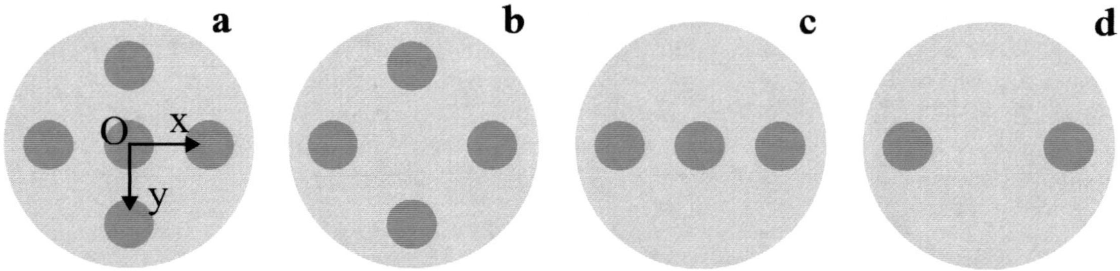

Figure 2. The scheme of distributed current application configurations

Gallium alloy GaSnZn is selected as a conducting medium, which has the following material parameters: kinematic viscosity $\nu = 3.1 \cdot 10^{-7}$ m^2/s, density $\rho = 6265$ kg/m^3, electric conductivity $\sigma = 3.56 \cdot 10^6$ S m^{-1}. The electric current has a fixed value of 1000 A.

The wavelet analysis was used in processing results [18, 19], since the numerical simulations showed the presence of non-harmonic oscillations in the process. The wavelet transform translates a one-dimensional time series $f(t)$ into the two-dimensional frequency-time plane (ν, t) is performed as:

$$W(\nu, t) = \sqrt{\nu} \int f(\tau)\,\psi\big(\nu, (\tau - t)\big)\mathrm{d}\tau.$$

The Morlet wavelet

$$\psi(t) = \exp\left[-t^2/2\sigma^2\right] \exp\left[i2\pi t\right],$$

where σ is an adjustable parameter, is one of the most popular wavelets. It was selected in this study as an analyzing wavelet, since it better suits for signals of complex nature. By changing the value of σ one can obtain optimal resolutions of time t and frequency ν. Low values of σ are used to achieve better time resolution, whereas large ones can provide better frequency resolution of the wavelet. However, an excessive increase of σ is undesirable, since it leads to attenuation of the analyzing function. The commonly adopted value of σ is 1 [20], while the limit $\sigma \to \infty$ corresponds to the Fourier transform. In this paper we assume $\sigma = 3$.

The map $W(\nu, t)$ has an easily interpreted structure only in case of the simple data analysis. However, in case of the real data, wavelet map interpretation can be a lot more difficult. The wavelet transform integral spectrum

$$\mathrm{M}(\nu) = \int |W(\nu, t)|^2 \mathrm{d}t,$$

which is one of its simplest characteristics, is calculated as a convolution of $W(\nu, t)$ along the time axis. It is a smoothed Fourier transform. The smoothing degree decreases with increase in σ.

3. Results

Several flow regimes that are realized with a localized and distributed electric current application, including up to 5 electrodes (see Fig. 2), are studied. The three-dimensional velocity fields of gallium alloy in the cell are obtained. To evaluate the flow characteristics, the time dependencies of the velocity z-component in the cell center are used, since it is precisely the component, which is measured by a Doppler anemometer in such laboratory installations [21,22]. The mean flow velocity in the cylinder center does not exceed 0.10 m/s, which corresponds to the Reynolds number determined through the cylinder diameter ($\mathrm{Re} = \langle V_z \rangle D/\nu$) $\mathrm{Re} = 64516$ for the examined configurations of localized and distributed current application (5 electrodes). The mean flow velocity in case of the localized current application is up to 6.4 times higher, than that in the multielectrode case.

A qualitative flow structure can be obtained using the visualizations of the velocity fields in the zOy section, shown in Fig. 3. Here, the color indicates the velocity magnitude, and the lines indicate the interpretation of the velocity vector field. As is evident from the figure, in the multielectrode case there is no pronounced large-scale single-vortex structure. In contrast, in the case of localized current application (Fig. 3a), one large-scale toroidal vortex is clearly distinguishable. In all the studied multielectrode configurations, the flow is represented by many small-scale vortices (Fig. 3b, c, d, e), the amount and size of which vary with the variation of the current application.

Fig. 4 shows the time dependencies of the velocity z-component in the center of the cell, for the multielectrode configurations only. It follows from the figure, that the observed EVFs are non-stationary, and differ both in the mean velocity and in statistical characteristics. Such a flow pattern is preserved over the entire time interval studied, which is significantly higher than the typical characteristic time scales in such problems [21].

The statistical characteristics of the process are studied using the wavelet analysis. In particular, the wavelet diagrams of the velocity z-component of the gallium alloy in the center of the cell were plotted (see Fig. 5), using using signal data presented in the Fig. 4. The wavelet diagrams show that the observed EVF in the cylinder are of non-stationary character, and the oscillations are non-harmonic.

The wavelet analysis of the data also makes it possible to calculate the wavelet integral spectrum, which is more convenient for quantifying the characteristic frequencies of the process. The values of these frequencies for each of the considered configurations are presented in Fig. 6. The figure shows the presence of two frequencies, one of which is higher than the other one. In the multielectrode case, the first frequency increases with increasing number of electrodes, reaching a maximum value of 0.10 Hz, which is 2.5 times lower than that with a localized current supply. With a further increase in the number of electrodes, this frequency decreases. The second (lower) frequency decreases monotonically with increasing number of electrodes.

Thus, the distributed electric current application reduces both the intensity and the characteristic frequencies of a non-stationary EVF.

4. Conclusions

The electro-vortex flow (EVF) of liquid metal in a closed cylindrical cell, with the height equal to its radius, is studied numerically. The direct electric current is applied at the cell bottom end face and is collected at the cell top end face. The the numerical simulations are carried out for configurations that include from 1 to 5 electrodes, placed on the bottom end face. The electric current value was fixed at 1000 A. The three-dimensional velocity fields of gallium alloy in the cell are obtained, as well as the statistical characteristics of the process.

It is shown that in all considered configurations the flow is non-stationary. In the case of localized current application, the flow is represented by one large-scale vortex. In the case of distributed electric current application the flow is represented by multiple small-scale vortices. The size and number of these vortices depend on the number of electrodes used. In the case of localized current application the mean flow velocity is 6.4 times higher and the process characteristic frequency is 2.5 times higher than in the multielectrode case.

Thus, for the same electric current value of 1000 A, the use of multielectrode electric current

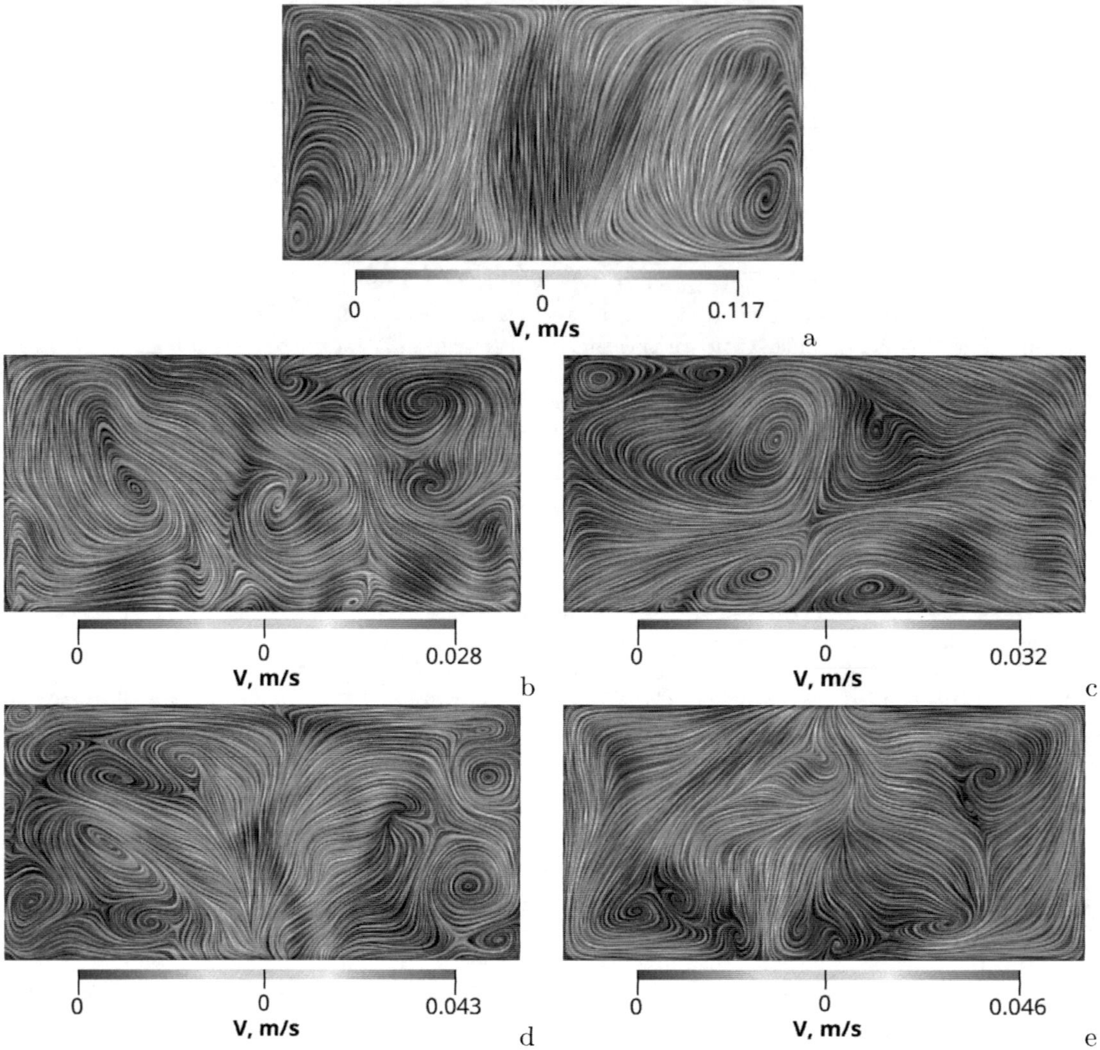

Figure 3. Visualizations of instantaneous velocity field in the zOy section. The color represents the velocity magnitude. The number of electrodes is: a — 1, b — 5, c — 4, d — 3, e — 2

application reduces both the intensity and the characteristic frequency of the EVF, compared to the case of the localized current application.

Acknowledgments

The study was supported by the Government of Perm Region (Program for the support of Scientific Schools of Perm Region, grant C-26/788).

IOP Conf. Series: Materials Science and Engineering **950** (2020) 012031 doi:10.1088/1757-899X/950/1/012031

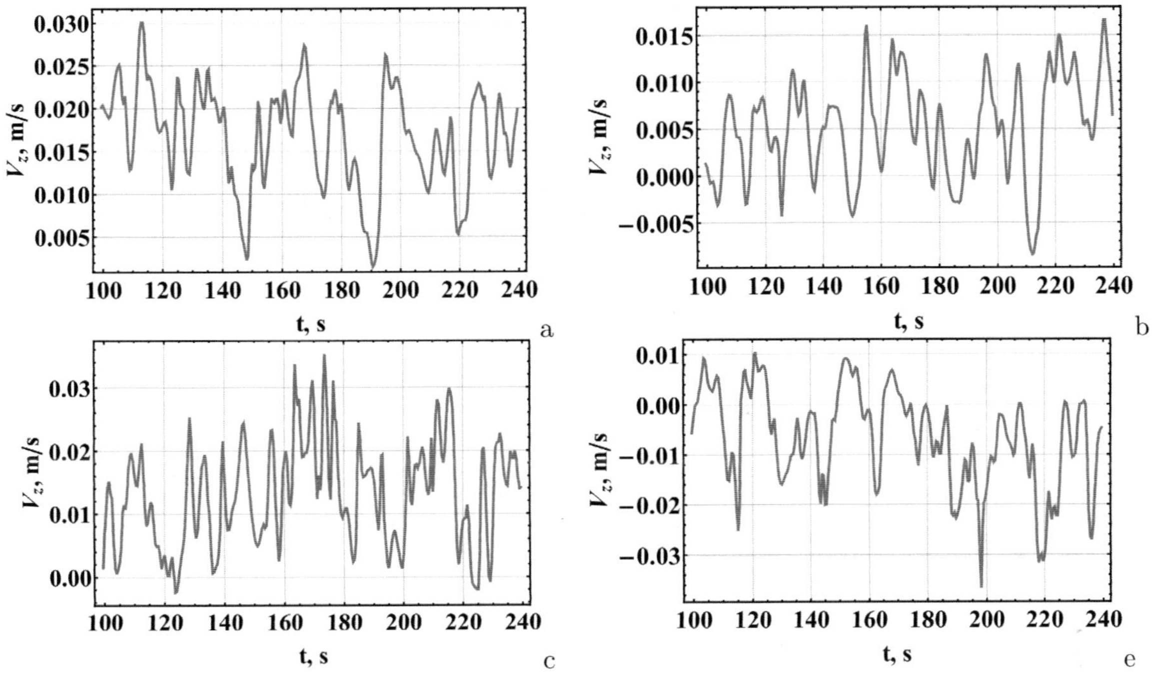

Figure 4. The velocity z-component as a function of time, in the center of the cell. The number of electrodes is: a — 5, b — 4, c — 3, d — 2

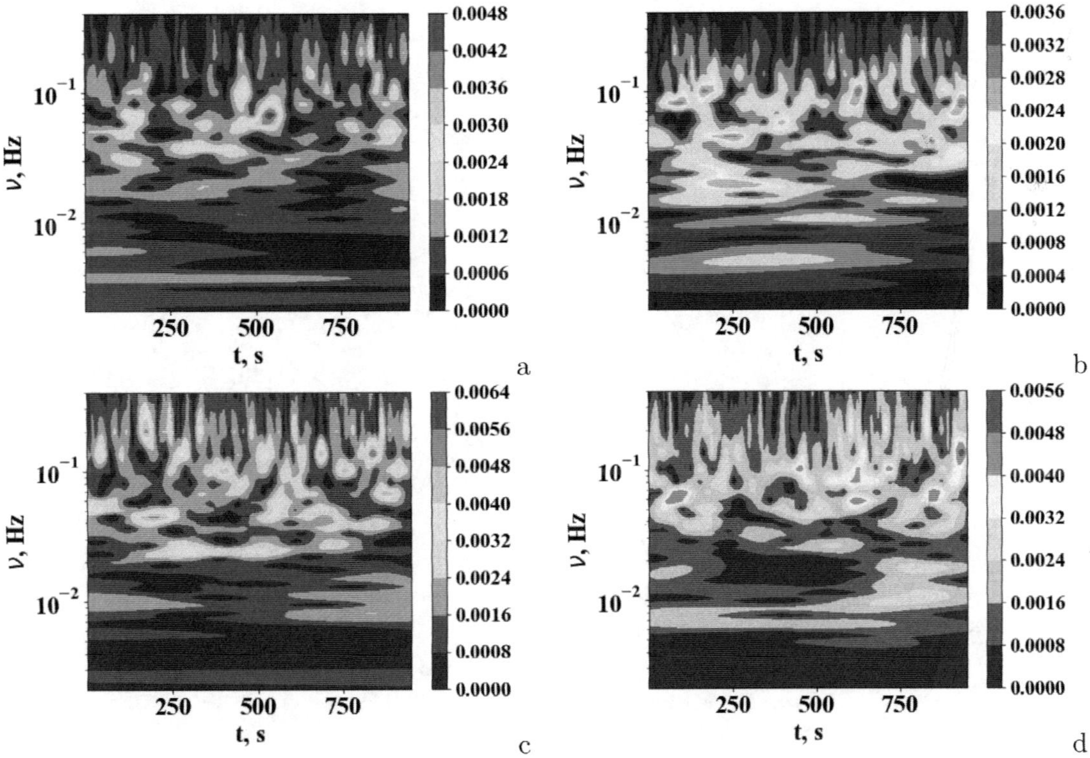

Figure 5. Wavelet diagrams of the velocity z-component in the cell center. The number of electrodes is: a — 5, b — 4, c — 3, d — 2

IOP Conf. Series: Materials Science and Engineering **950** (2020) 012031 doi:10.1088/1757-899X/950/1/012031

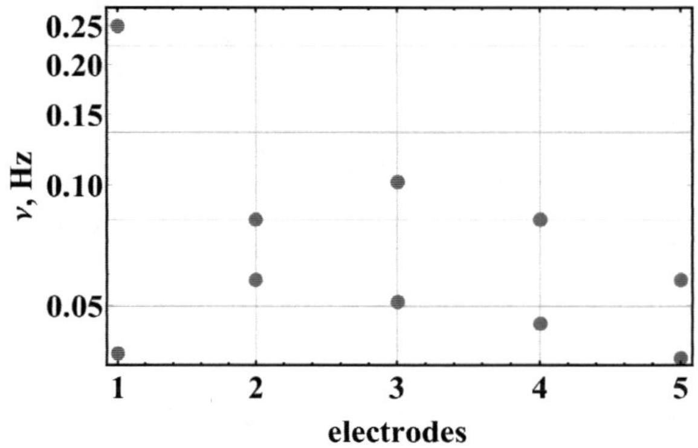

Figure 6. The EVF frequencies, obtained from the wavelet analysis

References

[1] Rabiger D, Zhang Y, Galindo V, Franke S, Willers B and Eckert S 2014 *Acta Materialia* **79** 327–338
[2] Kazak O 2013 *Metallurgical and Materials Transactions B* **44** 1243–1250 ISSN 1543-1916
[3] Sneyd A D and Wang A 1996 *Magnetohydrodynamics* **32** 487–493
[4] Weber N, Galindo V, Stefani F and Weier T 2014 *Journal of Power Sources* **265** 166–173
[5] Weber N, Galindo V, Priede J, Stefani F and Weier T 2015 *Physics of Fluids* **27** 014103
[6] Kelley D H and Weier T 2018 *Applied Mechanics Reviews* **70**
[7] Bojarevics V, Freibergs Y, Shilova E and Shcherbinin E 1989 *Electrically induced vortical flows* (Kluwer Academic Publishers, Dordrecht.)
[8] Vinogradov D A, Teplyakov I O, Ivochkin Y P and Klementeva I B 2017 *Journal of Physics: Conference Series.* **899** 082006
[9] Pedchenko A, Molokov S, Priede J, Lukyanov A and Thomas P J 2009 *EPL (Europhysics Letters)* **88** 24001
[10] Khripchenko S Y 1991 *Magnetohydrodynamics* **27** 77–83
[11] Kolesnichenko I, Khripchenko S, Buchenau D and Gerbeth G 2005 *Magnetohydrodynamics* **41** 39–51
[12] Weber N, Nimtz M, Personnettaz P, Salas A and Weier T 2018 *Applied Thermal Engineering* **143** 293–301
[13] Stefani F, Galindo V, Kasprzyk C, Landgraf S, Seilmayer M, Starace M, Weber N and Weier T 2016 *IOP Conference Series: Materials Science and Engineering* **143** 012024
[14] Davidson P A and Lindsay R I 1998 *Journal of Fluid Mechanics* **362** 273–295
[15] Weber N, Beckstein P, Herreman W, Horstmann G M, Nore C, Stefani F and Weier T 2017 *Physics of Fluids* **29** 054101
[16] Weber N, Beckstein P, Galindo V, Herreman W, Nore C, Stefani F and Weier T 2017 *Mangetohydrodynamics* **53** 3–13
[17] Zikanov O 2015 *Physical Review E* **92**
[18] 1999 *Wavelets: An Analysis Tool* (*Oxford Mathematical Monographs* no Paperback: pages Publisher: (July 29, 1999) Language: English) (Oxford University Press) ISBN 0198505213
[19] S M 2008 *A wavelet tour of signal processing, third edition: the sparse way* 3rd ed (New York: Academic Press)
[20] Kolesnichenko I, Pavlinov A, Golbraikh E, Frick P, Kapusta A and Mikhailovich B 2015 *Experiments in Fluids* **56**
[21] Mandrykin S D, Kolesnichenko I V, Losev G L and Frick P G 2018 *Bulletin of Perm University. Physics* **40** 20–27
[22] Mandrykin S, Kolesnichenko I and Frick P 2019 *Magnetohydrodynamics* **55** 115–124

The methodology of MMF calculation in magnetic circuit of induction motor by "ANSYS Maxwell"

O V Tikhonova, D D Mineeva, A T Plastun

Ural Power Engineering Institute, Ural Federal University, Yekaterinburg, Russian Federation

E-mail: olga_tihonova_91@mail.ru

Abstract. In the paper the methodology of MMF calculation in magnetic circuit of induction motor by "ANSYS Maxwell" is considered. The calculation is based on the use of the total current law. The magnetic circuit of the induction motor is divided into several sections: stator tooth, rotor tooth, air gap, stator yoke and rotor yoke. Each section of the circuit is analyzed separately. The calculation result obtained using "ANSYS Maxwell" is compared with the results of the classical methodology.

Key words: induction motor, MMF, magnetic calculation, ANSYS Maxwell.

1. Introduction

A magnetic circuit calculation is one of main stages in an electromagnetic calculation of an electrical machine. In this stage a developer determines the value of the magnetizing force, which is needed to create a magnetic flux in an air gap of an electrical machine to induce electromotive force (EMF) of a given value [1].

Also this calculation makes it possible to determine the value of the saturation coefficient of the stator and rotor steel; consequently it determines whether the geometric dimensions of the magnetic core of the electric machine are selected correctly. Generally developers determines those parameters of induction motors using standard calculation methods from benefits by I Kopylov and P Sergeev [2]. Such "ANSYS Maxwell" program cans be used to optimize this calculation and simplify engineering tasks.

In this work, the calculation of the magnetic circuit of a classic induction motor is made in two ways. The first one is using the classical calculation method [2] and the second one is using the application of "ANSYS Maxwell". The results of the research are compared with each other in order to verify the correctness of the definition of a particular value with "ANSYS Maxwell".

Using the concept of "the magnetic field tension" and the law of total current, this expression establishes the relationship between the magnetic flux and magnetomotive force (MMF) acting in this circuit and its geometric dimensions. This expression cans be obtained for any part of the magnetic circuit.

The law of the total current states that the line integral of some closed loop is the product of the magnetic field strength and the selected loop length and in other side is the total current flowing through the section bounded by this loop [3].

Mathematically,

$$\oint \mathrm{H}dl = \oint H\cdot\cos\alpha\cdot dl = I\cdot W$$

(1)

where H is the magnetic field tension, A/m; l is the magnetic line length, m; I is the current value, A;W is the number of wires in the loop.

The voltage drop between the points "a" and "b" of the magnetic circuit is equal the line integral of the magnetic field strength between these points [3]:

$$U_{\text{m a-b}} = \int \vec{H} \cdot d\vec{l}$$

(2)

The product of the electric current value and the number of wires is magnetomotive force (MMF):

$$F = I \cdot W$$

(3)

Using the above concepts, we can present the expression (2) in a form that is similar to the second Kirchhoff's law for electrical circuits.

$$\sum_{k=1}^{p} U_{\text{mk}} = \sum_{k=1}^{n} \pm F_k$$

(4)

Thus, the sum of voltage drops along the closed loop of the magnetic circuit is equal to the algebraic sum of MMF of coils in the loop.

2. The calculation stages

2.1. The magnetic tensors calculation in sections of the magnetic circuit by "ANSYS Maxwell"

The AO2-32-6 engine with a power of P = 2.2 kW, a voltage of U = 220/380 V, and a rotation speed of n = 1000 rpm was taken as the test machine [6].

The calculated engine model is shown in figure 1.

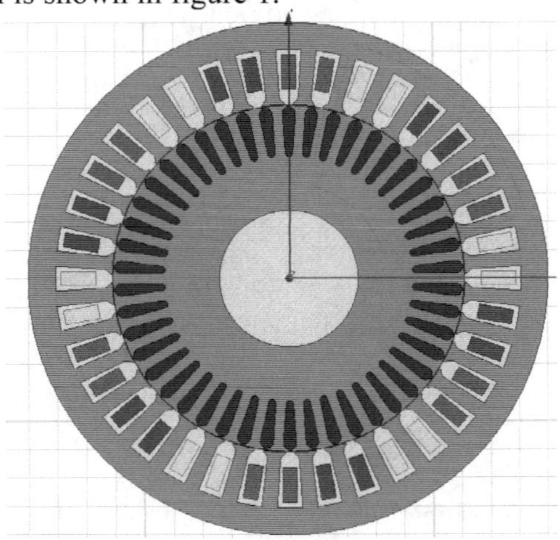

Figure1. The calculated AO2-32-6 engine model

Further it is necessary to select a closed integration loop (see figure 3). To find the integral over a closed line it is possible to divide this line into separate sections, along whose MMF is determined by the formula (2). Selected integration areas should be located along the magnetic lines.So it is necessary to provide a picture of magnetic field lines of the researched induction motor (see figure 2).

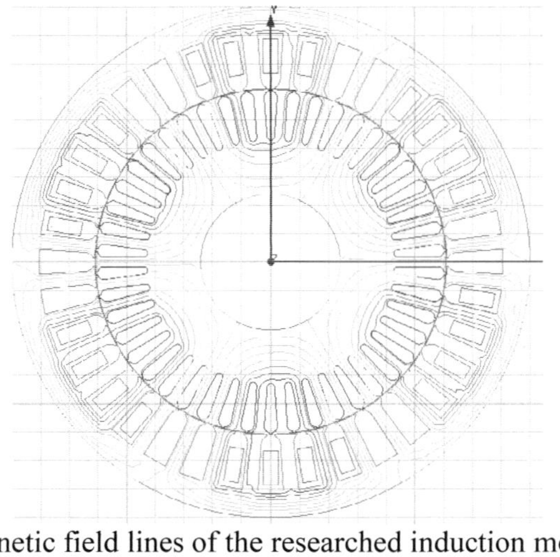

Figure 2. Magnetic field lines of the researched induction motor AO2-32-6

In figure 2 there is the picture of the distribution of the magnetic field for each pole. You can see that it is the same; therefore, the calculation cans be performed on one pair of poles. As in the classical technique, the magnetic circuit is divided into several sections. They are a stator tooth h_{z1}, an air gap δ, a stator yoke L_{a1}, a rotor tooth h_{z2} and a rotor yoke L_{a2}.

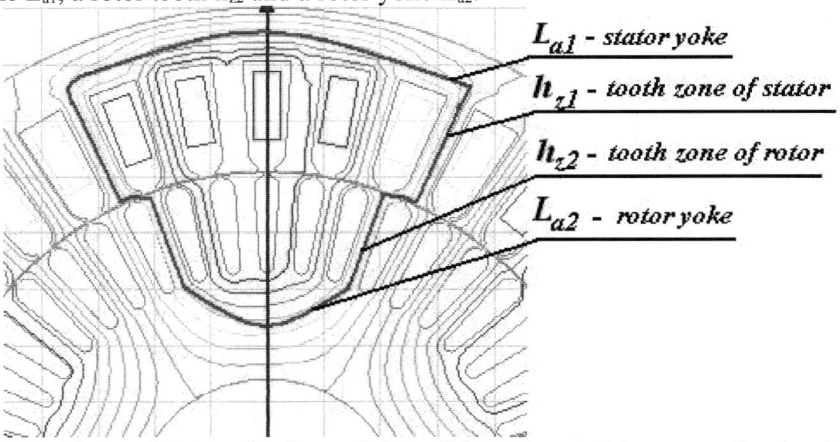

Figure 3. Lines for determining of the voltage of the magnetic circuit of the researched induction motor AO2-32-6

Closing in the magnetic circuit, the magnetic flux passes through the air gap and the tooth layers of the stator and rotor twice. A line is drawn along each section to determine the magnetic tension through integration (see figure 3).

The stator and rotor cores of the motor are made of laminated steel 2212.

The magnetic voltages are determined by the formula (2) using the calculator built into the program.

The current value is I = 9.3 A, the number of wires in the stator coil is W = 47. As is known, it is necessary to know the value of the tension H and the length of the integration line L to calculate the voltage. The value of the tension is taken according to the magnetization curve B = f (H) for steel 2212, the length of the magnetic line according to the geometry of the magnetic circuit. The values of MMF for the sections of the circuit, the length of the lines of integration are shown in Table 1.

Table 1. The values of magnetic circuit MMF calculated by "ANSYS Maxwell'

The area name	*The line designation*	*The line length, mm*	*The MMF value, A*

An air gap, l_δ (for one side)	δ	2x0,3	430
An stator tooth, h_{z1}	h_{z1}	19,6	36
An rotor tooth, h_{z2}	h_{z2}	18,25	51
A stator yoke, L_{a1}	L_{a1}	89,4	180
A rotor yoke, L_{a2}	L_{a2}	15,2	3,9

MMF of the magnetic circuit of an induction motor is equal to the sum of MMF values of all its sections:

$$F_\Sigma = F_\delta + 2\cdot F_{z1} + 2\cdot F_{z2} + F_{a1} + F_{a2} \tag{5}$$

According to the formula (3), the total value of MMF of the magnetic circuit is

$$F_{\Sigma 1} = 430 + 2\cdot 36 + 2\cdot 51 + 180 + 3{,}9 = 788 \text{ A.}$$

The saturation coefficient of the magnetic circuit is the ratio of the total MMF of the magnetic circuit F_Σ to the total value of MMF of the magnetic circuit in the air gap F_δ.

$$k_{\mu 1} = \frac{F_\Sigma}{F_\delta} = \frac{788}{430} = 1{,}83 \tag{6}$$

The program of "ANSYS Maxwell" [4] also make it possible to present zone patterns of the distribution of magnetic induction of an induction motor (see figure 4) and show the most saturated zones of the magnetic circuit using a color scale. As can be seen in the scale, blue corresponds to induction values close to zero, and red matches to maximum induction values close to two Tesla.

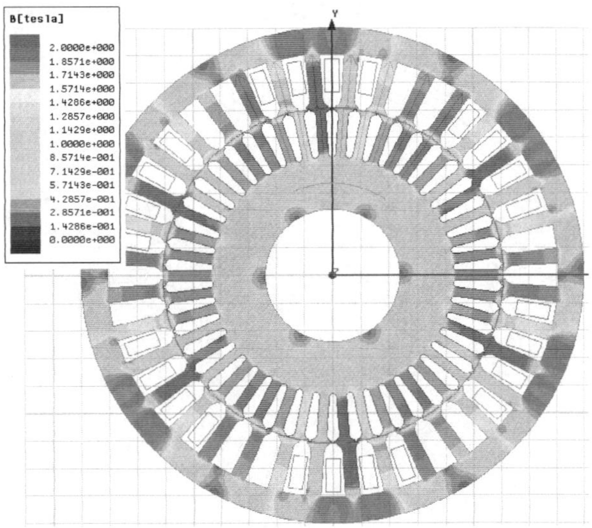

Figure 4. The zone pattern of distribution of the magnetic induction

Figure 4 shows that the most saturated areas are teeth and the stator yoke [5-6].

Based on the value of the saturation coefficient calculated according to the formula (6), the magnetic circuit of the induction motor is highly saturated.

2.2. The calculation of the voltages in the sections of the magnetic circuit using the methodology from the manual of I Kopylov

The following calculation was performed using the classical calculation methodology described in the manual by I Kopylov [2]. The calculation results are presented in Table 2.

Table 2. Values of the magnetic circuit MMF, calculated according to the methodology described in the manual by I Kopylov

The area name	*The line designation*	*The line length, mm*	*The MMF value, A*
An air gap, l_δ (for one side)	δ	2x0,3	417,4
An stator tooth, h_{z1}	h_{z1}	19,6	40,6
An rotor tooth, h_{z2}	h_{z2}	18,25	49,8
A stator yoke, L_{a1}	L_{a1}	89,4	179
A rotor yoke, L_{a2}	L_{a2}	15,2	3,9

The total value of MMF of the magnetic circuit was determined according to the formula (5), mathematically,

$$F_{\Sigma 2}=417{,}4+2\cdot 40{,}6+2\cdot 49{,}8+179+3{,}9=781{,}1\text{A}.$$

The saturation coefficient of the magnetic circuit was determined by the formula (6):

$$k_{\mu 2}=\frac{781{,}1}{417{,}4}=1{,}87.$$

2.3. Conclusions and the comparison of results

The study allows us to draw following conclusions and compare results:

- the total MMF value calculated using the application of "ANSYS Maxwell" is $F_{\Sigma 1}$ = 788 A and $F_{\Sigma 2}$=781,1 A, the latest was determined according to the classical calculation methodology. The relative error in determining the total MMF by two methods is equal to

$$\Delta 1=\frac{F_{\Sigma 1}}{F_{\Sigma 2}}=\frac{788}{781{,}1}\cdot 100\,\%\approx 0{,}88\,\% \qquad (7)$$

- the relative error in determining the saturation coefficient of the magnetic circuit is

$$\Delta 2=1-\frac{k_{\mu 1}}{k_{\mu 2}}=1-\frac{1{,}83}{1{,}87}\cdot 100\,\%\approx 2{,}1\,\%.$$

- thus, the difference in the results obtained by the two methods does not exceed 5%, from which it follows that the "ANSYS Maxwell" application software can be used to calculate the magnetic circuit of a classic induction motor.

IOP Conf. Series: Materials Science and Engineering **950** (2020) 012032 doi:10.1088/1757-899X/950/1/012032

References

[1] Sergeev P, Vinogradov N 1950 Designing of electrical machines Moscow
[2] Kopylov I 2001 Designing of electrical machines Moscow
[3] Bessonov L 1964 Theoretical Foundations of Electrical Engineering Moscow
[4] Tikhonova O Malygin I Plastun A 2017 Electromagnetic calculation for induction motors of various designs by "ANSYS Maxwell"Int. *Conf. ICIEAM*
[5] Bal S Tumay Demirdelen T Tumay M 2019 Lightning Electric Field Analysis of a Transformer Using ANSYS Maxwell 3rd *Int. Simp. ISMIST*
[6] Antik I 1938 Inductuion motors winding data Moscow

IOP Conf. Series: Materials Science and Engineering **950** (2020) 012033 doi:10.1088/1757-899X/950/1/012033

Prospects for the practical use of the Kaminskii thermovoltaic effect

P A M Napoleao, S N Costa and A A Vinogradov

University Agostinho Neto, 4th of February avenue, 71, Luanda, Angola

E-mail: vinluan@mail.ru

Abstract. The Kaminskii thermovoltaic effect, which is one of the new principles of converting thermal energy into electrical energy, is considered in the article. Samples of the rare-earth semiconductor samarium monosulfide (SmS), located in a uniform temperature field without any temperature gradient, generated an electromotive force (emf). The nature of the effect is associated with the concentration gradient of Sm impurity atoms in the sample, a combination of electronic and thermal processes at temperature about 200 °C, and the appearance of a concentration gradient of free electrons, which leads to the generation of emf. The following electrical characteristics of the effect were achieved on bulk samples: a 2.5 V emf in a pulsed mode with a pulse duration of 1.3 s and a 0.05 V emf in a continuous mode. The parameters of thermoelements based on the Kaminskii effect and based on the classical Seebeck thermoelectric effect were compared. The maximum values of the efficiency were 36 % and 10 %, respectively. The effect was observed on other semiconductors, however, its maximum values occur in SmS, which is explained by fundamental reasons related to the position of rare-earth elements in the periodic table. The thermovoltaic effect can be used for direct conversion of associated heat and utilization of waste heat as power sources for various devices.

1. Introduction

A qualitative theoretical analysis of the remarkable properties of samarium monosulfide and other rare-earth semiconductors was carried out on the basis of consideration of their features arising from the position of rare-earth elements in the periodic table of elements [1]. The value of the ionization potential *I* of the impurity element should be considered the most important parameter in semiconductors for the transition of an electron to a free state (conduction band). This follows from the fact that a rough estimate of the depth of the impurity donor level in the semiconductor material is proportional to *I*. Consideration of the second and third ionization potentials of the elements of the periodic table showed that rare-earth elements have their smallest values. We can conclude that the rare-earth element introduced into the matrix from any compound will most easily give its electron in comparison with all other elements. In semiconductors, an electron will fall into the conduction band and lead to the appearance of effects associated with such a transition. The reason for the smallness of ionization potentials in comparison with other elements may be due to the fact that the ions of rare-earth elements have the largest sizes of ionic radii. Large radii of atoms and ions of rare-earth elements are associated with the presence of electrons localized on 4f shells. These shells are located most closely to the nucleus of the element, and their electrons substantially screen the Coulomb potential of the nucleus, which leads to an increase in the ionic radius.

The described feature of the properties of rare-earth compounds allows us to determine the range of the most appropriate practical applications of rare-earth semiconductors. Since the rare-earth ion in these compounds most easily donate its electron under various external influences, these materials can be used as sensitive elements in the manufacture of sensors of various resistive type physical quantities (mechanical, thermal and gas sensors), as well as in converters of various types of energy into electrical energy. Samarium monosulfide is of greatest interest in this sense, since in this semiconductor the 4f levels of the samarium ion are localized in the band gap most closely to the bottom of the conduction band (~ 0.2 eV) among other rare-earth semiconductors. In particular, SmS has a record strain sensitivity among all semiconductors [2-4].

2. Thermovoltaic effect

The remarkable properties of samarium monosulfide revealed in a completely new effect, discovered by Kaminskii V. V. with colleagues in 1999 [5, 6]. The thermovoltaic effect is one of the new principles for the conversion of thermal energy into electrical energy. SmS samples located in a uniform temperature field without any temperature gradient generated an electromotive force. The nature of the thermovoltaic effect is associated with the presence of a concentration gradient of Sm impurity atoms over the sample volume, a change in the valence of defective ions located in the vacancies of the sulfur sublattice $Sm^{2+} \rightarrow Sm^{3+} + e$, electron transitions from 4f-levels to the conduction band, and the creation of large local carrier concentrations. Such electron transitions are collective. Electron transitions are accompanied by the appearance of pulses of electrical voltage and thermal processes synchronized with them. Generation was observed when the samples were heated up to about 200 °C. At higher temperatures, impurity levels with an activation energy of 0.04 eV are depleted, the concentration gradient of which over the volume of the sample causes the presence of a thermovoltaic effect. The following characteristics of the effect were achieved on SmS bulk samples: a 2.5 V emf in a pulsed mode with a pulse duration of 1.3 s and a 0.05 V emf in a continuous mode. The layout of the thermoelement [7] and the sample of a continuous signal generated for more than 5 hours are shown in Figures 1, 2.

3. Parameters of thermocouples and prospects for their application

A comparison was made of the values of the parameters of thermoelements based on the Kaminskii effect and based on the classical Seebeck thermoelectric effect (table 1). Maximum values of the efficiency were 36 % [8] and 10 %, respectively.

The thermovoltaic effect can be used for direct conversion of associated heat and utilization of waste heat as power sources for low-power electronic devices, autonomous sensors of physical quantities used for remote monitoring, in the automobile industry, when servicing pipelines and other facilities, complete with solar panels. The exceptional radiation resistance of SmS will allow using of generators in nuclear energy and spacecraft. The Kaminskii effect can be considered as an effective method of converting thermal energy into electrical energy, which reflects the current trend of the transition to environmentally friendly and resource-saving energy. The maximum effect is obtained on SmS. Such results cannot be achieved at the compounds of other elements for fundamental reasons related to the structure of the electronic shells of rare-earth elements, which are reflected in their location in the periodic table.

A further increase in the conversion characteristics based on the effect under consideration is associated with the working formula of the effect generation voltage [9] and can be achieved by increasing the maximum working temperature and developing effective technological methods for obtaining higher concentration gradients of impurity donor levels in the volume of the generating element.

Figure 1. The layout of the thermoelectric converter based on the thermovoltaic effect in SmS

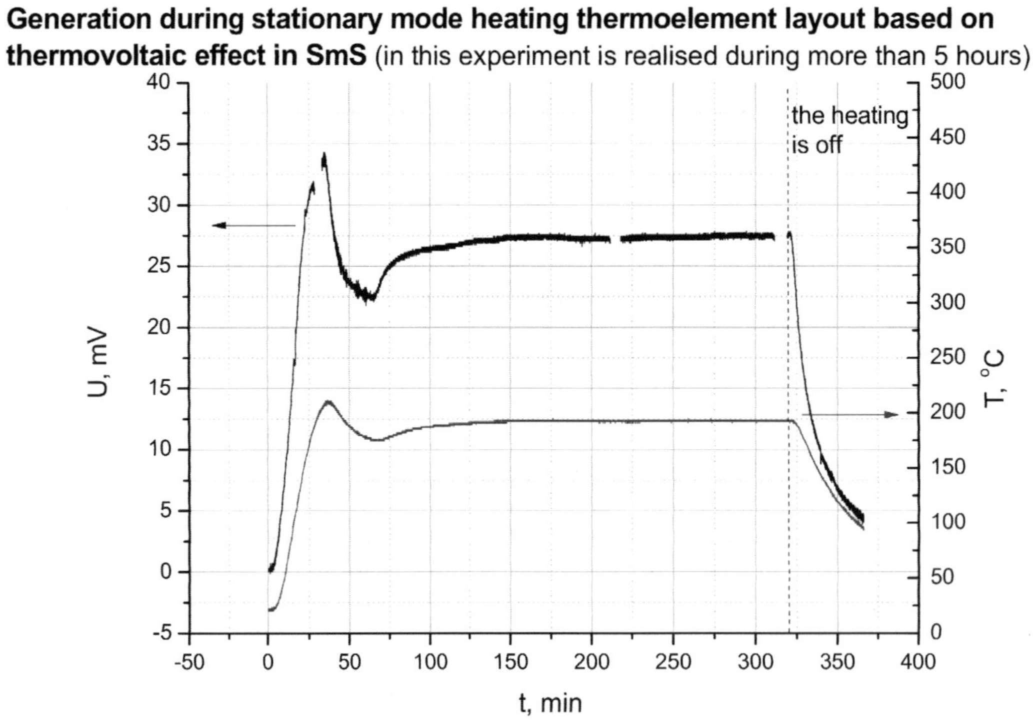

Figure 2. Dependence of the emf *U* and temperature *T* of the thermoelement based on SmS sample on the heating time *t*

Table 1. Maximum achieved results.

Characteristic	Thermoelement based on Kaminskii effect	Thermoelement based on Seebeck effect
Specific generating power, W/g	Up to 1,8	≤ 0,2
Voltage generated by a single thermoelement , V	Up to 5	≤ 0,1
Inner resistance, Ohm	0,2 ÷ 2	0,2 ÷ 2
Radiation resistance of electrical parameters (exposure up to the 1% change of the parameter, Roentgen)	10^{10}	10^{9}
Maximum power obtained from a single thermoelement	5 mW	Units of W
Transformation efficiency, %	Up to 36	Up to 10

Reference

[1] Kaminskii V V and Sharenkova N V 2019 *Semiconductors* **53** 150-2
[2] Kaminskii V V, Vinogradov A A, Kapustin V A and Smirnov I A 1978 *Sov. Phys. Solid State* **20** 2721-5
[3] Vinogradov A A and Kaminskii V V 2010 *Proc. Conf. on Physics of CPLP O Papel da Física e das suas Aplicações na Educação e no Desenvolvimento nos Paises de Língua Portuguesa* (Maputo: Eduardo Mondlane University) pp 249-55
[4] Kaminskii V V, Molodych A A, Soloviev S M, Vinogradov A A and Volodin N M RF patent No 2564698 10.09.2015 bul. 28
[5] Kazanin M M, Kaminskii V V and Solov`ev S M 2000 *Tech. Phys.* **45** 659-61
[6] Kaminskii V V and Solov`ev S M 2001 *Phys. Solid State* **43** 439-42
[7] Kaminskii V V, Kazanin M M, Kazakov S A, Grevtsev M A and Sharenkova N V RF patent No 2628677 21.08.2017 bul. 24
[8] Kaminskii V V, Kazanin M M, Solov`ev S M and Golubkov A V 2015 *Tech. Phys. Lett.* **4** 381-2
[9] Kaminskii V V, Solov'ev S M, Sharenkova N V, Hirai S and Kubota Y 2018 *Tech. Phys. Lett.* **44** 1087-8

IOP Conf. Series: Materials Science and Engineering **950** (2020) 012034 doi:10.1088/1757-899X/950/1/012034

High voltage electric power supply system with contactless power take-off

V Frizen, A Antipin, A Lapin, V Udintsev

Ural Federal University named after the first President of Russia B.N. Yeltsin

Abstract. Autonomous power supplies of light marking, monitoring, relay protection and emergency control systems of digital high voltage networks are considered, in particular, capacitive power take-offs from high-voltage three-phase power lines with voltage of 110 and 220 kV. Energy is collected in the immediate vicinity of the equipment, ensuring its operation without laying a separate power supply network and with minimal human involvement. The results of the analysis show a rather high technical and economic efficiency of the considered power supply system.

1. Introduction

The modern energy market dictates the use of increasingly sophisticated and complex electric power systems. Systems providing efficient, environmentally friendly production and reliable transmission and distribution of electrical energy. In turn, this requires the creation of highly automated electrical digital networks. They provide monitoring and control through digital communication systems and equipment, controlled in real time and monitoring the operating modes of all participants in the production process. They also transmit and consume electricity, which supports the functions of self-diagnosis and self-healing, which provides intelligent adaptation of operating modes and automatic synchronization of such networks with operating modes of consumers.

As a rule, such equipment is distributed along the entire length of electrical transmission lines (ETL) and it is not always possible to centralize power supply using step-down substations with power transformers. Currently, transformers of 110 kV and higher for a secondary voltage of 0.4 kV are not manufactured by industry, and the minimum power of three-phase transformers 110/10 kV is 2500 kVA. Obviously, the construction of 110/10 kV substations is very expensive, and in many cases transformers practically idle.

The problem of the transformation of low capacities from a medium voltage ETL has not been solved. Power transformers are produced for a voltage of 35/0.4 kV, starting with a rated power of 100 kVA and above. To transform power from the 10 kV line, transformers with a rated power of 0.63 kVA and higher are produced. At the same time, modern devices for monitoring, relay protection and emergency control of high voltage ETL, their light shields require significantly less power, often not exceeding 0.15 kVA. When using manufactured transformers, it causes unnecessary investment and overstatement of operating costs due to idling losses, which depend on the rated power.

For these reasons, and also in order to simplify the structure of monitoring devices and the light enclosure of high-voltage overhead electric lines, it is necessary to use autonomous power sources for these devices [1]. It is much more convenient and economical when energy is collected in the immediate vicinity of the equipment, ensuring the operation of the system without laying a separate power supply network and with minimal human involvement. Currently, autonomous power sources are known that use wind energy, solar radiation and electrochemical processes as an energy source. The disadvantages of the first two sources are the extremely unstable level of generated power, which depends on many external random factors. For this reason, the use of wind generators requires the use of intermediate energy storage devices of large capacity. The use of photovoltaic cells (solar panels) runs into their lack of reliability, especially in the winter time due to icing and contamination of the solar cells. A significant factor is the dependence of the output power on the level of irradiation, and degradation during the service life.

2. Unconventional solutions.

The solar battery (PV-panel) is a combination of series and parallel photocell circuits. A separate photodiode is characterized by a short circuit current − maximum generation current, and an open circuit voltage − voltage on the diode without load. In these two modes, the output power of the photodiode is zero. The open circuit voltage of the photodiode is in the range from 0.5 to 1.2 V, depending on the technology of its manufacture. The level of the short circuit current depends on the transition area and the radiation intensity and varies over a fairly wide range. As a result, the output power of the PV panel is a function of load resistance and light exposure (irradiation). Typical load characteristics (LC) for different irradiation intensities (200–1000 W/m2) of the PV panel are shown in figure 1. Each LC has a pronounced maximum. The geometrical location of the maximum power points forms the MPPT-curve of maximum power (Maximum Power Point Tracking). To remove the maximum power from the PV panel, the operating point must be on this curve.

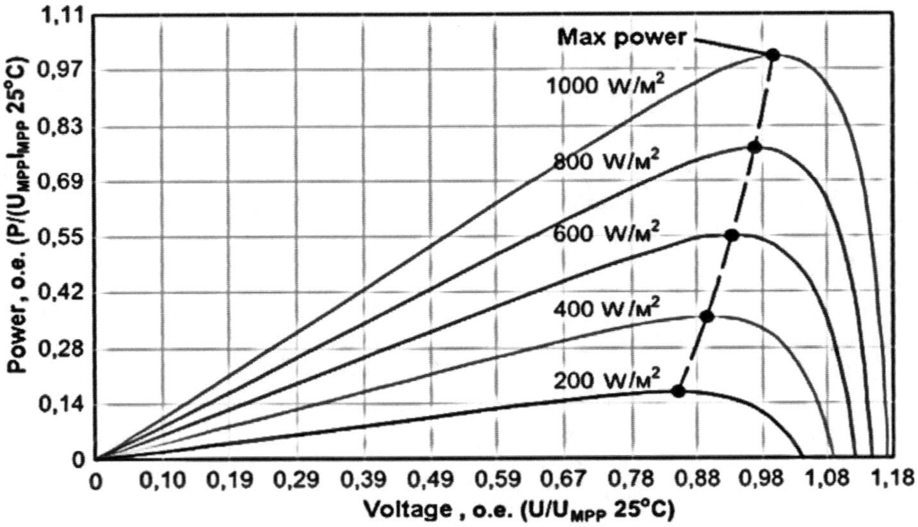

Figure 1. The dependence of the power of PV panels on the light and load resistance

The output power of solar panels depends on other factors: angle of incidence, spectral composition and intensity of solar radiation, air temperature, weather, season, fog, dust, etc. The output power of the PV panel can vary by an order of magnitude or more, depending on weather, season, fog, dust, angle of incidence, the spectral composition of sunlight, etc. Changing the load resistance also significantly affects the displacement of the operating point and the level of generated power. To ensure optimal operation of the PV panel, there should be a control system that will track the MPPT curve. All this requires a margin of installed power and large dimensions, and, as a result, causes greater windage and the inability to mount the structure on the ETL support.

Therefore, it is usually used electrochemical sources: batteries or accumulators for powering monitoring devices and light guarding high voltage ETL. Batteries require replacement or recharging, that is, periodic connection to an external power source and costly human intervention for their maintenance. Batteries require replacement or recharging, that is, periodic connection to an external power source and costly human intervention for their maintenance. The design of the device reduces operating costs for their operation, increases the duration of the overhaul intervals while increasing the reliability of operation, increases the safety of the ETL.

3. Design device.

Currently, several types of power sources are known that use high-voltage ETL as an energy source, for example, based on current transformers, providing operation in a certain range of measured currents [2]. They are conveniently used to power electronic components at high potential, since high-voltage isolation of the secondary windings is not required. As follows from the principle of their operation, the output power of such a source is proportional to the current in the line. At currents less than the minimum, this power source does not provide the necessary power for the system to work. This means that the system is inoperative in the current range from zero to minimum current, which is typical for small loads and idle power lines.

In this case, it is possible to use capacitive power take-off with the help of capacitors connected in series between the phase wire and ground [3, 4, 5, 6]. From the capacitors forming a capacitive voltage divider, lower voltage is removed, which is fed through the reactor to a step-down transformer. From its secondary winding through the electronic stabilizer is supplied to the load. Such devices can be used to power devices in the entire range of line currents, starting from zero. However, this device cannot be used to power devices with high potential. One lining of the power take-off capacitor is grounded and therefore, power supply of electronic units at high potential cannot be carried out from a capacitive divider. For this operation, appropriate high-voltage insulation of the secondary winding is required. Capacitors should be designed for the amplitude value of the ETL voltage, i.e. be high voltage and properly insulated. The reactor must be of a special design, which is not produced by industry (figure 2).

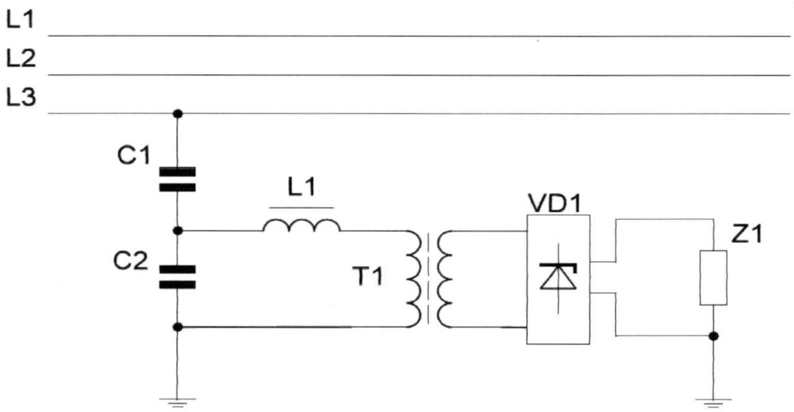

Figure 2. Capacitive Power Take Off

Also known is a device for capacitive selection power (CSP) from a high-voltage three-phase power line. The device contains an antenna located along the wires of the line at a safe distance from it and connected to an electromagnetic device. The electromagnetic device consists of a reactor, a single-phase step-down transformer and a damper. The reactor winding is connecting in series with the primary winding of a step-down transformer. The damper is connecting in parallel with the secondary winding of the step-down transformer. Damping conditions to suppress subharmonic oscillations that occur in the secondary circuit when the load or short circuit is disconnected. The reactor serves to compensate for the capacitive voltage drop and maintain a constant voltage value in the primary winding of the step-down transformer when the load changes. The design diagram of such a device differs from the circuit shown in figure 2 only in that the capacitor C1 will be replaced by three such capacitors connected to the three phase wires L1, L2, L3 [7].

An antenna suspended along a power line, for example, in the form of a cable with a length of at least one span between the supports, is a fairly simple and cheap solution. It does not require high-voltage capacitors; it allows you to have a large power margin in the entire range of line currents. There are several problems with this solution. Firstly, the volume of energy received in this way is usually quite modest, which necessitates the proximity of the antenna to the phase wire. Secondly, the most effective mode of operation of the antenna is resonant, which leads to the need to tune the connected electromagnetic device to resonance. An electromagnetic device must be bypassed with an expensive high-voltage capacitor (C2) or a large reactor (L1), which is under a voltage of about 5–10 kV. There is the possibility of the appearance on the reactor of the full phase voltage of the line when the antenna cable is whipped together with the line wires and the antenna breaks when the cable is iced up.

When supplying consumers with particularly low power, for example, the above-mentioned measuring, control or control systems, high-voltage ETL light-shielding systems, etc., such CSPs can be very effective, especially when the antenna is design as a small-sized insulated device. According to the output characteristics, such devices correspond to a current source. With fluctuations in the load resistance, the voltage at the output terminals of the source will change. The most effective solution in this case will be a combination of CSP with an electronic voltage stabilizer or battery charger [7]. Their main characteristic will be the output power and its dependence on the load resistance and external factors: air temperature, weather, season, fog, dust, etc.

4. Experimental data

In order to study this dependence, a CSP prototype was create and test results were obtaining. A positive result can be considering that it was not possible to establish the dependence of the output power on weather, temperature and humidity, season, snow, ice, fog, dust and other environmental factors. The measurement results differ by 2–3%, which can be explain by the natural measurement error. The dependence of the power on the secondary winding of the transformer on the load resistance during operation of the CSP in the resonant mode is show in figure 3. As follows from the graph, when the load resistance varies from 200 to 500 Ohms, the power changes from 20.52 to 19.24 W, i.e. within 5%. In this case, the voltage on the secondary winding of the transformer varies from 72 to 104 V, and the current from 285 to 185 mA. This allows us to conclude that when working with an electronic voltage stabilizer, high output voltage stability will be ensuring.

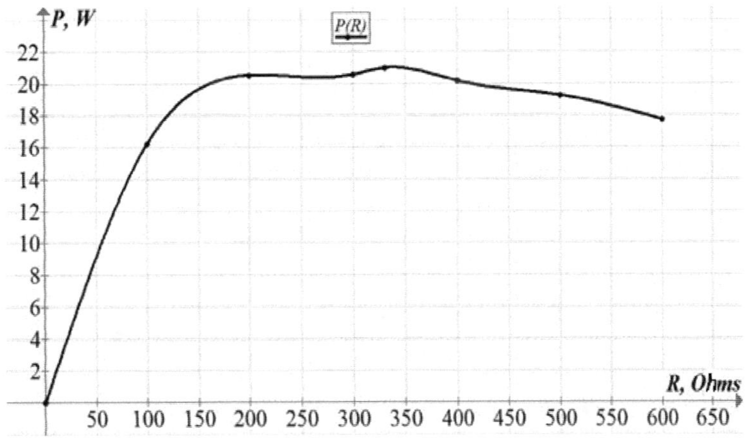

Figure 3. Dependence of the output power of the CSP on the load resistance

As a result of the research, it can be concluded that non-contact power take-off from a high-voltage power line using an antenna is an actual solution. As a result of the research, we can conclude that non-contact power take-off from a high-voltage ETL using an antenna. The proposed solution gives a good result compared to other methods of power take-off from high-voltage power lines and allows you to replace expensive chemical current sources.

5. References

[1] Mao, X., Huang, Q., Lan, S., Lin, M. & Lu, W. 2017, "Contactless Powering Via Magnetic Resonance for On-line Monitoring Wireless Sensor of High-voltage Electrical Equipment", *Gaodianya Jishu/High Voltage Engineering,* vol. 43, no. 11, pp. 3718-3725.

[2] Patent USA № 4794318, IPC H01F 27/42; G01R 15/22; 15/07, published 27.12.1988.

[3] Nefed'ev A I Emkostnyi delitel' vysokogo napryazheniya na gazonapolnennykh kondensatorakh / Nefed'ev A I, Isaev A V and Azaryan D K // Energo- i resursosberezhenie: promyshlennost' i transport. – 2017.- №2(19). - S. 43-46.

[4] Patent for invention № 2381585 Russian Federation, IPC7 H01F27/42, G01R15/04. Emkostnoi istochnik pitaniya / Arson A G, Ostapenko E I and Chursinov A M. - № 2008139248/09; stated 03.10.2008; published 10.02.2010, newsletter № 4.

[5] [4] Patent for invention № 2558690 RF, IPC7 N02J3/00. Ustroistvo otbora moshchnosti ot linii elektroperedachi / Korolyuk Y F, Buryanina N S and Lesnykh E V. - № 2013129389/07; stated 26.06.2013; published 10.08.2015, newsletter № 1.

[6] Bulashevich D N and Yurenkov V D. Emkostnyi otbor moshchnosti ot linii elektroperedachi – Gosenergoizdat, Moskva, 1959.

[7] Frizen V E, Antipin A S, Nazarov S L and Udintsev V N *The contactless power take-off from the high voltage power line* // XX-th International Symposium on Electrical Apparatus and Technologies SIELA 2018 3-6 June 2018, Bourgas, Bulgaria.

AUTHOR INDEX

Institute of Physics
Dirac House, Temple Back
Bristol BS1 6BE UK

ISSN: 1757-8981
ISBN 978-1-7138-3198-3

CURRAN ASSOCIATES INC.
proceedings.com